*Chief Engineer*

# By the Same Author

---

*First Light: A Celebration of Alan Garner* (Editor)

*Pas de Deux/Concert of Stories*
(with Abbi Patrix and Linda Edsjö)

*Seizure*

*Ariel's Gift: Ted Hughes, Sylvia Plath and the story of
'Birthday Letters'*

*Gravity*

# Chief Engineer

## Washington Roebling:
## The Man Who Built the
## Brooklyn Bridge

## Erica Wagner

BLOOMSBURY
LONDON · OXFORD · NEW YORK · NEW DELHI · SYDNEY

Bloomsbury Publishing
An imprint of Bloomsbury Publishing Plc

50 Bedford Square
London
WC1B 3DP
UK

1385 Broadway
New York
NY 10018
USA

www.bloomsbury.com

BLOOMSBURY and the Diana logo are trademarks of Bloomsbury Publishing Plc

First published in Great Britain 2017

© Erica Wagner, 2017

British Library Cataloguing-in-Publication Data
A catalogue record for this book is available from the British Library.

ISBN:   HB:   978-1-4088-4377-2
ePub:   978-1-4088-3775-7

2   4   6   8   10   9   7   5   3   1

Typeset by Westchester Publishing Services
Printed and bound in Great Britain by CPI Group (UK) Ltd, Croydon CR0 4YY

MIX
Paper from
responsible sources
FSC® C020471

To find out more about our authors and books visit www.bloomsbury.com. Here you will find extracts, author interviews, details of forthcoming events and the option to sign up for our newsletters.

For Francis and Theo

*two strong towers—*

Faith—is the Pierless Bridge
Supporting what We see
Unto the Scene that We do not—
Too slender for the eye

It bears the Soul as bold
As it were rocked in Steel
With Arms of Steel at either side—
It joins—behind the Veil

To what, could We presume
The Bridge would cease to be
To Our far, vacillating Feet
A first Necessity.

—Emily Dickinson

# Contents

*Foreword*  ix

1. "No one ever does just the right thing
   in great emergencies"                                       I

2. "The finest place in the world"                            19

3. "Something of the tiger in him"                            36

4. "I was not a chip off the old block"                       61

5. "It is curious how persons lose their heads
   in times of excitement"                                    77

6. "The urgency of the moment overpowers
   everything"                                                89

7. "I am very much of the opinion that
   she has captured your brother Washy's
   heart at last"                                            113

8. "All beginnings are difficult, but don't
   give up"                                                  130

9. "I will have to go to work at
   something"                                                151

10. "Good enough to found upon"                              171

11. "I have been quite sick for some days"                   189

12. "Now is the time to build the Bridge"                    197

13. "Trust me"                                               219

14. "She goes everywhere and sees everything"                247

15. "The image of his wife floats before him"     266

16. "You can't desert your job"                    288

17. "Time & age cures all this"                    303

Epilogue: Cold Spring                              310

*Acknowledgments  313*

*Notes  317*

*Bibliography  347*

*Index  355*

# *Foreword*

IN THE ARCHIVES OF Rutgers University is a photograph of two gray-haired gentlemen. They are sitting in a grand room filled with fine furniture and there are oil paintings on the wall. The clean-shaven gentleman is sitting beneath a large portrait of the man who wears a beard; easy enough to guess it was taken in the elegant home of the latter—Washington Augustus Roebling. He was seventy-three when this picture was taken in 1910, a man who appears ready to retire from active life, though as it happened this was not the case at all. He had been trained as an engineer, to follow in his famous father's footsteps. He had fought with the Union Army all the long, bloody years of the Civil War; at the end of that war he had found himself a clever and handsome wife. When the fighting was over and the reconstruction of the country had begun, he and his father together had set out to aid that rebuilding in the most practical—and yet metaphorical—way possible: by building bridges. That father, John Augustus Roebling, had bridged Niagara when Washington was a boy; together they spanned the Ohio at Cincinnati; and when, in 1869, his father had died, suddenly and tragically, the position of Chief Engineer of the great East River Bridge—the Brooklyn Bridge—had been taken up by this thirty-two-year-old veteran of several different sorts of wars.

The story of his life holds "all the fascination of an interesting novel," as one who knew the Roebling family well once wrote. But this little-known photograph, tucked into a folder in a basement in New Brunswick, New Jersey, raised the hairs on the back of my neck. For the man seated next to Washington Roebling is Admiral Jacky Fisher.

"We are today entertaining Admiral Lord John Fisher of the British navy," Washington wrote to his own son, John, in the letter that corresponds to the date of the photograph. Fisher was in the United States to see his

son married in Philadelphia—near enough to Trenton to feel he might pay a call on the famous engineer at the mansion that, for the last two years, Washington had shared with his second wife, Cornelia. "He is a jolly, jolly, British Tar," he told John. "The admiral is 69 and acts like 40—he danced all around the parlor with Cornelia and told a hundred stories . . . He is a man of wonderful vitality with a temperature of 98° all the time—Has to stand in the open door to keep cool," Washington wrote buoyantly.

A high compliment indeed, for not all company pleased Washington Roebling. But somehow I was not surprised that Lord John Fisher's had. One of the touchstones of my decision to write about Washington's life was an extraordinary biography of Fisher written by Jan Morris. *Fisher's Face* is an account of the life of Lord John "Jacky" Fisher, born in 1841 in Sri Lanka (making him just shy of four years younger than Washington), who rose to be commander in chief of the British Mediterranean Fleet in 1889, and eventually First Sea Lord, head of the whole Royal Navy. He was a man with a "genius for enjoyment," as Morris writes—a genius visible even in

Washington's brief encounter with him by the banks of the Delaware River. Fisher died in 1920—six years before Washington Roebling's death, and six years before Jan Morris was born—but he had been, she wrote "one of my life's companions." She kept a photograph of him tacked up inside her wardrobe: "Apart from those of my family it is the face I know best, far more familiar to me than the features of statesmen, actors, artists or even old friends."

For many years, I too had seen an old photograph each and every day, a battered image from a book, made in the New York Public Library when I was nineteen years old. Taken at the beginning of the Civil War, it's a photograph of a young man in the slouched forage cap of the Union Army, his pale, wide-set eyes looking directly at the camera, his mouth unsmiling, a faint mustache just visible on his upper lip. I no longer recall exactly which book it was I took to the library's photocopier to make a souvenir, my own *carte-de-visite*, of that face. My picture is only about an inch square. I put clear tape over it to protect it, and I made a tiny envelope to hold it, on which I wrote "W. A. R." We share a middle name. I have never been without this little image since.

Growing up in Manhattan, I could hardly avoid an awareness of the Brooklyn Bridge, but I can't say that it ever impinged on my consciousness until I was in my teens. I acquired an English boyfriend, a little older than I was, who had just qualified as a civil engineer. He said he wanted to cross the Atlantic to visit me one Christmas—but now I realize it wasn't really me he wanted to visit; it was that big work of stone and steel over the East River. So one day we walked out of the subway into bright winter sunshine, east into Cadman Plaza Park before hooking west up the steps leading to the pedestrian walkway that rises, elevated above the traffic, up toward the towers, through the cradle of cables, right over the glittering river. In those moments, as we strode back toward Manhattan in the cold, I had the experience the bridge's original designer, John Roebling, intended I should have— just as many had before me and will have long after I am gone. I wrapped my cold fingers around one of the vertical suspender cables and felt the steel vibrate under my touch as if this structure, this place, were a living thing. I could feel in my body the words that had been spoken when the bridge had opened to the public in May 1883, that the structure "looks like a motionless mass of masonry and metal; but, as a matter of fact, it is instinct with motion. There is not a particle of matter in it which is at rest even for the minutest particle of time. It is an aggregation of unstable elements, changing with every change in the temperature, and every movement of the heavenly bodies. The problem was, out of these unstable elements, to produce absolute stability; and it was the problem which the engineers, the organized intelligence, had to solve, or confess to inglorious failure."

The boyfriend went back to England, but something had changed in me, for good.

I wanted to know more. I read David McCullough's magisterial account of the construction of the Brooklyn Bridge, *The Great Bridge*. He first took the measure of the materials available on both the construction of the bridge and the history of the Roebling family; reading through Washington's notes and letters, reports and sketches gave him, he wrote, "the odd feeling of actually having known the Chief Engineer of the bridge." The wry, stoic, practical and yet elegant voice that I could begin to hear in McCullough's book was speaking, I felt, directly to me.

I began to haunt libraries and archives, listening for that voice, wondering what more Washington had to say. He was a man who would seem to have been born with a silver spoon in his mouth. By the time he went to study engineering at the Rensselaer Institute (later renamed Rensselaer Polytechnic Institute) in 1854, his father, a German immigrant who had come to the United States to be a farmer, had become one of the country's most innovative engineers; his 1842 patent for wire rope would not only make him a staggeringly wealthy man, it would lay the foundation for an industry that would make the building of modern cities possible. But John Roebling's children paid a high price for his brilliance—Washington, as the eldest, most of all. His story drew me to him—and the sound of his own words. Here was a real writer's voice, "genuine literature, which is Arcadian and original." I heard it in my head as clearly as the voice of anyone I knew. It belonged wholly to him: it was nothing like his father's. For all his long life, Washington suffered in comparison to his father, that powerful presence. Indeed John Roebling's death in 1869 means it has often been thought, and said, that Washington Roebling only completed the plans for the Brooklyn Bridge his father had drawn up.

Nearly a century and a half after it was begun, the Brooklyn Bridge remains a wonder, its image displayed on tourist brochures, on film posters and drugstore wrapping for toilet paper, on packets of Italian chewing gum. The story of its construction, one of the greatest emblems of "progress" in the nineteenth century, is such a dramatic tale of vision, innovation, and endurance in the face of extraordinary odds that it overshadows—understandably, perhaps—the rest of its builder's life. At the same time, if you stop a few dozen pedestrians walking across the bridge and ask them who brought it into being, they won't be able to give you an answer. The architectural historian Lewis Mumford—who wrote of the "clean exaltation" engendered by a walk over the Brooklyn Bridge, also admired "the tradition of anonymity" cloaking the identities of most engineers.

Washington's life is, in itself, a striking one. Born in 1837, he lived to be nearly ninety, and so his years spanned the most American of centuries. He was an extraordinary witness to his times; his own recounting of those times is vivid, entertaining, drily observant—and often wonderfully cantankerous. However, there is one quality that stands over all: his tenacity. His father was a genius, certainly, endowed with the kind of imaginative gifts rarely given to men or women; the more one reads about John A. Roebling, the less he seems a mere mortal. In contrast his son Washington was one of us: no genius, not near it, but willing to put his shoulder to the wheel—whatever the cost—until the job was done. "It's my job to carry the responsibility," he told a journalist near the end of his life. "And you can't desert your job; you can't slink out of life, or out of the work life lays on you." That he never did.

The British engineer David Blockley writes of a quality he calls "practical rigour," which is vital for a civil engineer. This quality, he says, "requires wise foresight to anticipate what can go wrong and put it right before the consequences are serious . . . Every practical possibility must be considered and every reasonable precaution must be taken." The life of Washington Roebling offers an enduring demonstration of that quality, which is reinforced by his own cogent expression of what it means to be tested, what it means to persevere. And so whenever possible in this book, I have allowed Washington to speak for himself. There is a steadfastness in his writing, in his voice, that makes him an admirable companion; but there is anxiety, intemperance, anger, and prejudice in his character, too. But flaws as much as virtue make a man. Biography is an argument as much as it is a conversation.

And biography is also an act of imagination. Some time into this process of researching Washington's past, I found myself at lunch with the writer Colm Tóibín, in Dublin. I had just returned from a trip to the archives at Rutgers University, where for the first time I had held in my hands Washington's little pocket notebooks, kept during the early years of the construction of the Brooklyn Bridge. They are aide-mémoires and contact lists combined: the necessary information for a working engineer out on the job, lists of things to do, people met, queries to follow up. Tóibín and I had been talking of the elusiveness of Henry James—who, in 1904, saw the Brooklyn Bridge as part of "the bold lacing together, across the water, of the scattered members of the monstrous organism" of the whole city. I said to Tóibín that holding those notebooks made me feel, absolutely, as if I were *with* Washington Roebling as he walked the streets of Brooklyn and New York. He smiled at me across the table. "But then he put the notebook in his pocket, and you don't know where he went," he said.

And this, of course, is the problem. Biography promises truth where, to put it plainly, there is none—or at least, not very much. Facts can be ascertained and agreed upon—the Battle of Gettysburg took place on these dates, the opening ceremony of the Brooklyn Bridge on that day—but truly the past vanishes into air, despite our efforts to pin it down with notebooks, with letters and diaries. Furthermore, biography is inherently transgressive in nature, as Janet Malcolm so keenly perceived; the biographer is "like the professional burglar, breaking into a house, rifling through certain drawers that he has good reason to think contain the jewelry and money, and triumphantly bearing his loot away. The voyeurism and busybodyism that impel writers and readers of biography alike are obscured by an apparatus of scholarship designed to give the enterprise an appearance of banklike blandness and solidity." A biographer wants to know her subject's secrets, to discover the real character behind the public face. But that "real character" is a creation of imagination, too.

Not all biographers are interested in banklike blandness and solidity: neither description applies to Washington Roebling's biography of his own father. That text is also, to a large degree, a memoir of his own life, and I refer to it as such. The Roebling family's first chronicler, Hamilton Schuyler, remarked that if Washington Roebling could have been persuaded to write an autobiography, the resulting book would have been "a production of rare merit, for not only was his memory for persons and events of unusual clearness and accuracy, but whatever he wrote bore the impress of his strongly marked personality. In his case the style was the man. He wrote as he talked. He had a gift for racy description and his comments were always pertinent and characterized by a certain caustic, not to say sardonic, humor, and a native shrewdness of observation that were most illuminating. Moreover, his perfect candor, his tendency to use direct language, always to call a spade a spade, and not an agricultural implement . . . proves that he could have written of himself with the same detachment from personal issues as he was wont to do in the case of others." Schuyler—whose account of the Roebling family was published in 1931—was spot on regarding Washington's character, directness, and style; but he was not aware that Washington had, in fact, written an account of his own life. That manuscript, however, had vanished.

That Washington was writing a book was no secret, in his later life. In 1910 he wrote to Henry S. Jacoby, a fellow engineer who shared his passion for geology, in reply to an inquiry as to whether he might one day write an autobiography. Thanks to the crippling illness brought on by his work on the Brooklyn Bridge, Washington said, he had found it impossible to consider any such task for many years; but then, in the mid-1890s he commenced making notes about his early life in Saxonburg, and his service in the Civil War. He returned to the manuscript in the early years of the twentieth century, picking up where he had left off. "Some of the matter might best be called irrelevant," he told Jacoby; he allowed that his son John, his only child, might edit it after his death. But he would show it to no one—it remained, for the time being, "his own exclusive property." Henry Dodge Estabrook, a lawyer who gave the address when John A. Roebling's statue was unveiled in Trenton's Cadwalader Park in 1908, had read it—and called it "one of the most remarkable books I have ever read" while admitting that it "may never be published." It was "honest and unlacquered," Estabrook said. It was "remarkable for its analysis of men and events and for an acidulous humor that is almost styptic; but chiefly it is remarkable for the frank revealment of the *intime vitae*, the qualities and inequalities of the extraordinary man who was his father." He noted too how revealing it was of its author, John's eldest son. "Unconsciously to himself, perhaps, the biographer has given us a study in evolution, with the factors of heredity, environment, the struggle for existence, and all the rest of it, plus a psychic something that Darwinism might consider negligible."

Every so often Washington's son, John, would write and ask how the book was coming along. Washington was reticent, calling what he'd written "scraps," or claiming they had been lost.

They were not lost; but when David McCullough went hunting for what Washington called his "biography of J. A. Roebling," there was no trace of it in the archives at Rutgers University, where the family's personal papers were deposited and remain. But then, in the 1980s, Donald Sayenga, a historian of the Roeblings and the former general manager of Bethlehem Wire Rope, which acquired the Roebling name in 1973, turned up the manuscript, which had been hiding in the archives all along. It is an extraordinary source. While Washington intended to write only of his father, he was often—very often—distracted by the happenings of his own life up until 1869. Here is his childhood in Saxonburg, the town his father had built from the wilderness in western Pennsylvania; his difficult years at school and college; his service in the war; his service alongside his father, who is painted as a figure of shocking brutality. The strength of his feeling is the "psychic something" which Estabrook found so striking. There is no doubting that the manuscript is damaging to the reputation of the man who was once Trenton's first citizen; while making faithful accounting of John Roebling's achievements, his son's book was by no means the celebration Washington Roebling had planned.

It is not a complete account of Washington's own life, because it ends with John A. Roebling's death, in 1869. That incompleteness can be linked, perhaps, to another ellipsis in Washington's life—the record of his marriage to Emily Warren Roebling, and their life together. There is a plaque on the Brooklyn Bridge to Emily Roebling, "whose faith and courage helped her stricken husband, Col. Washington A. Roebling, complete the construction of this bridge"; it is dated 1951 and was sponsored by the Brooklyn Engineers' Club. It is there with good reason: during the long years of his often mysterious illness—precipitated certainly by his work in the pressure of the caissons of the bridge's towers—she was very much more than his invaluable assistant, and she was, clearly, a remarkable person, who rose to a challenge rarely presented to a woman of her age and station. "Mrs. Roebling was a woman of strong character and rare cultivation," Hamilton Schuyler wrote, adding that she had "an almost masculine intellect"—which may be taken to mean a very good one. But Washington kept very little of their correspondence, although his letters remain in archived manila folders. The letters she wrote to him are not preserved; and she is, altogether, an elusive presence in the Roebling archives. Given the story of the discovery of Washington's memoir, I know I should be cautious in arguing that Washington destroyed much of her correspondence—but as he burned the letters she sent him before their marriage, getting rid of much more pertaining to Emily might not have troubled him. He knew he was a public figure; but he was a private man, who, in his lifetime, felt he had been ill served by the busybodyism of journalists and writers.

That said, a great deal of his writing survives—and all of it, even the most technical reports, is accessible and engaging. During the Civil War, Washington's letters from the front provide an extraordinary personal record of one man's conflict—as do his later recollections of that time, pages and pages of which are shoehorned into his account of his father's life. In his later life the conflation, in the public mind, of his own life with his father's was the source of no small frustration, despite his recognition of the magnitude of John Roebling's achievements. But in this section of his account he cannot resist the pull of his own story, his participation in the war that would alter his country forever. Ball's Bluff, Antietam, the Second Battle of Bull Run, South Mountain, Gettysburg, Spotsylvania, the Crater at Petersburg— the battles that Washington witnessed, and in which he participated, were some of the most terrible of the war. And in one of them, Gettysburg, he did no small service in turning the fight to the Union's advantage—and this was a battle that, most agree, was a crucial turning point of the war.

And so I have done my best to serve him, this flawed and fascinating man, this bridge builder. He was a son, a soldier, a husband, a father, an engineer, a businessman. His life—like all lives—was intricately made; solving the problem of how to give an account of such a life is, to some extent, a problem of design, one Washington might have recognized. A book, like a bridge, must have firm foundations, a strong structure, and effective working if it is to endure; in that way writers and engineers are alike.

Bridges are always more than a way to get from one shore to the other: from the rainbow bridge of the old Norse gods, to Pontifex Maximus in

Rome, bridges are symbolic of the desire for connection, the possibility of connection. Where there was nothing, now there is something, arcing miraculously through the air: and this is particularly true of suspension bridges, it seems to me. The curve of the cables echoes the shape of a simple cord you can hold between your hands; and when you cross the water you are indeed suspended, held safe in the bridge's cradle. You are in a place that is no place at all, that is in itself *between*: you belong, quite simply, to the bridge. And then you keep walking, and reach the other side.

# "No one ever does just the right thing in great emergencies"

Death as destruction is impossible, it is only a change of all parts, whose composition on earth is adapted to local purposes but only to local purposes, and as long as they fit together and work together in accordance with the laws of nature, so long do they exist in the formation of a human being . . .

—John A. Roebling in a letter to his father, 1844

DECADES LATER, WASHINGTON ROEBLING would blame himself. In describing the event that would alter the course of his life forever, he would return to its details, almost as if by conjuring them he could call back the past. "Nothing happened to me," he wrote of that dreadful time—but he was wrong. Nothing, for him, was ever the same again.

In 1869, the year in which he became Chief Engineer of the East River Bridge, Washington Augustus Roebling was thirty-two years old. His childhood had been spent in Saxonburg, Pennsylvania, an industrious community of German immigrants founded, a few years before his birth, by his father, John Augustus Roebling. In his lifetime the elder Roebling was often referred to as "a lesser Leonardo"; he had been born in Saxony in 1806, and his mother had scrimped and saved to send her clever son to Berlin, where he would study architecture, bridge construction, hydraulics—and philosophy with Georg Wilhelm Friedrich Hegel. But, with a party of fellow pilgrims, he had escaped the restrictions of Prussia for a new life in the new world, a

*John A. Roebling's drawing of the tower of a proposed bridge
over the East River, 1857*

life in which he would swiftly establish himself as one of the greatest engineers and visionaries of his day.

John Roebling had proposed a bridge over the East River as early as 1857,
setting out his stall in a letter to the *New-York Tribune*: "The plan in its general features proposes a wire suspension bridge crossing the East River by
one single span at such an elevation as will not impede navigation . . ." Precisely ten years later, his dream had become a reality: in April 1867, a charter authorizing a private company to build and operate an East River bridge
had been voted through the state legislature in Albany; a month later, John
Roebling was named the company's Chief Engineer. He was the one man
alive, it was generally reckoned, capable of building a span across the river—
which is, in fact, a congested, turbulent tidal strait. Such an endeavor was
an unprecedented task: in the middle of the nineteenth century, suspension
bridge technology was still in its infancy. In 1854 the bridge over the Ohio
River at Wheeling, West Virginia—built by John Roebling's great rival, Charles
Ellet—was nearly destroyed in a storm; it had been standing a mere five
years. But the *Brooklyn Daily Eagle* put its faith in John Roebling, announcing at the end of June 1869: "The East River Bridge having now been finally
approved by the Federal Government as well as by the board of U.S. Engineers and the eminent civil engineers to whom the plans were submitted, the

work has now been placed in actual progress, and the erection has become a fixed fact." Should that fixed fact cause any anxiety, the article's subtitle read "The Bridge Six Times as Strong as the Greatest Possible Load Upon It." Now the work could begin.

And then, just three days after the *Eagle*'s boast, John Roebling was injured in what most took to be a minor accident. "The circumstances were as follows," his son wrote of that otherwise ordinary day. "We had gone down after lunch to inspect the site of the Brooklyn Tower in the spare ferry slip— In order to see better he climbed on a heap of cord wood, and from that on top of the ferry rack of piles—Seeing a boat coming, and fearing that the heavy blow would knock him off I cried to him to get down—(I was up there also some 20 feet off) In place of getting down all the way he stepped from the fender rack down on the string piece of the prominent outside row of piles—the blow from the boat was severe, sending the fender rack so far in that its string piece overlapped the other one at the same time catching the toe of his boot on the right foot and crushing the end of his toes—When he uttered a cry I did not realize at first what had happened." Washington wrote to his younger brother Ferdinand that the piles had clipped his father's foot "like a pair of big shears would do."

Washington was his father's right-hand man. He and his young wife, Emily Warren Roebling—sister of Washington Roebling's commanding officer during his distinguished service in the Civil War—had, not long before, returned from an extensive tour of Europe, which they had taken at John Roebling's behest. It was no honeymoon, but a tour of steel mills in Germany and Great Britain, and the study of pneumatic caissons, which would be the foundations of the bridge. Washington Roebling had used his time at Rensselaer —then the United States' only engineering college, and still one of the most prestigious institutions in the United States—to qualify in his profession, then honed his trade building bridges for the Union Army to cross before Confederate troops could blow them up. He had volunteered in the spring of 1861, first in New Jersey and then again in New York: he entered the Ninth New York State Regiment a private, but by the war's end—by which time he was a veteran of Second Bull Run, Chancellorsville, Gettysburg, the Wilderness, and the Crater at Petersburg and had been a witness to the Battle of Hampton Roads, the first great clash of the ironclads *Monitor* and *Merrimack*—had risen to the rank of colonel. He had then supervised the work on his father's Covington–Cincinnati bridge over the Ohio, completed in 1867 and today known simply as the Roebling Bridge. Yet for all his expertise he was still his father's lieutenant; John Roebling, at sixty-three, was a vigorous man.

At the time of the accident the elder Roebling hadn't moved from his home in Trenton, New Jersey, to the work site; according to Washington he came over once or twice a week, and kept a couple of rooms at a Turkish bath on the corner of Cranberry Street and Columbia Heights in Brooklyn. Washington and Emily had taken a house on Hicks Street, a few blocks away. "As quickly as possible I got him into a carriage and took him up to the Turkish bath where he was staying. As Dr. Sheppard did not feel like keeping such a sick man I took him to my house—With great difficulty we got him up the stairs where I undressed him and laid him on the bed which he never left. Dr. Sheppard recommended a Dr. Barber as surgeon. When he arrived he proved to be a young man of not much force. But he trimmed the wounds, cut away the crushed tissues and put on the first dressing all right—the mistake I made was in not taking Mr. Roebling to a hospital at once—but I had been brought up to look upon hospitals as the abode of the devil and upon a doctor as a criminal, perhaps I am excusable as no one ever does just the right thing in great emergencies."

Nothing more is known of Dr. Sheppard or Dr. Barber. What is known is John Roebling's violent dislike of any medical regimen other than one he had himself devised. Recalling his youth in Saxonburg, Pennsylvania, Washington wrote of his father's "especial animosity" toward doctors; "the average kindly family physician was held up to his children as a monster in human form." In the city's summer heat, Washington was unable to withstand his father's will.

"He did not rally from the first shock until the next day—Then when Dr. Barber came he told him that he would take command of his own cure himself and would take no orders or treatment from him. Dr. Barber shook his head with a dubious smile—A Tinsmith was ordered—He fixed up a big tin dish like a scale, supplied with a hose and running water—into the dish Mr. R. put his foot—the stream of water playing on it all the time—When Barber saw that he exclaimed you are inviting sure death for yourself—nature in endeavoring to cure such a severe wound must have recourse to pain, to fever, in order to supply the increased vitality necessary to the healing process . . . He ordered Barber out of the room in a violent manner."

TETANUS REMAINS DEADLY in developing countries. In most Western nations, the incidence is less than one case per million population per year, following the development of a widely available vaccine in the 1920s. *Clostridium tetani* is a bacteria that produces a toxin affecting the brain and central nervous system; it is not contagious, but can be acquired through contamination in a wound such as the one John Roebling suffered. Untreated, the symptoms are dreadful, even when described clinically in the modern day. "Trismus (lockjaw)—the inability to open the mouth fully owing to the rigidity of the masseters [jaw muscles]—is often the first symptom . . . Generalized tetanus is the most common form of the disease, and presents with pain, headache, stiffness, rigidity, opisthotonus [the severe, rigid arching of the back, head and neck], and spasms, which can lead to laryngeal obstruction. These may be induced by minor stimuli such as noise, touch, or by simple medical and nursing procedures . . . The spasms are excruciatingly painful and may be uncontrollable leading to respiratory arrest and death."

Washington Roebling kept watch over his father. "After three or four days I noticed an inability to eat or speak. Barber ventured up and at once pronounced it lockjaw, incurable at that—Dr. Kissam was called in, in consultation he confirmed it . . . Tetanus antitoxin was then unknown. Now came ten terrible days. As the jaws set, eating and swallowing became impossible. With feverish haste he started to write all kinds of directions about his treatment about the bridge about his financial affairs as his powers waned his writing became more and more illegible, nothing but scrawls at the end. As there was no trained nurse, I assumed that function, with an occasional friend to sit up nights—Dr. Barber being dismissed I telegraphed to Dr. Brinkman of Philadelphia a water cure doctor whom my father knew—he tolerated him—but Brinkman knew it was too late, coming only as a matter of form and to write the death certificate."

Washington wrote this text nearly forty years after the event, decades beyond his father's death; but despite the passage of time it is clear from his swift cursive hand that in his mind the images from those days remained

clear, haunting, violent. The scene crowds in on him; his father's feverish
haste becomes his own, and his self-reproach is the more poignant for his efforts to refute it—even as he returns to the dawn of his father's death, his
final stillness as the sun rose over the East River, over the ferries, over the
unbridged stream.

"Daily and hourly," he wrote, "I was miserable witness of the most horrible tetanic convulsions, when the body is drawn into a half circle, the back
of the head meeting the heels, with a face drawn into hideous distortions—
Hardened as I was by scenes of carnage on many a bloody battlefield, these
horrors often overcame me—When he finally died one morning at Sun rise
I was nearly dead myself from exhaustion.

"We all have to die—It is useless to tax one self as to whether life could
have been prolonged or death hastened by this or that treatment—Criticisms
are all in vain, we should be thankful that we know not what the future has in
store for us." His father's death was the pivotal moment in Washington Roebling's life; the criticisms he feared were not made by others: they were the
voice in his own head. Thinking back to that moment on the pier he wrote:
"I have often taxed myself that if I had kept still and given no warning nothing might have happened—But the experiences of a long life teach me that
such self criminations [sic] are futile."

JOHN ROEBLING HAD set out his vision for his great work in a letter to the
New York Bridge Company dated September 1st, 1867.

Gentlemen.

On the 23rd of May last, I accepted the appointment as Chief Engineer of
the Bridge proposed to be erected over the East River, between the two Cities of New York and Brooklyn, under the provisions of your charter, with
the understanding that I should proceed to make the necessary Surveys, to
determine upon the best location, to make out Plans & estimates, and to
report upon the subject at as early a day as practicable. I commenced this
task without delay, and have been engaged on it ever since.

The following Report & accompanying plans are the results of my la-
bors & are respectfully presented to your consideration.

The contemplated Work, when constructed in accordance with my designs, will not only be the greatest Bridge in existence, but it will be the great Engineering Work of this Continent & of the Age.

Its most conspicuous features, the great Towers, will serve as landmarks to the adjoining cities, & they will be entitled to be ranked as national monuments. As a great work of art, & as a successful specimen of advanced Bridge Engineering, this structure will for ever testify to the energy, enterprise & wealth of that Community which shall secure its erection.

Respectfully Submitted/John A. Roebling

The report that follows runs to fifty-two pages of John Roebling's bold, clear, sloping hand. Nine days later, in the pages of the *Brooklyn Daily Eagle*, Roebling's plans were reproduced for all to read and approved in a long editorial essay that spoke to the necessity of the work. Roebling's report, said the paper, "will attract great interest, for it may be accepted as the first practical step towards the realization of one of the most remarkable enterprises of our time." There could be no doubt of the need for the bridge as the cities of New York and Brooklyn flourished in the years just after the Civil War. "Is the bridge necessary?" asked the *Eagle*. "We have nearly reached the accommodation the ferries can furnish on the routes of travel. The population of Brooklyn has increased four-fold within fifteen years . . . If the ferry companies cannot more than accommodate the travel of our present population, how would it be if three times the number pressed upon them? . . . Last winter on several days, and for hours each day, ferry travel was interrupted for hours. We assert, without fear of contradiction, that it would be better for Brooklyn to sacrifice an amount equal to the whole cost of the bridge rather than have it established as a fact that, in winter time, no resident of Brooklyn could count on getting to New York to do his business." (On the same page the paper noted under "Topics of To-day" that "the Spiritualists' proceedings were duller than usual at the Cleveland convention. The feminine delegates, as described by a reporter, are not specially attractive . . . Women care less for blossoming in Bloomers than ever for voting.")

But—largely thanks to the corrupt, wholly male world of politics of the city of New York, still a separate metropolitan entity from its as yet more rural neighbor across the water—by the time of John Roebling's death little evidence of a bridge could be seen.

That said, New York and Brooklyn were growing and changing at such a rate in the years after the Civil War that they were in a state of constant flux. In a process that had begun long before the conflict between North and South, these great cities were undergoing dramatic shifts in function and form. Where once homes and businesses had existed side by side, now the rapid expansion of business districts meant that the more affluent sought peace and quiet away from their places of work and the middle-class habit of commuting began. From 1837, the New York and Harlem Railroad offered regular service to 125th Street—at the time a full six miles north of the built-up area downtown. Lines reached Westchester County by 1844: the *New-York Tribune* predicted that "the line of this road will be nearly one continuous village by 1860." Between 1810 and 1860 the steam ferry, the omnibus, the commuter railroad, and the horse-drawn streetcar were all introduced into city life, changing the pattern of that life forever; cities were being turned "inside out."

By the third quarter of the nineteenth century, wrote William Stone in his *Centennial History of New York City*, New York had become "the national city": "cosmopolitan, European as well as American, and obviously one of the few leading cities of the world—the third city of Christendom." (He declined to name the first two, allowing other grand conurbations to flatter themselves, perhaps, that they remained ahead of New York.) Stone claimed that the real value of property in New York City was, at this time, one trillion dollars—or a thirtieth part of the entire property of Great Britain; and, he wrote, "35 tons of mail-matter are received here for our citizens, and 55 tons are sent out daily." The city could boast 2,621 blacksmiths, 6,307 boot- and shoemakers, 9,501 dressmakers, 5,978 merchants, 1,232 lawyers, and 855 piano makers.

Then, as now, the gulf between rich and poor was great indeed. In September 1869 one of the century's great robber barons, Jay Gould, attempted to corner the market in gold and nearly bankrupted a nation, in an event that came to be called "Black Friday." Gould, born in upstate New York almost a year to the day before Washington Roebling, had risen from surveyor to tannery owner, to director of the Erie Railroad, finally rising to become one of the wealthiest men in American history. At the time, U.S. "greenback" currency—paper money—was still backed by gold; Gould hoped to drive up the price of gold and then sell off what he had bought, thereby raking in huge profits. By September 24, the price of gold was 30 percent higher than it had been just six months earlier; but when the government flooded the market by releasing four million dollars in gold, panic ensued. Investors who had borrowed to jump on the gold-buying bandwagon could not repay their loans; thousands were ruined. Gould's partner in the scheme was "Diamond Jim" Fisk, another of the city's shady financiers—who was himself a close associate of the corrupt "boss" of New York, William M. Tweed. But despite the panic, for the top tier of society, life carried on as usual. Later that year, in November, a "committee of fifty" gathered together to raise $250,000 for a museum that would make the new world worthy to stand alongside the old: the Metropolitan Museum of New York.

But away from Fifth Avenue—where Gould kept a mansion at Forty-Seventh Street—things were very different. "A glance at New York City, embracing the entire of Manhattan Island, will show that its geological position, its advantages for sewerage and drainage, in fact for everything that would make it salubrious and healthy, cannot be surpassed by any city in this or any other country," wrote the *Catholic World* in 1869 in undertaking a survey of "the sanitary and moral condition of New York City." "And still," the report continued, "with its bountiful supply of nature's choicest gifts, many of our readers will be surprised to hear that our death-rate is higher than that of any city on this continent, or any of the larger cities of Europe." A table showing "the relative per annum mortality" in various cities around the globe reveals that one in every thirty-five New Yorkers was liable to end up dead in any year; in London the ratio was one in forty-five.*

The living conditions the report goes on to describe retain the power to shock. The sanitary superintendent of the city, a Dr. E. B. Dalton, counted 18,582 "tenant-houses," or tenements, home to more than half a million people—well over half the city's population. "Such an economy of space was never known to be displayed in sheltering cattle as is here shown in the houses, if they can be so called, of the laboring classes." Two houses, seven or eight stories high, would stand on a lot just twenty-five feet wide and

---

*Parisians were worse off than Londoners, with one in forty perishing; and Copenhagen was only slightly less noxious than New York, at one in thirty-six.

one hundred feet long; the "yard" of the two buildings had stalls for privies—hardly any of which were connected to any kind of sewer or drain. Filth and human waste simply accumulated where it was left. As many as 126 families, more than eight hundred people, could be packed into every house. Furthermore, Dr. Dalton reported that of this housing, "52 percent [are] in bad sanitary condition, that is, in a condition detrimental to the health and dangerous to the lives of the occupants, and sources of infection to the neighborhood generally; 32 percent are in this condition purely from overcrowding, accumulations of filth, want of water-supply, and other results of neglect."

Cholera, typhoid and typhus fevers, smallpox, measles, and scarlet fever were rife; scurvy was as common as it was aboard ship. Because so many tenement houses were damp, tuberculosis was a particular scourge: in the fifteen months ending December 31, 1867, 4,123 people were reported as having died of the disease, most of them aged between twenty-five and forty. One out of every four babies born in the city died before reaching the age of twelve months—in 1868, that was 7,494 infant deaths. And the discrepancy in death rates between rich areas and poor areas was very clear: in 1863, in New York's Fifteenth Ward—the wealthiest at the time, in the area around Washington Square—death rates were one in sixty; in the grim Fourth Ward, down by the East River waterfront, where the tenements described above stood, the rates were one in twenty-five. Strikingly, the *Catholic World* saw a solution on the city's opposite shore, in Brooklyn, where—with plenty of open space available—"neat and comfortable cottages can be built for the laboring classes"; the proposed East River Bridge "will form the first of the links which are to bind our Island City to the surrounding rural districts . . . Over the East River Bridge it is intended to run cars by an endless wire rope, worked by an engine under the flooring on the Brooklyn side. The minimum rate of speed is put down as twenty miles an hour. It is such traveling facilities as these structures will afford which are necessary to enable the workingmen to reach healthful and salubrious homes outside of the metropolis."

The city of Brooklyn had been incorporated in 1834—despite the opposition of some on that other, more crowded shore in Manhattan, who "foresaw that the incorporation of Brooklyn as a city, would give a new impetus to her growth and population; and that Brooklyn lots would soon become formidable rivals to their own," wrote Henry Stiles in his history of the city, published between 1867 and 1870. Brooklyn—which took its name from Breukelen, a township in the province of Utrecht—had its original founding in 1636 when William Adriaense Bennet and Jacques Bentyn purchased from the native inhabitants a tract of 930 acres of land at "Gowanus." "The occupation of this farm, over a portion of which the village of Gowanus subsequently extended—and which comprised that portion of the present city lying between Twenty-seventh street and the New Utrecht line—may be considered as the first step in the settlement of the city of Brooklyn."*

With its beginnings in farmland, Brooklyn was always of a much more rural character than New York. By the end of the Civil War, the grandest houses in Brooklyn stood on the Heights—though it had a greener, more rural past. "The face and brow of this noble bluff were covered with a beautiful growth of cedar and locust, while its base was constantly washed by the waves of the East river. From its summit, the land stretched away, in orchards, gardens and pasture, out to the old highway (Fulton Street). The red men, who first roamed over this spot, named it in their expressive language 'Ihpetonga,' or 'the high sandy bank,' and it must have been a favorite place of resort for them, if we may judge from the large quantities of stone arrows and other implements . . . which used formerly to be found here after the washing of the river banks by storms, or heavy rains. To the early villagers, it was known as 'Clover Hill' . . . in the memory of some yet living among us, the brown freestone glories of these latter days can never eclipse the simpler natural beauties of the 'Clover Hill' of their boyhood."

By 1869 Brooklyn was the third largest city in America, and more than simply "a sleeping apartment" for its neighbor, New York; yet the overflowing prosperity and greatness of that neighbor was crucial to Brooklyn's growth and success. An obstacle in the way of that success was the river keeping the cities apart. A ferry across the river had been established by the Dutch by 1642; in 1870 the *New York Times* estimated that seventy million crossings were made every year—over half of those on the fourteen boats

---

*No one seems very sure where the name Gowanus comes from; it may have its origin in the name of a native Canarsee sachem, Gouwane—but then it may also come from the Dutch word for a bay, *gouwee*.

belonging to the Union Ferry Company, then the wealthiest ferry company in the world. "Crowds of men and women attired in the usual costumes! how curious you are to me!/ On the ferry-boats the hundreds and hundreds that cross, returning home . . ." Walt Whitman wrote in *Leaves of Grass*, first published in 1855. But in the winter, especially, when ice clogged the river, crossings could be unreliable; in January 1867, the river froze up completely, as it had done in 1852 and 1856. Rush hour was disrupted by a frost fair, as the *Brooklyn Daily Eagle* reported, the river's frozen surface transformed into an icy recreation ground. A boat "struggling out of Fulton ferry, in the New York slip, found itself imprisoned, and the chilly dawn brought the discovery that she was frozen in as effectually as was ever Dr. Kane in the Arctic ocean—differing only in degree." Furthermore, the ferries were not always safe. On November 14, 1868, two of the Union Ferry Company's boats—the *Union* and the *Hamilton*—collided on the New York shore; twenty people were injured and a boy named George Bower was killed instantly. One of the results of the catastrophe was perhaps unsurprising: "The accident had no inconsiderable effect in forming the public mind towards the building of a bridge across the East River."

In the spring of 1869, Congress authorized the construction of the bridge, "provided that the said bridge shall be so constructed and built so as not to obstruct, impair, or injuriously modify the navigation of the river." Only then did the Roeblings, father and son, take the stockholders of the bridge company—along with a party of consultant engineers—to see the other bridges that had made the elder Roebling's name, and that qualified him for this work. An elegant traveling circus journeyed to the Smithfield Street Bridge and the Allegheny River Bridge in Pittsburgh; to the bridge over the Ohio between Cincinnati and Covington, Kentucky; and finally to the dramatic railroad span across the gorge at Niagara Falls, called simply Suspension Bridge, which had been completed in 1855.

Evidence of the work on the Brooklyn Bridge that was just beginning in 1869 can be found in a sequence of little black pocket notebooks now held in the collection of Rutgers, the state university of New Jersey. The center line of the bridge had been established, Washington Roebling working on the survey with Colonel William Paine, a self-taught engineer from New Hampshire whose work during the Civil War had been commended by no less than President Lincoln. "Record of the various marks & pins on the Center Line of the East River Bridge," John Roebling wrote, "as finally established on June 7th 1869." The notebooks are beautifully illustrated with delicate, accurate drawings, showing exactly where the survey lines are to go. The property at 170 South Street was part of "Mark on New York Side," with a note that "There is a mark above the letter O in the name Cornford, made by leaning out of the 3rd story window. This mark is a little east of the true line, about 1/2"." The drawing shows the building, and where the mark is above the names "Cornford & Collins." A few blocks away Water and Front Streets are shown, and note is made of where a little carving would show where the surveyors had been: "The Water St marks are two pins as located, the measurement from No 290 to 1st one being at right angles. The triangulation line through Water St. strikes 1" south of the north pin or 11'-7 1/4" from building & point of crossing is marked by a crowfoot . . ."

The first of five books is headed "John A. Roebling: Brooklyn Br. 1867"; and up until page 91 the notes contained therein are indeed in John Roebling's handwriting. That his son was not a wasteful man is revealed by the change, in the course of a page, to his own small, neat writing. That any change had occurred is noted in the most practical manner: "John A. Roebling's salary commenced May 22nd 1867 $8000 per annum and continues at that rate. If the bridge is abandoned he was to have nothing but if it continues he was to have it. As the bridge was continued he was entitled to the above from May 22/67 to July 22/69."

But none of this was evidence to any outside eye that a bridge would—or could—yet be built.


AT THE TIME of his father's death, Washington Roebling was thirty-two years old. Years later, recalling the toll exacted on himself—and, it might be inferred, on his family—by his father's behavior, his attitude, his unbending will, Washington wrote that "whatever a man gets he has to pay for in some shape." In July 1869 the cost was clear.

Washington was the oldest of John Roebling's surviving seven children; a little daughter, Jane, had died in infancy in 1838, and a son, Willie, had died of diphtheria at four the year before the Civil War began. His younger brother Ferdinand was working with his father at the wire rope business in Trenton; Charles was also being schooled as an engineer at Rensselaer

Polytechnic Institute in Troy. His youngest brother, Edmund, was only fifteen at the time of his father's death. Laura, the oldest girl, was married and living on Staten Island with her schoolteacher husband; Josephine was married to Charles H. Jarvis, a brilliant pianist of the day; Elvira had married only weeks before John Roebling's death.

The Trenton manufactory of John A. Roebling's Sons would keep its name, according to John Roebling's will; the business was the cornerstone of the family fortune, the wire rope it produced the centerpiece of the engineer's bridge designs. Also included in that will as a partner in the business was Charles Swan, who Roebling had first encountered in his earliest days in America, when Swan was working in a carpenters' gang. Washington Roebling called him "a bright eyed black haired red cheeked chap" distinguished by "his daring, his handiness, adaptability to all kinds of work, and the good natured honesty which characterizes the German." Swan had taken charge of the wire rope business, as John Roebling was so frequently away. "The mutual respect and esteem and the implicit confidence and trust which my father reposed in him grew with age and only ended with his life." John Roebling left Swan twenty thousand dollars in his will, and instructions that he should be a full partner in the business. His will had been generous to charities and good causes; he was rather less generous where his children were concerned. The residue of his estate was left to his sons, after taking away "such amount or amounts as have been or shall be advanced by me to my children in my lifetime on account of their respective shares of my estate, which amounts will appear from the entries and accounts made by me against them in my private ledger."

He trusted Charles Swan; and he trusted that the wire business would continue as it had gone on. But the question of trust might also have been raised in regard to John Roebling's successor. The bridge was planned, but not begun; it was a work the like of which the world had never seen. As to whether Washington Roebling was confident that the New York Bridge Company would see him as his father's rightful heir as chief engineer, his own accounts differ. In a letter written in 1916, he set out his case strongly and clearly.

"First—I was the only living man who had the practical experience to build those great cables, far exceeding anything previously attempted, and make every wire bear its share.

"Second—Two years previous I had spent a year in Europe studying pneumatic foundations and the sinking of caissons under compressed air. When the borings on the N.Y. tower site developed the appalling depth of 106 feet below the water all other engineers shrank back . . .

"Third—I had assisted my father in the preparation of the first designs—he of course being the mastermind. I was therefore familiar with his ideas and with the whole project—and no one else was."

Yet in his memoir—in a section probably composed some eight years before the letter above—he sounds much less certain. His father's funeral was a very grand one, as the *New York Times* noted. "Mr. Roebling had for many years been a resident of Trenton, and the circle of his friends and acquaintances in the city and throughout the state was very large. The attendance upon his obsequies yesterday was numerous, thousands of the citizens of Trenton as well as a large number from this City, Brooklyn, Jersey City, Newark and Philadelphia being present. A special train over the New-Jersey Railroad carried the friends of the deceased from this vicinity, and special arrangements were made upon other roads in the state to accommodate those who desired to attend." The *Brooklyn Daily Eagle* reported in detail on the appearance of the late Mr. Roebling: "The body arrived shortly after seven o'clock," the paper said of the night before the funeral, "in charge of the undertaker, Mr. T. W. Barnum, of Pierrepont Street in your city, who was called in immediately after Mr. Roebling's death. Owing to the precautions taken by Mr. Barnum, not the slightest trace of decomposition was perceptible. Emaciated it certainly was, but that is accounted for from the fact that for nearly a week before his death Mr. Roebling was unable to eat anything. In other respects the features are perfectly natural . . . The body had been attired in a suit of black clothes and then placed in a casket of solid rosewood lined with white satin. The ornamentations are of solid silver and in exceedingly good taste, consisting of massive bar handles at the sides and ends . . ."

With his father in that solid rosewood casket, a great burden had shifted to Washington Roebling's shoulders. "After a week I had become sufficiently composed to take a sober look at my own situation," he wrote in his memoir. "Here I was at the age of 32 suddenly put in charge of the most stupendous engineering structure of the age! The prop on which I had hitherto leaned had fallen—henceforth I must rely on myself—How much better when this happens early in life, before we realize what it all implies . . ." He was, more-over, the head of the family now; his obligations did not only concern the East River Bridge. "As the principal executor of my fathers [sic] will much responsibility was placed on me in dividing his considerable estate." Even after substantial charitable donations to the Union Industrial Home Asso-ciation for destitute children of Trenton and the Marlbury Orphans' Farm School near Mount Vernon—and several others—the remainder of the estate was valued at $1,250,000—worth many, many millions of dollars by any reckoning done today. More than money was at stake: he was also left as the guardian of his youngest brother, Edmund, who had been badly affected by their mother's death in 1864. But Washington knew that despite the burdens now placed upon him, he at least was exceptionally fortunate in the partner he had found for his life: Emily. "At first I thought I would succumb but I had a strong tower to lean upon—my wife, a woman of infinite tact and wisest counsel." At the time of his father's death he could not have known just what a singular helpmeet she would prove herself to be.

The Sunday funeral was an extraordinary affair. The men who worked in Roebling's mill, along with their wives and families, passed through the par-lor of the Roebling house, where the body was laid out, to "gaze with tearful eyes on the inanimate form of the man who had so often proved himself their friend." The industrial and financial titans of the cities of New York and Brooklyn also paid their respects, as did key officeholders: Brooklyn's con-gressman, Demas Barnes; Andrew Green, comptroller of Central Park; C. C. Martin, chief engineer of Prospect Park; men without whom the notion of a bridge across the East River might never have got as far as it did, such as Julius W. Adams, chief engineer of the Brooklyn Water Works—who had himself proposed a plan for an East River crossing. Henry Cruse Murphy, publisher, lawyer, politician, and Brooklyn patrician, had worked with his fellow mourner, the redheaded, energetic William C. Kingsley, to drive the bridge plan through. Kingsley—loathed by Washington Roebling, who was only four years his junior—was Brooklyn's leading contractor, a builder and political operator who had much to gain by such a grand construction. In the afternoon, the cortege was prepared; as the funeral carriages passed through Trenton, "Along each side of the carriages the workmen from the different iron works marched two and two, while the rear of the procession, which was upwards of a mile and a half in length, was brought up by the Fire Companies. It is estimated that there were about fifteen hundred persons in the procession. No such demonstration had ever been witnessed in Trenton before, and all along the line of march . . . the sidewalks were crowded with people. Every window was occupied, and nearly one-half of the population of the city appeared to have turned out to do honor to the remains of their beloved fellow-citizen, John A. Roebling, as they were being carried to their last resting place." He was laid to rest in the family plot in Trenton beside his wife Johanna, who had died five years before.

The press were cleanly confident in the young engineer's ability to take over the job. "Not long since, before the accident which led to his death, Mr. Roebling remarked to us that he had enough of money and of reputation, and he scarce knew why, at his age, he was undertaking to build another and still greater bridge," ran an editorial published in the *Eagle* on the day of his death. "His son, he added, ought to build this Brooklyn bridge—was as competent as himself in all respects to design and supervise it; had thought and worked with him, and in short, was as good an engineer as his father. Little did we imagine then that a time was so near at hand when these words might be important to record as evidence that the loss the Brooklyn Bridge scheme has sustained in the death of Mr. Roebling, though great, is not wholly irreparable."

The "us" in question translates to the person of Thomas Kinsella, who had taken up the editorship of the *Eagle* at the age of just twenty-nine, in 1861, and served until his death, a year after the bridge whose construction he so helped to promote was completed, in 1884. Like John Roebling, Kinsella had been born across the ocean, in County Wexford, though to all intents and purposes he was, like Roebling's son Washington, an American through and through. He recalled nothing of Ireland, only that when he arrived in the new world as an orphan boy "New York Bay opened before him like a vision of heaven."

Kinsella and the *Eagle* were quite correct in noting John Roebling's complete confidence in his eldest son; and they were wise to reassure the reading public—including, of course, the project's financiers—that this unprecedented structure would rise to completion. John Roebling, the *Eagle* said, "lived not to see his work finished, or even actually begun. It exists, happily, complete in his plans and drawings. Not only the outlines, but the details and dimensions, the working plans and quantities, have been calculated and completed."

But while Washington Roebling was, most likely, grateful for the *Eagle*'s support, he knew that John Roebling's plans were very far from complete. This would be Washington's bridge.

2

—

# "The finest place in the world"

"THE CIRCUMSTANCES WHICH SURROUND the life of a boy leave their marks which last through manhood into old age," Washington Roebling wrote. He was writing about his father—and how the Saxony of his father's boyhood had been ravaged by the French armies of Napoleon. The elder Roebling had been raised to hate old Bonaparte. But while in so many ways Washington Roebling was a different kind of man altogether than his father, the adage holds true. Washington, a young man in charge of the greatest engineering project the world had ever seen, was very much what his boyhood had made him.

When we think of "the frontier" in the history of the United States, we think mainly of the west. And yet in America's earliest days all was the frontier. The town of Saxonburg, in Butler County, Pennsylvania, had existed for a mere five years by the time Washington Roebling was born there in 1837. It had been built by the hard labor of a group of German immigrants on what had been—to European eyes—virgin land. "One hundred years ago the territory lying north and west of the Allegheny river was a wilderness," wrote a historian of the county in 1895, "inhabited principally by wild beasts and Indians. The solitude of nature was yet unbroken by the advancing tide of civilization, and the wisest statesman did not dream of the wonderful changes which the progress of a century has produced."

Before Europeans arrived, the land had been peopled by the Seneca, the Delaware, the Shawanse, the Muncey. The French were the pioneers of western Pennsylvania. In 1749 the governor of what was then New France, the Marquis de La Galissonière, organized an expedition to the region under the command of Captain Pierre Joseph Céloron, Sieur de Blainville, "a fearless and energetic officer." They came down through Lake Erie and Chautauqua Lake and headed for the Allegheny River, taking possession of the land by burying plates of lead in the ground as evidence of ownership. Buried slabs of lead or no, the English were not best pleased by the French claim; conflict over the territory would lead to the French and Indian War—and the final expulsion of the French from the territory in 1759.* The native people, however, didn't cede their land without reluctance—or violence. Just forty years before the planting of Saxonburg, a young woman of Butler County, Massy Harbison, was abducted by native people; two of her children were killed—her five-year-old son murdered and scalped before her eyes—and she was driven across country with a babe in arms, her own life saved only, it seems, because one of her attackers "claimed her as his squaw." It was nearly a week before she was able to escape with her baby.

But by the time of Washington Roebling's birth, Saxonburg was beginning to be a neat little town—just as it remains today. The place was orderly, Washington Roebling would say, "in true German style." "One broad main street running exactly east and west, flanked by lots which were from 100 to 200 feet or more in width, and ran back to Water Street, almost half a mile, so that each man had a little farm to himself as is the custom in Germany. Water Street, so called, ran parallel to Main Street, and was considered the poorer quarter. Much of the land was still virgin forest; mostly black oak . . . Deer and bear were still met with and much smaller game. As late as 1845 a black bear walked down Main Street—he got away." The land on which it stood had been purchased by John Roebling in 1831. He and his brother Carl had made the decision to emigrate to America after a group of German-American citizens "who had spent several prosperous years in the United States" had returned to the land of their birth to spread the gospel of the New World; the pair were convinced, and with a group of fellow colonists sailed for America on May 11, 1831, aboard an American ship, the *August Edward.*

---

*One of the first attempts to oust the French was made in 1753, at the behest of Governor Dinwiddie of Virginia: in charge of the English troop was a promising young major named George Washington.

John Roebling would never see Germany again. The brothers arrived in Philadelphia on August 6 and began to consider where best to set down roots. When Washington came to tell the story of his father's arrival in America, many years later, he had the great good fortune to come into a cache of letters he had previously never seen. Written in German, and bound together into a collection, the letters were addressed to an old friend in Germany who later brought them to the United States; they revealed John Roebling's determination to build a wholly new life. "He looks at everything with the eye of a man who is going to be a farmer," Washington writes. "You would never dream that he had been educated as an engineer." He weighed up the possibilities offered by America's vastness with what seems like a practiced eye—his youth being no impediment, just as it would not be to his son in 1869. "Considering that my father was only 27 years old then he understood the situation very well," Washington wrote. "He didn't like the idea of a farmer trading his products to a country storekeeper for goods, the latter making all the profits . . . Ohio, Ind[iana] & Ill[inois] are condemned unequivocally as unhealthy & full of miasma . . . Missouri he thinks the worst state of all. New-Orleans is the natural market on account of Water transportation—but New Orleans can't take it all so the surplus goes to New York by sail and by that time freights have eaten up profits . . . The merits of Western P[ennsylvania] are well set forth."

And so Butler County was settled on for the home of this "Dutch" plantation—in those days Dutch, from *Deutsch*, being the adjective applied to German natives and their kin. Mrs. Sarah Collins, a widow, owned a great tract of land, nearly twenty thousand acres, which she was anxious to sell a part of for cash. "The lands were inspected, enthused over and finally purchased, and the future site of the great city of Saxonburg, my birthplace was in his possession." John Roebling saw, Washington said, the beginnings of a kind of Utopia; there would be not only farming but the life of the mind, with "a modern academy which shall rival the academic groves of ancient Athens, with a Mühlhausen Socrates as director." Saxonburg would never be a great city; but even today, with the neat frame houses built by John Roebling and his fellow colonists still standing, it remains a monument to his industry and vision. The brothers bought 1,600 acres from the good widow, Washington reports, "at $1.75 per acre. One thousand dollars down and $750 in a years [sic] time—with the refusal of 2000 acres more at the same price and 3000 acres more at $1.00 per acre." Johann and Carl first settled at a farm not far from what would become the center of town; it served as their base as land was settled and homes were built.

"I am constantly amused at his enthusiasm on the land purchase and the commencement of farming," Washington wrote of reading his father's letters home. "No doubt it was excellent training—he acquired the language and became thoroughly selfreliant [sic]—but later on he thought it a mistake after all." Several attempts were made to get rich off of the Pennsylvania soil. Sheep farming was the first; friends back in Mühlhausen were enjoined to bring over German shepherd dogs with them when they made the voyage to America. Washington wrote of his father's disappointment in a vivid present tense. "Goes to Economy [Pennsylvania] & buys sheep—Finds that sheep bit off the grass close to the ground and spoil the pasture for cattle and sees to his horror that American sheep will roam wild in the woods and won't mind a Dutch shepherd." Furthermore, the sheep would eat the neighbors' crops, leading to no small argument. The raising of rape for oil was tried; sunflowers, too—and then even silkworms. "One of my earliest recollections is the plantation of several acres of young Mulberries (*Morus multicaulis*)—where I had to pick leaves and carry them up stairs to the silkworm room—where they ate the leaves and finally spun their cocoons. With what interest I used to catch the silk worm moth, pin him to a card where it deposited its eggs! The cocoons were roasted in our bake oven—then mother and aunt pulled the silk off the cocoons and after working all week, accumulated perhaps a pound or two—at any rate my father took the first premium for silk at the Butler Co. fair and that ended it, except that mulberry sprouts came up for years afterwards."

In 1836 John Roebling married Johanna Herting, whose father, Ernest, had been a well-to-do tailor in Mühlhausen before emigrating to Saxonburg. The boy was close to his mother's family: "My grandfathers [sic] house was always my favorite abode and haven of refuge," he wrote. "I could write pages of ravishing reminiscence about my grandfathers [sic] humble abode and what real enjoyment I had there. He loved to get off a good joke and indulged in the quaint sayings of the Harz mountaineers—There was so much to attract a boy—His mangy cur Molly—the cat Buckel [sic] that always sat on his shoulder at meals—the Antwerp raspberries behind the house—Early oxheart cherries—black, red & white currants—delicious rye bread smeared with thick layers of cream *butter* & honey—Early & late apples shaped like a pear which kept until spring—white and red mulberries—the coolest spring in Saxonburg—a great hop field and the frolic when picking time came—the . . . excitement when the white muley cow fell into the cellar through the trap door in the hall way, and how the neighbors worked to get the beast out . . ."

Washington recalled how the walls of one room in his grandfather's house were papered with newspaper brought from Germany—there was a serial story printed on the pages, but since they had been hung any which way "I had to stand on my head to read the tale." In the garret of the house were the boxes and crates in which the family had ferried their possessions over from Bremen: "the lids were adorned with lithographs of stirring battle scenes such as Blücher at the battle of Katzbach—The battle of Navarino—Waterloo etc—I could look at these for hours—His tailoring always interested me—He could cut out a suit without measuring—an immense mystery to me . . ."

Washington Roebling's intention in sitting down at his desk in 1893 was to write a biography of his father. And yet in a strange echo of the manner in which the identities of the two men would become confused in the public mind following John Roebling's death, Washington's own life intruded upon his task again and again. He meant to set down the story of his father's development of the wire rope that would found the Roebling fortune, and yet he was drawn back into his own childhood, calling up a list of those who had settled in Saxonburg along with and after the Roebling brothers. "The people my father had hoped would come there, came not—in their place came men of a humbler walk of life. Mechanics, people with a trade, small farmers, all hard workers and much better adapted for a life there than more educated people." Washington, as a boy, knew the place inside out. After 1841, when his father began to give up farming for manufacture, a workforce was required—and it was the Roebling boy's job to gather that workforce. "In the process of ropemaking, seven strands are first manufactured, by a regular force working daily—But when the strands are laid up into the large rope a force of 20 men was required for one or two days; this force had to be summoned from all the neighborhood and I was the little messenger who did the running. As the men were only too glad to get an occasional job with cash pay and plenty to eat, the little harbinger was very welcome. In that way I became intimate with every family and knew every path and byway for miles around."

All those years later his recollection of the German men and women who peopled Saxonburg in his youth still took a geographical shape from street to street, as if he were re-running in his mind routes of his boyhood. "Saxonburg was full of idiosyncracies [sic] of character, the result of maintaining German customs in a strange land," he wrote; and indeed, the society described is wholly and completely "Dutch." There is Ferdinand Bäer, his father's great friend, whose wool-carding business was a source of delight because "wool bags are soft and nice to sleep on." There is a tailor called Haase from Leipzig; and a store owned by the Vogeley family, from Cassel, in Hessen. Kunze, a weaver, lived in a comfortable log house. Bernigau was a cabinetmaker, "excelling in coffins," another of his father's great friends. "The vogue was still German," Washington recalled. "Lager beer had not arrived; ale was not to the German taste; wine was too dear; hence Monongahela rye-whiskey became the staff of life. Entertainments at home, small parties and dances were frequently held. Bernigau played the violin; Wickenhagen the violincello; Neher the cornet; Roebling the flute and clavier." Aderhold had been a baker in Germany; in Saxonburg he was the "host and innkeeper" who also kept a very jolly little theater—lest anyone think Saxonburg life lacked culture. A great favorite, apparently, was Schiller's *Die Räuber* (*The Robbers*), a tale of rivalry between two brothers that caused a sensation when it was first performed in 1782: "Vice is here exposed in its innermost workings," the playwright said of the drama. Perhaps unsurprisingly, Washington noted that the version used was "much expurgated." The Saxonburg impresario, he notes a few paragraphs down, eventually "fell dead while shoving a loaf of bread into the oven." (When he was in his late fifties Washington would return to Schiller, especially *The Robbers*. He was enraptured again, but in a different manner: "I read it now from a different standpoint than in my youth," he wrote to his son, John. "Then it was the plot, the action, the events that fascinated me, now it is the whole motif— I put myself in his place and think what I would say next.") He himself took up the violin, and his love of music never left him.

There were eagles in those parts, then; and passenger pigeons, too, extinct before many more years had passed. For a boy the wilderness was a cornucopia, full of "wild gooseberries, plums, hazelnuts, raspberries, and all the nuts that boys delight in." Toward the very end of his life, he would write to his son of Saxonburg that: "To me as a child it was the finest place in the world—after I went to school in Pittsburgh I always went home with the greatest delight—The village in reality was not as bad as it is painted, and the people were jovial and contented, having preserved many of their German customs . . . Those early recollections are as vivid in my mind today as if they had happened yesterday—I have kept away on purpose so as not to have the early impressions disturbed."

THE TOWN THAT John Roebling had created in the wilderness couldn't hold his interest for long. Sheep, sunflowers, and silkworms: he had attempted to impose his will upon the land, but his ambition and energy needed a broader, more worldly canvas than a rural community in western Pennsylvania.

It's no wonder he had not stayed in the old country. By the time Johann August (he would anglicize his name to John Augustus in America) was born there on June 12, 1806, the years of Mühlhausen's glory had long since vanished. Today there are some 35,000 inhabitants in the town, smack in the middle of modern Germany, in the federal state of Thuringia; pleasingly it is twinned with Saxonburg, Pennsylvania, these days. Fortified around the tenth century, and a significant place in the centuries before the Reformation, by the eighteenth century it was a sleepy little place. John's father, Christoph Polycarpus Roebling, was a tobacco merchant, "content to pass his days in the familiar surroundings of his native town, enjoying its excellent beer and withal an inveterate smoker of his own tobacco." Washington's interpretation of the character of the grandfather he never met was that he was simply "a very peculiar man, what nowadays would be called a crank . . . Every evening he would spin the most delightful yarns about his travels in Brazil—Africa & India, whereas in reality he had never left Mühlhausen."

But Johann August's mother, Friederike Dorothea, was cut from a different cloth. By all accounts a forceful and ambitious woman, she determined that her children would find success beyond Mühlhausen's thick stone walls—especially when it came to her youngest boy, the one who most resembled her. Even strangers remarked on his intelligence and his energy, it was said; according to Washington, her plan was clear. "My father's mother had the opposite characteristics from her husband and my father inherited many of her traits," he wrote. "She was of a very positive temperament, had much executive capacity, made everybody work, managed her household, family, the business and her quarter of town besides. It was her ambition that her youngest son should become a great man." His mother arranged for him to study in Erfurt, a town sixty miles away, first under the well-known mathematician Doctor Ephraim Solomon Unger (whose father had been the first Jew accorded the rights of citizenship in Erfurt, just a few years before). This was in 1820; Johann August was not quite fifteen years old. In 1824 he went to Berlin, to the Bauakademie, or Berlin Building Academy (later assimilated into the Polytechnic Institute in Berlin). He studied "the Construction of Roads, Bridges, Locks and Canals," among many other subjects; the course included the study of suspension bridges. Teaching—as it was when Washington would come to study—consisted almost entirely of lectures, hours and hours of lectures; John Roebling's notes from Berlin make up ten bound volumes. "To one who has not been through such a course, the amount of laborious application seems incredible," Washington would write, "and simply shows how much can be accomplished by steady work from early morn till midnight. Thirty years later I went through the same task myself at Troy and know what it means. I do not consider the lecture systems a success. Good text books are better and are of use later on in life. The notes are always taken in haste and written in a crabbed hand so that later on you cannot read your own handwriting. So much time is taken up in writing that the mind gets tired and cannot properly understand the subject. That was my father's testimony and I second it."

It was always said in the family that, while in Berlin, he was the favorite pupil of the philosopher Georg Wilhelm Hegel—who was chair of philosophy at the university at that time. There is not much evidence for this, though it does seem that Roebling attended lectures given by the great German Idealist, and certainly what might be called the rational theology that characterized Hegel's work was a great influence on Roebling: for all his practicality, throughout his life he would remain close to what he thought of as a spirit world. Later in his life he would fill page after page with his metaphysical investigations of the conjunction between matter and the ideal condition of that matter; there existed, in Roebling's mind, an absolute perfection external yet available to the human world. A family friend would drily call John Roebling's metaphysics "his dissipation," remarking that "his son has a manuscript volume of thousands of pages written by his father, called 'Roebling's Theory of the Universe.' I have not read this book— Heaven forbid that I should ever be asked to do so . . ." And indeed, Washington confirmed that "the manuscript of this ponderous work is in my possession and numbers thousands of pages . . . The love for the abstract thus early implanted, grew with age and finally became his master passion." In the manuscript copy of Washington's memoir, "religion" has been crossed out, replaced with "master passion."

And yet, as his son remarked, "Philosophy was only the shadowy side of his nature. The practical side was always dominant. The time that most people waste in reading works of fiction, newspapers, at cards or amusements, he spent in earnest thought—he never permitted card playing at home & forbid the newspaper craze."*

That practical side enabled him to earn his living: after Berlin he worked for the Prussian government, building roads in Westphalia. Around this time, he first saw a suspension bridge—a chain bridge that had been built over the river Regnitz, at Bamberg. "Of course he sketched it examined it & made it the subject of a thesis—It was the memory of this structure which turned his attentions later on in that direction." In 1828, he proposed the building of a suspension bridge over the Ruhr at Freienohl, with the main cable composed either of linked iron rods or strands of wire laid parallel. The planning department for the Ministry of the Interior was impressed—but not impressed enough. "The project for this suspension bridge must be rejected, because the chains and especially the rings that connect the chain elements, as well as the wire cables and the anchoring of the chains in the earth are surely too weak. As well, the non-perpendicular form of the suspenders . . . do not have our full approval. Nevertheless, the industriousness with which the engineer Roebling has completed his suspension bridge project deserves our

---

*And it was a craze: in the latter half of the nineteenth century, there were hundreds of newspapers in New York and Brooklyn alone: twenty-nine morning daily papers were published, plus more than 250 weeklies and well over 150 with issues every month.

full recognition, and we wish to convey to him our encouragement." Those "non-perpendicular suspenders"—or diagonal stays—can be seen today on the Brooklyn Bridge.

But prosperity was hard to achieve for a young, ambitious man like John Roebling in those days. The July Revolution in France in 1830—when the fiercely royalist Charles X was ousted in favor of the Duke d'Orléans, Louis-Philippe, called "the citizen king"—caused unrest throughout Europe and led the Prussian and Austrian governments to restrict freedom of the press and what could be taught in universities. Between 1831 and 1840 more than 150,000 people left what we now know as Germany for the United States— fleeing not just political unrest but the economic stagnation such turmoil brings with it. Washington put the choice that his father had faced in stark terms: there could be no real opportunity for a man like John Roebling in Prussia. "Being now an engineer the question of his future life confronted him. Should he remain in the fatherland, tied down to the strict rules of semi official life—a perpetual subordinate with no opportunity to gratify a laudable ambition or to follow the bent of his own genius." The memory of that letter of rejection remained always with the elder Roebling. "My father often told me when referring to the Pittsburg Suspension aquaduct [sic] that he never would have been allowed to build such a structure in Prussia. The dignity and pride of the supervising engineer would have ground down the ambitious attempt of the young engineer in even proposing such a structure which had no precedent . . . America was the goal which all young men aimed to reach then as well as now." Indeed, John Roebling never graduated from the Bauakademie: while he was one of only six out of 126 students in his year to win an award for the diligence of his studies, he never took the final examination that would have allowed him to be called Baumeister, or master builder. The exam itself, which took days to complete, was as good an example as any of the oppressive quality of Prussian bureaucracy; and perhaps the young man's sights were already set on a very different kind of life. His son would later remark, too, how violently John Roebling would react whenever his opinion was challenged; perhaps the knowledge that he had not completed his course stayed with him even in his new, American existence.

So Johann August and his brother Carl set off for America—and a remarkable insight into the mind of Washington Roebling's father is provided by the journal he kept of his voyage across the ocean. "Diary of My Journey from Mühlhausen in Thuringia via Bremen to the United States of North America in the Year 1831" is a document of 150 pages.* His excitement—and his scientific bent—is palpable when the *August Edward* finally left Europe behind and came into the deep swell of the open sea. "We have now left the Channel and have entrusted ourselves to the ocean, which is recognized by its more darkly shining water and by its heavier waves," he wrote on May 29. "The water appears almost entirely black, but if one looks at it in a glass, it is as light, transparent and clean as the finest crystal, and seems to offer a splendid cordial to the thirsty. But woe to him, who ventures to swallow or even to taste! Already in washing the face one takes care that the lips are not rubbed too much with it. One cannot imagine a more abominable taste than that which seawater has . . ." He described the winds, and how they blow, and marveled at the ocean currents they follow. "If the magic of the magnet did not guide us, who would show us the way?" he wrote on June 12. "Who would entrust himself to this immeasurable body of water, now smooth as a mirror, now a boisterous, roaring mass of lofty billows, whose depths are not to be sounded in several thousand feet? And yet this waste of water is nothing but life; and life may exist down to its furthest depths . . . A single seal pushed its way snorting by our ship yesterday; also we saw today, but at a greater distance, a whale spouting streams of water into the air."

Even as he blended scientific observation with poetic reflection, his practical mind was never at rest. "Up to now," he observed, "there has been no system in the distribution of water, something which ought to be established and strictly observed." He knew how things should be run. He and his brother Carl were not the tired, poor emigrants that Emma Lazarus hymned in the poem that—written the year the Brooklyn Bridge opened—would be engraved at the base of the Statue of Liberty. The brothers—traveling first class—were berthed in a comfortable cabin, well lit and spacious, and food was plentiful and healthy. (There were, to be sure, certain things one had to watch out for at sea: "For constipation a small dose of rhubarb is to be recommended, when it is very severe. The frequent use of rhubarb can only be harmful and weaken the stomach. On the other hand, brisk motion on the deck, frequent use of plums, beer and *bierkaltschale* [a cold beverage made

---

*It was privately printed, in German, and translated into English in 1931.

of beer, with grated bread, sugar, lemon or fruit], commonly induce a good movement.") Those in steerage were not so lucky—so Roebling did his best to ameliorate the situation, insisting to the captain that a roof be constructed over the steerage quarters so the hatches could be left open when it rained—thus allowing in a bit of fresh sea air. There was no proper toilet aboard for those in steerage: Roebling saw to it that one was built. Even though he was only twenty-six years old, the force of his personality was felt. Already obsessed with his own health and that of others, he felt that the life aboard a well-run ship could work wonders for her passengers: he noted that the skin condition of a fellow passenger disappeared after a few weeks at sea.

Flying fish skip along beside the ship, their fins glinting in the sun; they pass by a schooner that the captain fears may be the home of pirates; they keep their distance. The young man considered just what he was undertaking to do. "The lonely, isolated stay on the ship during a long sea-voyage from one part of the world to another has in a certain respect an entirely well-founded utility," he wrote on June 27, after six weeks aboard. "Thus, as we see the last of the shores of Europe vanish from our sight, we separate ourselves at a single stroke from the Old World that is known to us, and to which by our birth, by our education, by the entire youth that we have lived there, by the nation to which we belong, by its history and our own, by our relatives, and by many other circumstances, which have a bearing on our life, we are bound and continue to be bound, as long as we draw breath." He would make an entirely new beginning, knowing it would not be easy. "Much [in the New World] may be exaggerated and set forth in too bright a light, yet everything is also very simple and natural, and hence well-founded also. The greatest art will indeed consist in shaking off the old European prejudices and fitting one's self to the New World; and every people demands of a foreigner, who wants to settle permanently amongst it, that he will assimilate himself to his environment as soon as possible."

By the end of July Roebling, like all the passengers, was longing for the sight of land. On July 30 the distant New Jersey coast was spotted, and by August 5 they were moving through Delaware Bay—and Roebling, unusually for a man of his time, could not help but consider what its earliest inhabitants had lost. "Three hundred years must have passed since the sheltered loneliness of these wild surroundings was interrupted by the all-disturbing European," he wrote.

This extraordinary journal shows a restless, unsleeping intelligence—one that, ultimately, couldn't possibly be satisfied by life in Saxonburg, the very definition of a one-horse town. Following his unsatisfactory agricultural initiatives and his marriage to Johanna, in 1836 John began to look for employment as an engineer, writing to the legislators in Harrisburg—Pennsylvania's capital—offering his services to the state. The Erie Canal had opened, after eight years of construction, in 1825. This 363-mile-long channel, with its 83 lift locks, cut through fields, swamps, and forests and ran over cliffs and rivers, connecting the port of New York with the Great Lakes and so the whole interior of the country—and served as a school for a nation's engineers. More than that, however, it cut the cost—in money and time—of passenger travel and freight enormously. What had been an uncomfortable, two-week-long stagecoach journey for passengers could be accomplished in a few days aboard a comfortable canal boat; freight costs fell by 90 percent. But the canal threatened the prosperity of Philadelphia, so the state of Pennsylvania began to construct its own system of canals connecting up to the Ohio River and to the rich country that lay beyond. Travel by canal boat had none of the romance of the steamboat. "There is something mysterious, even awful, in the power of steam," wrote one traveler, Harriet Beecher Stowe, a little over a decade before *Uncle Tom's Cabin* made her the most famous novelist in America. "But in a canal boat there is no power, no mystery, no danger . . . one sees clearly all there is in the case—a horse, a rope, and a muddy strip of water—and that is all."

It was on these unglamorous canals that John Roebling restarted his professional career, building dams and locks along the Sandy and Beaver Canal, which ran from Ohio to Pennsylvania, before surveying a route for the railroad over the Allegheny mountains. Before long he was appointed as principal assistant to the chief engineer of the survey, Charles L. Schlatter. His salary was four dollars a day. Around this time John Roebling made friends with another young engineer, called Washington Gill, after whom John's firstborn would be named. "I have always understood that my father named me 'Washington' after him, and not directly after the father of his country," his son said. Washington was born the same year John Roebling became an American citizen, 1837.

While on survey work John Roebling had the technological encounter that would change his life and make his fortune. He had to travel on the portage railway, a system of planes—like long slides—built to haul freight along the canals and over the mountains. Washington Roebling takes up the story. "These surveys caused my father to make many journeys over the Allegheny mountains across the Portage R. R. which extended from Hollidaysburg to Johnstown, forming a series of 20 or more inclined planes. Every Canal boat was built in 3 or 4 Sections. In order to cross the Mountain these boats were dismembered, each section placed on a car and hauled over to Johnstown or vice versa and then put together again in the water." Another traveler on the portage railway—in 1842—was a famous visitor to the young United States, Charles Dickens; he described the experience vividly in *American Notes.* "On Sunday we arrived at the foot of the mountain which is crossed by a railway. There are ten inclined planes, five *a*scending and five *de*scending; the carriages are dragged up the former and let slowly down the latter by means of stationary engines; the comparatively level space between being traversed, sometimes by horses and sometimes by engine power as the case demands. Occasionally the rails are laid upon the extreme verge of a giddy precipice, and looking from the carriage window the traveller gazed sheer down without a stone or scrap of fence between, into the mountain depths below . . ."

The problem with the portage railway was the rope used to haul the cars up and down the planes. Great hawsers made from Kentucky hemp were used, some of them over a mile in length. They could cost three thousand dollars apiece, a fortune at the time—and they were prone to fraying and breaking; inevitably the resulting damage to loads and goods—and life—was costly and dangerous, too.

John Roebling believed that rope might be made of something other than Kentucky hemp. "He knew that wire rope had been made in Europe some years before," Washington wrote, "first at Clausthal in the Harz by Herr Albert, Supt. of the mines in 1832." Wilhelm Albert was a German engineer working in the silver mines of the Harz Mountains, which rise from the North German plain. His were simple inventions, to be used vertically in the mine hoists: three lengths of wrought-iron wire were twisted around each other by hand; then three or four of these strands would be twisted together to make a rope—the method was no different than that for making ropes of hemp. Around 1840 John proposed a trial of wire rope to the proprietor of one of the smaller planes of the portage railroad, at Johnstown, Pennsylvania; he would make a rope four hundred feet in length and three quarters of an inch thick. He had, as his son wrote, "Nothing but the rough meadow to work on"—an ordinary ropewalk on the ground at Saxonburg. This first rope, made of wires laid parallel and wrapped with an outer skin of wrapped wire, was "a complete failure": the wrapping wire was tightened by an ordinary serving mallet, such as was used with ordinary hemp ropes—"Of course as soon as the serving broke the whole rope went to pieces."

John Roebling, however, was undaunted, even though he had already drafted a patent application for his idea, which would now have to be completely revised; "the experience gained by one failure is often of the greatest value in pointing the way to the right path," as Washington wrote. "About the particular method of manufacture he knew nothing. He had everything to learn, everything to evolve from his own brain." And so he did: a new and more sophisticated method was devised, with seven "strands" arranged (or "laid up") into a rope with a central core of wire supporting the outer strands. The strands, each composed of nineteen individual wires, were parallel to each other, rather than twisted like ordinary rope. He received a contract for this new rope on April 5, 1842; made in three pieces and spliced together on site, this rope was a great success—and remained in use for a full five years. U.S. Patent No. 2,270, "Method of and machine for making wire ropes," is in John Roebling's name, and was granted in July that same year. By the beginning of 1844 he proposed to supply ten inclined planes of the portage railroad with ropes over a period of five years, for the sum of seventeen thousand dollars a year. Immediately and entirely confident of the enduring success of his invention, he stated that he was happy to supply "ample security for the fulfillment of the contract on my part." When he had written to the patent office in 1841, he received a letter back that expressed some surprise at his assurance: "You state that you are proposing to put up an experimental cable on the Portage R. R. Would it not be well for you to defer taking a patent until this essay has been made?" Roebling's iron determination enabled him to overcome, as Washington wrote, "the reluctance with which the powers that were came over to his views [and] the bitter hostility of the hemp rope interest. But he carried it through with undaunted energy and thus laid the foundation for his fortune." His life as a surveyor was over.

Washington beat the bounds of the town, calling the men from farm and field to this new, industrial work. "They commenced work in summer at 5 a.m.—came to breakfast at 6.30—at 10 a.m. I carried down a basketful of ryebread and whiskey—at 12 I blew the horn and they came to dinner—at 4 p.m. I carried down a basketful of ryebread with <u>butter</u> and more whiskey and at 7 p.m. They came up for supper . . . This the way Germans work in Germany." An ordinary rope would take a day to lay up, Washington wrote; a longer one would take a day and a half. "When rope making commenced, as many as 20 people were daily fed at our house—The wonderful American stove had not yet appeared in Saxonburg and my poor mother was ground to the dust in trying to cook for so many on a primitive stone hearth, with wood fire and one pot hook and iron Kettle—Mühlhausen savagery."

It is as good an image as any for the America in which Washington Roebling was growing up: men making one of the great industrial advances of the age while a woman cooked for them without even a stove, all in a land that was still half-wilderness. His first ropes were made on a "rough meadow," using a force of local farmers gathered from the town by his young son; he built a "simple twisting machine" at one end of the meadow to twist the iron wires, which had been ordered from a local mill; to build his business "he took stock in pay for many of the ropes and this proved a profitable investment."

John Roebling was trying to build a business in a nation that had yet to decide whether economic growth and stability would come from individual states or from the capital. The dissolution of the Second Bank of the United States by President Andrew Jackson, and his mandate, in the Specie Circular of 1836, that all government land be paid for in either gold or silver, led to the Panic of 1837, which threw the country into a depression that would last until the 1840s. Jackson believed that a nation's financial strength should not be concentrated into a single institution. When the Second Bank lost its charter, the nation's 850 banks were each able to issue their own currency; but ten months later banks refused to redeem their notes in specie, bringing commerce to a shuddering halt. "The Administration is in a most awkward predicament," noted the London *Times* on June 13, 1837. "These troubles have come upon them like a clap of thunder, notwithstanding they have been warned of their approach by the Opposition for two years past."

John Roebling, however, weathered the storm. In 1837, his son said that was mainly because "he had little money to lose." That said, "he gained financial experience which enabled him to weather the next panic twenty

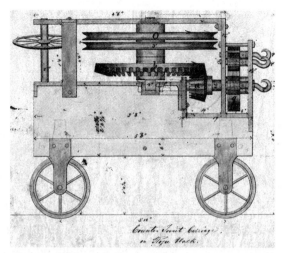

*John A. Roebling's drawing of a counter-twist carriage
on a rope walk, 1848*

years later, without losses of any account." His habits were those of caution and prudence; the foundations upon which he built his great fortune still stand as a lesson today. He always operated, Washington recalled, "with saved capital, not borrowed capital, the secret of most success—when a small capital has been got together by pinching economy and a process of saving akin to penury, it forms habits which last through life and cannot be laid aside when they are no longer necessary."

Washington's words appear approving: but they also hint at what was inflicted on John Roebling's family as his business grew. "Pinching economy, a process of saving akin to penury"—life in the Roebling household could be very grim indeed, and left Washington Roebling with scars he would carry to the end of his days.

# *"Something of the tiger in him"*

AT THE TIME OF his death in 1869, not only John Roebling's genius was remarked on, but his kindness and generosity, too. Hundreds of his workers
had gathered around his funeral cortege; he had left substantial bequests in
his will, not only to the Industrial Home Association for destitute children of
Trenton and the Marlbury Orphans' Farm School near Mount Vernon, but
also to the Widows' and Single Women's Home of Trenton and another to
the Orphan's Farm School at Zelienople, Pennsylvania—not to mention the
Pittsburgh Infirmary. Some of those bequests were for as much as thirty thousand dollars, a huge sum in those days, equivalent, at the most conservative
estimate, to a half million dollars today. "In his charities he was exceedingly
liberal, and what he bestowed was given without ostentation," it was said
of him. "A plea for aid from the poor and humble, although his time was
precious, always received attention, and a share of his bounty. His social and
domestic relations were most agreeable. In conversation he was earnest, in-
structive, and exceedingly entertaining. His sympathies for the working man
were large, and he lost no opportunity to promote harmony and good feeling between the employer and the employed." This is no falsehood; but the
difference between the public and the private man is striking. In letters and
public recollections, his son remained respectful of his memory—and while
it was clear that John Roebling was a stern father, this was nothing out of the
ordinary, nor did his family seem to perceive it as such. Only in Washington's
memoir of his father can his son's true feelings—and the father's private
actions—be observed. His account is a shocking revelation.

Even in his later years, with his father long dead, Washington was hesitant
to speak his mind. It seems that his plan was simply to set down for posterity

the story of his father's coming to America, the growth of the community he built, his success as an engineer and a businessman. And yet the truth of the matter could not be avoided—certainly not if he was to do justice to his beloved mother. "In 1836 my father married my mother, Johanna Herting a handsome young woman of 18 or 19, of an amiable and gentle disposition, entirely too good for him," Washington writes. The run-on sentence seems freighted with feeling. "Her life was a most unhappy one—she died at the early age of 45, a martyr to brutality." And here it is worth remarking that, in the typescript version of the memoir which is held at Rutgers University, someone—we will never know who, only that they were keen to protect, however feebly, John Roebling's reputation—has deleted Washington's carefully chosen word, *brutality*, and substituted the words *hard treatment*. There is a difference. "I have debated with myself whether it is proper for me to put into writing or allude in any way to the dark sides of my fathers [sic] character—There is a superstition that only good should be spoken of the dead—How that arose I cannot say—We all have to die and render an account of ourselves."

Washington Roebling was fifty-five years old when he wrote those words, ensconced in his ornate new mansion in Trenton, with a fine study in which to work; but he knew what it was to consider one's legacy. Some years thence John Roebling's statue would be erected in Trenton's Cadwalader Park; at its dedication, in 1908, the speaker of the day—Henry Dodge Estabrook, a lawyer and orator with "an attractive style and very forceful delivery"— called him "a man of iron. Iron was in his blood, and sometimes entered his very soul; a man of iron with the virtues of iron and peccancies of iron to his account." Estabrook had seen what Washington had written. The son wrote bluntly of the father: "His domestic life can be summed up in a few words—domineering tyranny only varied by outbursts of uncontrollable ferocity—His wife and children stood in constant fear of him and trembled in his presence."

When John and Carl Roebling first settled Saxonburg they owned nearly all of the village, selling it off in parcels to other incomers as they went along. "Selling land in small parcels, giving deeds, mortgages, making agreements, contracts, etc., made my father very familiar with legal papers and with the common law of the country, which was of great use in later life." The Roebling house was built flush to the street, with a hallway between two rooms on the ground floor, one a sitting room and the other his father's office, with kitchen, storeroom, and dining room off the back. Bedrooms were upstairs—more often than not occupied with German waifs and strays given houseroom by the benevolence of John Roebling. "All his lifetime my father was burdened with people of this sort—While he would treat his own family with the utmost rigor and a penurious parsimony, yet he always found room for these people." But that benevolence was part of his public face. His father's office Washington called "the sacred room"—which was also "an execution room where I was beaten nigh unto death—None of us children ever dared enter it."

The idyllic image of a frontier boyhood vanishes as his father's true nature is revealed. At suppertime his wife and children, by Washington's account, sat in silent terror. "The table was a gloomy feast—not a word was spoken or sound heard. The cravings of hunger satisfied, each poor wretch would slink away, fearing to hear some harsh reproof before he escaped from the room—

"When anyone was sent for to his room, it was with fear and trembling, as it was a grave question whether he would return alive from the execution chamber . . . As he grew older he realized what he had done and tried to control himself but it was impossible."

No wonder young Washington thought of his grandfather Herting's house as a "haven of refuge." For what Washington described was a violence that was not an occasional outburst but an onslaught, a constant danger. To the world at large John Roebling appeared the very definition of the rational man, whose endeavors to tame the wilderness of America first by farming and then by construction might be seen as perfect expressions of enlightened thinking. But he was Janus-faced. Having resolved to tell the truth as he lived it as a boy, Washington is unsparing.

"There was something of the tiger in him—the sight of blood following the strokes of the raw hide, brought on fits of ungovernable fury—To fell my mother with a blow of the fist was nothing uncommon." Once again some person unknown has removed this last sentence entirely from the typescript version: it can only be found in the holograph, handwritten version of the memoir, in Washington's tiny, penciled hand. An interesting editorial decision, revealing the notion that it would be more acceptable to beat one's children than to beat one's wife. Yet condemnations of wife-beating were published as early as the seventeenth century. William Heale's *An Apology for Women*, produced in England in 1609, said there is nothing in the law to suggest that a man has any right to strike his wife; it is unnatural, Heale wrote, to do so. The Massachusetts Bay Colony had outlawed wife-beating as early as 1655: "No man shall strike his wife nor any woman her husband on penalty of such fine not exceeding ten pounds for one offense, or such corporal punishment as the County shall determine." In the nineteenth century many states already had statutes against such behavior on their books.

Washington Roebling's tormented account of what he had to witness and experience in his home makes the father's brutality horrifying. Yet Washington preserved a note of wry humor, making what he says all the more dismaying. After describing his mother's treatment, he went on to say that "my good old grandmother"—his mother's mother, called Adelaide—"once saved my life by knocking him down with a fence rail as I lay writhing on the floor in my last agonies—It was a fortunate thing that his engineering engagements kept him away for prolonged periods, otherwise his children would have all died young." The children who spent their earliest years in Saxonburg were Laura, three years younger than Washington; Ferdinand, born in 1842; Elvira, born two years later, and Josephine, born in 1847. All of them, by Washington's account, suffered indiscriminately at their father's hand. "A huge cowhide always hung by the clock, ready to come down without warning," he recalled. "All our ingenuities were bent to one purpose, to keep out of his sight—The result was he brought up a set of sneaks." Washington didn't mince his words when describing the life-long effects this upbringing had on all of John Roebling's children. "Such a bringing up could only humble the pride and crush all manly independence in his children— His 4 sons and daughter now living are simply examples of warped— misanthropic cranks, peculiar, odd and unhappy—Fortunately not one is his father exactly over again—Charles is the nearest," he wrote. Charles, born in Trenton in 1849, just after the family moved there, inherited his father's genius for building and management, and was largely responsible, in later years, for the vast expansion of the John A. Roebling's Sons Company, and the development of the town of Roebling, New Jersey. But he was taciturn to the point of complete silence in company, apparently almost completely incapable of ordinary conversation; "he had a positive genius for reticence," as Hamilton Schuyler put it. No one—no creature—escaped John Roebling's terrible wrath. "These outbursts of violence were not confined to his family—he preferred to trample a rattlesnake to death than waste powder and shot on it—a disobedient dog or horse had the alternative of life or death presented to it at once," Washington wrote.

Washington's strength of feeling, as he recalled those days, must not be underestimated. It was not merely violence he feared; it was death itself. Fiddle-playing Bernigau—his father's friend the cabinetmaker—"was fond of kidnapping me for a few hours to accompany him on short excursions"; once, however, "this proved the cause of the most terrible beating I ever got in my life." Again it was his grandmother who intervened, his own mother "already helpless by the blows of a brutal monster whose greatest pleasure was to murder helpless little children, and yet the world called him great." When Bernigau died, aged only forty-two, Washington wrote that he felt "great regret"—then crossed that out, substituting "inexpressible grief." Even when allowance is made for the vagaries of memory, the portrait Washington paints overall is a dreadful one.

But he did at least have the consolation of his mother's presence. Johanna Roebling's characteristics, her eldest son wrote, had "some tempering effect" on himself and his siblings; "I venture to say that my mother was the only woman in the U.S. with whom my father could have lived, any other woman would have been murdered or else driven to suicide." Her husband's cruelty was expressed not only in violence, but also in the stern economy by which he insisted his household must be run. "After my father had become a wealthy man, a millionaire, my mother received a weekly allowance of $50 with which she had to support and feed a large family and keep up a big household. Of course she had to slave it in the kitchen and over the wash tub—She never owned jewelry and the one solitary silk dress was looked on as an unpardonable extravagance—No wonder she succumbed early—My features resolve themselves into a sardonic smile when I see my own wife squander $500 a week on herself alone—Was it for her that this miserly parsimony was exercised? A curious world indeed, is this."

Surely, in a small, closely knit community such as Saxonburg—founded by people who mostly came from the same place, who were not strangers to each other, and who would have lived in a tight web of interdependence—not to know what was going on in the Roebling household would have been impossible. But this was the household of the community's patriarch, its leader and founder, a man whom none dared question.

In these circumstances, Washington's efforts to understand his father, to empathize and project himself into the older man's mindset, are extremely moving. He considered the source of his father's miserliness, his unwillingness, finally, to provide for his family in a way that—as soon as he began to get steady work as a surveyor and engineer—would have been perfectly possible; and he seems to forgive it. That meanness, he sensed, "was inherited from both sides, for many generations back—Thüringen of all the German states is one of the poorest—a yield of 15 bushels of wheat to the acre is exceptional, and other crops in proportion . . . You might say that my father sucked in economy with his mother's milk. All the Saxonburgers are the same way—unfortunately as he became wealthier and his circumstances easier his views did not broaden—He early realized that wealth meant power and so he cherished it—His tendency was rather to conceal from his children the fact that he had wealth—How much better it would have been if he had taken his sons into his confidence as they grew up, taught them the wise use of money, how to take care of it, how to invest and how to avoid the snares and pitfalls that surround all young people with inherited wealth." But such a father was not his own. "Well, I suppose he acted according to his convictions and did it for the best—I myself feel poor to the present day—and it grows on me."

The constant threat of physical violence and savage economic strictures would be enough to make any child's life, any family's life, a misery. But these weren't the only methods by which John Roebling tyrannized his household. "There were two professions which were subjects of his especial animosity—namely a respectable intelligent physician and a theologian," Washington recalled, and painted an especially vivid picture of both his father and the unlucky clergymen who came into John Roebling's orbit. "His encounters with the poor clergy were harrowing spectacles—in a few minutes he would be worked up to a fever heat, the spit would fly, his arms gesticulated violently—The amazed opponent became terror stricken, finally seeking safety in flight, even sacrificing his hat or coat umbrella or overshoes," he wrote. "The last encounter of the sort that I witnessed was at Covington, Kentucky," Washington says, referring to the time when father and son were at work on the bridge over the Ohio that, just after the Civil War, was the last John Roebling built before beginning the great work in Brooklyn. The theological target of John Roebling's rage in this case was a not insignificant one—"a fellow called Cramer, a methodist preacher and brother in law of Genl. Grant."

Michael J. Cramer, a graduate of Wesleyan University, married Grant's younger sister during the Civil War; during which time President Lincoln had appointed him hospital chaplain.

Washington didn't say what brought about the argument with Cramer. However, in John Roebling's philosophical writings is evidence of his appreciation for the work of Henry James Sr.—the father of novelist Henry and philosopher and psychologist William—who disdained organized religion as a conduit to true spirituality. "Religion is now become the idol of men's impure devotion," James Sr. wrote, "the one conventional decency which more effectively separates man from God." Cramer, a man of standing, attempted to put up a fight; but it was, apparently, all to no avail. "Although he stood his ground well he was finally routed with great slaughter," Washington records. Since he worked closely with his father on the Covington bridge, he may have witnessed the rout himself. "The Tendencies of his were largely the result of his university life in Berlin," Washington noted. Hegel had something to answer for, then.

Also subject to the "great slaughter" of John Roebling's temper was any physician he encountered. "No epithet was too vile, no treatment too harsh" for such a man. Why? Because he was "an ardent devotee of the water cure"—a treatment that would finally contribute to his premature death.

"When one of his children fell ill, the first impulse was to conceal it from him, because the torture of his water treatment was worse than death," Washington remembered. "If the patient was still able to walk the doctor was visited surreptitiously or else was secretly smuggled into the house, then my poor mother was afraid to present the doctors [sic] bill as it might cost her her life, so the amount was slowly laid aside out of the $50 a week, twenty five and fifty cents at a time." This subterfuge came at high cost to a family already struggling—needlessly—against severe financial restraint. "All this meant so much less to eat—This kind of treatment injured his children for life in many ways—Every household ought to have its family physician, in whom the various members have confidence, who by wise counsel in time can head off serious maladies, which sometimes are as much mental as physical in origin. These priviledges [sic] were denied to us, and yet the poorest of the poor can have them."

The belief that water has not only the ability to cleanse but also to heal is deeply rooted in human nature. "Everywhere in many lands gush forth beneficent waters, here cold, there hot, there both . . . in some places tepid and lukewarm, promising relief to the sick," wrote Pliny the Elder in his *Natural History*, which appeared not long after the birth of Jesus. The well-preserved baths in both Herculaneum and Pompeii offer fine examples of the kind of amenity any proper town would be expected to have, with a series of rooms—for both men and women—graded by temperature and by the facilities they offered. Water has long preoccupied both the sick and the medical profession. "This concern has taken a multitude of forms," according to social and scientific historian Roy Porter. "Physiological interest traditionally centered on water as an 'element,' macrocosmic and microcosmic, constitutive of the balance of fluids in the economy of nature and the human body alike, and drawing upon the religious and mystical cleansing properties of the cleansing 'waters of baptism' and, perhaps, the unconscious associations of amniotic fluids." Water, even more than food, is the first necessity of life; we emerge into life from the waters of our mothers' wombs.

In the nineteenth century water treatment was fashionably mainstream—not at all an aberration. Charles Dickens and his wife, Catherine, took treatment at England's Malvern baths, as did men of science, most notably Charles Darwin, who took his beloved eldest daughter, Anne, there in 1850. "I feel certain that the Water cure is no quackery," he wrote to Joseph Hooker—one of his very closest friends and another giant of nineteenth-century British science. But Anne died at Malvern early in 1851, succumbing to a wasting fever. She was buried there, and such was Darwin's distress, he was unable to attend her funeral. Despite all this, he never lost faith in the water cure.

Neither did John Roebling, though as a follower of the ruthlessly simple teachings of Vincent Priessnitz he had no need of expensive spas—though such things did exist in Pennsylvania even before the American Revolution. The Bristol Chalybeate Baths—chalybeate indicating mineral springs containing iron salts—had been established in Bucks County as early as 1760. But Roebling didn't care for special waters, iron-infused or otherwise. Plain fresh water, and plenty of it, would do the trick—that was the Priessnitz way. "Priessnitz was at that time the idol of the water fiends," Washington recalled drily. Roebling's guru was a simple man, a farmer's son, born in Graefenberg, Silesia, in 1799. Some said that Priessnitz's father had a neighbor who had practiced a "water-cure" on his cattle; it was that which inspired the boy—that, and taking his fate into his own hands. In a detailed account of his methods published in 1842 in English, we learn that one day as a young man during hay-making season he was kicked in the face by a horse; if that wasn't bad enough, when he fell a cart ran over him and broke two of his ribs. A local surgeon announced that he would never be fit to work again; but the boy was determined to cure himself. An early admirer described how he took matters into his own hands: "To effect this, his first care was to replace his ribs"—presumably, jam them back into place—"and this he did by leaning his abdomen with all his might against a table or a chair, and holding his breath so as to swell out his chest. This painful operation was attended with the success he expected; the ribs being thus replaced, he applied wet cloths to the parts affected, drank plentifully of water, ate sparingly, and remained in perfect repose. In ten days he was able to go out, and at the end of a year he was again at his occupation in the fields."

Clearly Priessnitz was just as tough a nut as John Roebling. In 1840 a Dr. Engel of Vienna reported that the system would cure indigestion, jaundice, gout, rheumatism, scrofula, "diseases affecting women, such as hysterics &c.," and "Typhus, putrid and scarlet fevers." From dysentery to cholera to ringworm, from smallpox to cancer ("For this awful complaint, cold water is a certain remedy"), the Priessnitz method had the answer to everything.

Usually the cure began with the patient being swaddled for hours in heavy woolen blankets to induce sweating; this would be followed by a cold bath and a system of "douches"—cold water poured from a great height. There were eye baths, foot baths, half-baths, sitz baths—and the water was not just to be sat in or under. Drinking twelve glasses of water per day was the absolute minimum for the Priessnitz method; twenty to thirty glasses was much more like it. Cold water "clysters"—enemas—were also crucial. No orifice escaped unwatered.

To keep up-to-date, Roebling maintained a correspondence with one Georg Westermann, a publisher in Braunschweig, back in the old country: Westermann was to send every new book on water treatment to Saxonburg. "Many were lost or given away," Washington recalled, "but they still fill a long shelf in my library—Each new luminary claimed that he alone was right and his predecessors all wrong. One would cure typhoid fever with a wet rag up side down—the next man would put it down side up." Not only John Roebling himself and his long-suffering family were made to partake of the water cure; no one who came near him, apparently, escaped. "The worst of it was that he insisted on treating every poor wretch that came within the sphere of his influence—he was packed and douched and steamed, plunged into cold water and hauled out by his heels à la Achilles, sitzbathed, hip bathed, Sprayed, water bagged, fomented and revulsed and God knows what all—If the victim would not respond with a good reaction it was looked on as a personal insult." Among John Roebling's papers are a group of little notebooks keeping careful track of each sitz bath, steam bath, drenching, every cold wet sheet he wrapped himself up in—even what came out of his nose. Of his "chronic catharr [sic] and sliming" he wrote: "My sliming proceeds certainly from the Brain, the slime is the excrement of the Brain . . . that drawing water through the nose has a good effect is evident, because the discharge of slime immediately after it is always much increased."

It might seem funny—except for the effect it had on those who were forced to take part. "How I pitied my poor mother—it killed her in the end," Washington remarked brutally. "His children would run away and would rather die than acknowledge they were sick—This sort of business lasted with unabated vim until he killed himself with it—No doubt there is some merit in the judicious practice of water cure, but to apply it to the exclusion of all other treatments is simple murder . . . I am now nearing 60 and have been a quasi invalid for over 20 years—by force of habit and tradition I gravitated at first towards water cure and only made myself worse—I will not maintain that any other treatment would have cured me (I was beyond cure) but water was the last thing—I don't think anyone can be perfectly cured. It is prevention that is worth everything."

The power of John Roebling's mind cannot be doubted. What is striking, for the modern reader, is the manner in which, during the nineteenth century, scientific practicality—exemplified by John Roebling's skills as an engineer, or indeed Charles Darwin's observations of the natural world—sat happily alongside what seems now to be nothing more than damaging fantasy. "The bizarre aspects of such procedures did not escape the eye of the satirist or the unmasker of quackery," Roy Porter observed, "yet they lasted—at least until the era of modern scientific, professional medicine—because they satisfied a deep desire that the healing enterprise should proceed within frameworks essentially sociable in their nature, and suffused with symbolic cultural meanings."

The cost, to his family, of John Roebling's attitudes and beliefs was great. For the most part he seemed entirely unaware of what damage he wreaked on those close to him; very rarely was there a glimmer of self-doubt. "Existence of evil in human life is a force too potent to be ignored or to be denied," he wrote in one of his philosophical meditations. He had been reading Henry James Sr.'s *Substance and Shadow, or Morality and Religion in Their Relation to Life*, published in 1863. "There is evil & plenty of it in the world now, but all this evil may be traced in its origin to man's transgression of the laws of his being. Man is the Origin of his own Evil, not God; neither is nature to be charged with it. On the contrary, Nature is constantly endeavoring to wean us back to her true meaning." Reading the son's writings, reading the father's: in so many ways, the two men could not have been more different. Washington is particular, specific, amused and amusing; the father, in contrast, often seems hardly human in his striving to reach a higher plane of existence, never noticing the misery of those around him. Or almost never, in any case. "A man may be content with the success of an Enterprise; he may have succeeded in overcoming all obstacles; in vanquishing his adversaries & Enemies; in achieving a great task; solving a great mental problem, or accomplishing a work, which was previously pronounced impossible," he wrote in the spring of 1863. "The hero is admired, is proclaimed as a public benefactor; the observed of all observers, he feels himself elated, and in his own estimation a great man, if not a true man. Retiring in a calm moment within the recess of his own inner self, he assesses his past deeds, his thoughts & acclaim. And before the judgement of his own conscience he stands condemned, an untruth, a lie to himself. But no body knows! Does he himself not know? Who can hide me from myself? Vain conceits! I am my own judge, and from my own judgement either happiness or misery results!" No one's opinion, other than his own, mattered to John Roebling.

On May 24, 1844, Samuel F. B. Morse transmitted the first telegraphic message over an experimental line stretching from Washington, D.C., to Baltimore, Maryland. The text had been suggested by Annie Ellsworth, the young daughter of a friend, and was taken from the Bible—Numbers 23:23: "What hath God wrought?" That year, too, Elias Howe, a young man who had lost his factory job as a machinist following the Panic of 1837, invented the sewing machine. Those couple of years were momentous for John Roebling, as well; his achievement in that time was as significant for his adopted country, in its way, as any of those developments.

Saxonburg had become a little manufacturing town, busy with the work of supplying wire ropes for the portage railroad. And then John Roebling took on what his son would call "the greatest feat of his life"—though it hardly seems that now. Yet the Pittsburgh Aqueduct, built in nine months over the winter of 1844–45, marked the real beginning of his career as an innovative engineer. An aqueduct had first been built over the Allegheny River in 1829; it gave the merchants of Pittsburgh their primary access to the east via the Main Line canal. But it had fallen into disrepair; and following the Panic of 1837, the State of Pennsylvania and the city of Pittsburgh were struggling to find the cash to repair it. So the city burghers offered a prize for the best solution to the problem and got the state government to agree to endorse the repairs. Robert Townsend—an influential Pennsylvanian who early on supplied John Roebling with wire for his ropes—encouraged Roebling to enter the competition.

It was, his son would write, "an untried problem, without a precedent and undertaken in the face of violent opposition, raided by the press, by rival contractors, engineers, canal men, merchants, etc. . . . He conquered everything with that wonderful personal force, a power which only fed on opposition and knew no defeat—I consider the effort of merely getting started fully as great as that of actually executing the work." For a grand total of $62,000 (which included the removal of the old structure and repairs to the piers) John Roebling built a great flume, wide enough for a single boat to pass through. There were seven spans of 162 feet each, consisting of a wooden trunk to hold the water, and supported by a continuous wire cable on each side, seven inches in diameter. The length of the aqueduct was 1,140 feet, the cables 1,175 feet, and the total weight of water carried by the aqueduct 2,200 tons.

For the first time the wire cables were made—just as they would be, decades later, on the Brooklyn Bridge—in place. In the old French system of cable-making the strands were made on shore, and then lifted into place and tightened. But, as Washington wrote, "the location here was such that it was impossible to make the strands on shore in line with the bridge—hence they had to be made in place—Their consolidation into one compact cable was a great step forward!" In the manuscript holograph is Washington's neat drawing of how each cable was made from seven strands of wire laid up parallel: two along the bottom, three above, two more at the top. Individual wires were spliced together on great reels, which then made, as Washington puts it, "practically one endless wire," the wires passing around a cast-iron "shoe" at the ends into which they were laid. The cables were then wrapped around tightly with wire—the wrapping machines were like two halves of a cast-iron barrel slid on to the cable, around which turned a loose reel of wire that passed over and under the cable; this was tightened as the "barrels" slid forward. That first patent filed by John Roebling in 1842 for making wire ropes also included this wrapping tool.

It was an unprecedented construction—not least because it was built in a mere nine months, and in a severe winter with snow on the ground and ice on the river. "Everything was new, everything required an invention," Washington recalled. "There was no time for failures . . . without the Saxonburg wire experience and the skilled men, J. A. Roebling could not have made the cables in time." Seven men were killed on the job over that winter—mostly drowned, the son reported. And it was on this aqueduct that John Roebling first worked with the bright-eyed, black-haired Charles Swan, who would become such an integral part of the Roebling family and business. "J. A. Roebling" the son called his father—as if the man who built the aqueduct was a different man from the tyrant who ruled his family life at home.

The memory of this breakthrough work was clearly imprinted on young Washington's mind. "I myself saw the aquaduct [sic; this was a word he often struggled to spell] for the first time in May 1845, just before the opening having been brought down from Saxonburg to go to school in Pittsburgh—Although I was still only 8 years I remember its appearance very well and shall never forget the disagreeable smell left on my country nose by the coal tar with which everything was drenched." At the time coal tar was used as a wood-preservation treatment; and while the scent might have left a strong memory, even more traumatic to the boy—despite the sometime horrors of home—was being sent away to school.

Washington's earliest education had come under his father's roof. "Being the first child, I was the first to be experimented on and could read German when I was about 4 years old—entirely too soon—an age when only an unhealthy precocity can be engendered," Washington remembered. He would have no need of English until much later in his boyhood. "Owing to the fixed purpose of making Saxonburg a German settlement only—Americans had no object in coming there and the English tongue was seldom heard. I myself did not learn to speak English until I was along towards 10 or 11 years of age, and most of the older settlers never learnt it. My father however made the most determined efforts to speak nothing but English. Unfortunately his strong Saxon accent usually betrayed him, and as soon as he opened his mouth some fellow would cry out Dutch in derision, then came the usual fit of rage and that fellow had to fly for his life." Washington remained enough of a native German speaker that for the rest of his working life he would make calculations using the German method, rather than the English.

When Washington was six, things took a rather eccentric turn when an extraordinary character arrived in Saxonburg. "One day in 1843 a new character appeared on our front steps—a short, stout man with a rifle and shotgun slung over his back, a carpetbag in one hand and leading a hunting dog, Rolla by name, by the other," Washington recollected. "A huge pair of spectacles adorned his face. He had walked up from Freeport, was exhausted and done up. My father took him in and practically took care of him the rest of his long life . . . Having no profession or trade, it was hard to place him, so he was made a sort of tutor for me, teaching me the three R's in German." Julius Riedel was the ninth son of a Lithuanian baron and had grown up near the Russian frontier. Since he had had few prospects at home, at the age of sixteen he had joined the Prussian Army and served under his father's cousin, Field Marshal von Blücher. Along with the Duke of Wellington, Blücher would lead the allied forces to victory over Napoleon at the Battle of Waterloo in 1815; Riedel was a young attaché on his staff, and was present not only at the battle but at the later occupation of Paris, too.

Later on, John Roebling made Riedel the town preacher—more evidence that the former's word held sway over all. The appointment was a great success; the church was always packed, and "young couples yearned for his services." But "no one knew the catastrophe that was impending. One day a delegation arrived from the Lutheran Synod in Pittsburgh and demanded his removal on the ground that he was no preacher, nor the graduate of a theological seminary, had no license to preach; that the marriages that he had performed were illegal and must be performed over again; likewise the baptisms; but the dead could remain in their graves. The congregation was in dismay, finally gave in and bade a silent farewell to poor Riedel."

Julius Riedel's departure had an unfortunate consequence for Washington, too. When he departed with his family, "I, poor wretch, was dumped into that same wagon and trundled off to Pittsburg [sic] to go to school— This was in many respects the worst thing that ever happened to me in my life." Separation from his mother was hard to bear. "Worse than that [my father] believed that a mother was not the proper person to bring up children, and so I was torn away from a happy home, a little strippling [sic], and sent away among strangers to be educated some more." The distance of years had not diminished the pain of this alteration in his circumstances when he wrote his memoir decades later. "I have always looked upon this act as a crime and the evil consequences of it have affected my whole life."

Johanna Herting Roebling's life has a shadowy quality—as did the lives of so many women in the early years of the nineteenth century. Washington's lively descriptions of life in his grandfather's house are not matched by any real descriptions of that grandfather's daughter, his mother. A surviving image shows an unsmiling woman with dark hair and handsome, wide-set eyes like those of her eldest son, who saw her suffering so clearly. In describing being torn from her at the age of eight, the words *crime* and *evil* are especially strong. He couldn't let go of the subject; a few pages later, while allowing that his father's desire to further his education had been quite correct, "the question was when should a boy be removed from the care of a mother and the happy surroundings of his own home—I claim that 8 years is entirely too soon . . . I know that it has injured me and warped my character and existence for my whole lifetime—The little advantage it may have given me amounts to nothing in comparison."

In those years Pittsburgh was developing into an industrial powerhouse. It had done well out of the War of 1812, when British blockades had meant great opportunities for American manufacturers; it had been incorporated as a city in 1816. Three years later the Union Rolling Mill—the first complete rolling mill for iron in the United States—was built there; in 1818 the first bridge in Pittsburgh—a covered wooden toll bridge—was built over the Monongahela River. The journey between Pittsburgh and Saxonburg was an all-day affair. While the trip could be "a great pleasure in summertime," the return trip, "especially in the wintertime" was dreadful. "Uphill all the way, snow or mud ankle-deep . . . The road through Sharpsburg was a bed of mud, churned up by the many iron wagons. The hill at Deer Creek was so steep that the load had to be transferred. Finally Saxonburg was reached late at night—half frozen and hungry." Washington was allowed back home to Saxonburg twice a year: "I counted the days for a month in advance," he said.

The boy was sent to a German-American school run by a man called Henne, who had begun life in Germany training for the Roman Catholic priesthood "but abjured the faith and ran off to America, where he fought starvation by trying to teach little wretches like myself." The schooling he began to receive—as the school moved, in the course of the four years of Washington's attendance, peripatetically from one ramshackle house to the next—was more German than American: "German Latin of the most useless sort, which I had to learn all over again when I went to an English school." His favorite reading, by his own account, were German translations of the *Iliad* and the *Odyssey*; he learned whole passages by heart. John Roebling, realizing that this German environment wouldn't aid the boy's language skills, arranged for a weekly visit to two local families, the Townsends and the Caldwells, "where the respective young misses were to bawl nothing but English into my dutch ears."

A certain bravado pervades Washington's recollections of his earliest schooldays. When not at his desk he was free to roam the streets and wharves of Pittsburgh at will, keen to escape "the Garret where I slept & cried—close quarters, no servant—and but little to eat." His expeditions were motivated in part simply by hunger: Henne's rations were meager. But he discovered that the steamboats berthed in town were full of generous folk. "I soon found that after dinner there was plenty of rice pudding left, which could be had for the asking, especially on the New Orleans boats—So I became a rice pudding beggar." He was allowed to help a local sign painter with his work; next door to the sign painter's was a bookbinder's, run by a Belgian "with a deep groove in the top of his skull, deep enough to lay your finger in, the result of an encounter with the Prussian dragoons at the battle of Ligny"— Napoleon's last victory before his final defeat at Waterloo.

But strange, dark notes also sound through his recollections. Once, when Henne and his scholastic establishment had moved yet again, Washington found himself living above a piano maker—who then shot himself. "The pallid dead face with the blood stained hole in the forehead haunted me so that

I was afraid to pass the door in the dark." Later on, the boy who lived next door died of dysentery. The memory of poverty haunted him—not least because it was a poverty unnecessarily imposed by his father. And it was, too, "genteel poverty—the worst kind." He recalled the condescending arrogance of a well-dressed boy who scorned Washington in his shabby clothes—and yet the boy's father, Washington noted acidly, "had not half the means my father had nor half the position in society . . . I am not a believer in my father's theory that economical habits can only be acquired by being poor, acting poor, and living poor—He always took care to live in the best of hotels and wear the best of clothes himself."

Just at this time, catastrophe in Pittsburgh brought opportunity for John Roebling. In the early days of April 1845, a fire broke out in an old shed down by the water, on Ferry Street, that would devastate the city. PITTSBURGH IN RUINS!! read the headline of the *Pittsburgh Daily Gazette and Advertiser* on April 11. Eventually nearly a third of the city would burn, more than one thousand buildings across fifty acres. "The fire absolutely appeared to dance from roof to roof," the paper reported, and within a very short space of time the whole area down by the waterfront was "a sea of flame." Water was no barrier: "The Monongahela Bridge has taken fire," the paper's editor wrote, "and is entirely consumed."

John Roebling got the contract to rebuild the bridge—his first proper bridge. Work was begun in May of that year, and the bridge was finished by February 1846—and cost, according to Washington, a mere $46,000: "and even at that fig[ure] a little money was made." The bridge was 1,500 feet long, consisting of 8 spans of 188 feet, and its cables were made in the old French style, alongside the construction rather than being laid in place. The bridge would be superseded in 1881, but it had a quality that would be significant to the success of John Roebling's work, and the work of his son. "The peculiar construction of the Monongahela Bridge was planned with the view of obtaining a high degree of stiffness, which is a great desideratum in all suspension bridges," John Roebling wrote. "This object has been fully attained. The wind has no effect on this structure, and the vibrations produced by two heavy coal teams, weighing seven tons each, and closely following each other, are no greater than is generally observed on wooden arch and truss bridges of the same span."

As 1848 became 1849, the Hennes departed for the west, as did so many in those days; Washington was sent to live and study with Professor Lemuel Stephens, a New Englander in his early thirties with a wife and a new baby. Life in this new situation was even less congenial than it had been with Henne and his family. It wasn't simply that he couldn't bear New England cooking—"the constant smell of boiling codfish was nauseating"—or that Stephens was an amateur taxidermist, and so the garret in which Washington slept was festooned with a stuffed porcupine, an alligator, a wild cat, a wolf, a raccoon, two snakes, and a coyote, "which filled my heart with terror as they swayed back and forth." There was worse to come.

"Satan induced the Rev. Mr. J. McLeod from Mississippi to send his son & daughter to us to board," he wrote. "I was moved down from the garret to room with young David McLeod. The young devil ruined my peace of mind for life. Would to God I had never seen him, he was a lunatic. No other boy would have stood it—but as my spirit had been crushed by my fathers [sic] cowhide I was afraid to say anything, until after a few months they found it out themselves & he was moved to an asylum where he died shortly after. I never got over the wretched association—That is one of the evil results of sending young boys away from their homes to learn manners—"

Not so many years later his service in war and his work on the Brooklyn Bridge would nearly cost him his own life; yet it was not those events, he said, that ruined his peace of mind for life. What can David McLeod, a boy from Mississippi, have done to Washington in that room they shared? Whatever it was, when it was discovered, the perpetrator was banished not only from the school but from his own family, too, locked up, it seems, to meet an early death.

In many respects it might be said that Washington Roebling had a privileged childhood. His father, while an immigrant, had never been poor; Washington's life as a boy was worlds away from the crowded, filthy city slums where children of his age grew up without education or sanitation, without heat or light or running water, with hardly any hope of a better life. Young Washington, in contrast, could listen to his grandfather's tales of old Prussia, hear the story of the Battle of Waterloo from one who had been there, read the *Iliad* and the *Odyssey*—and delight in them. He could run for miles in the woods and watch his father's first bridge being built over a great American river.

But each man's sorrow is unique. Washington Roebling's boyhood cast a long, dark shadow over the rest of his life. His father's savagery could be implacable, near murderous; and he had been removed from his mother's love and sent to a coal-darkened city where mysterious torment awaited him. And yet, as his first biographer, Hamilton Schuyler, wrote: "Neither the buffets of circumstance nor those of parental infliction had availed to break his spirit or spoil his disposition, which always remained generous and kindly."

JOHN ROEBLING'S WIRE business was too big for Saxonburg, now. As the 1840s drew to a close he began work on four aqueducts for the Delaware and Hudson Canal—which had been conceived by two Pennsylvania businessmen, brothers, in the 1820s to transport anthracite coal from land they owned in northeastern Pennsylvania to the markets in New York City. They had been prescient; the canal was so heavily used that by the early 1840s a complete program of rebuilding was undertaken. The enlargement, as Washington recounted in his memoir, "necessitated the reconstruction of its large aqueducts, 4 in number—crossing the Lackawaxen at its mouth, the Delaware close by, the Neversink above Port Jervis, and the Rondout at High Falls." The aqueducts were needed to carry the artificial waterway over natural rivers in two states: the first two were in Pennsylvania, the second two in New York. And so, "it was therefore a matter of course that he should move east, when his engineering engagements demanded it and the product of his rope making was also consumed."

John Roebling uprooted his family from western Pennsylvania and moved to Trenton, New Jersey. "The summer of 1849 was spent in getting ready to move. The children were glad, but my mother wept and the old folks went under compulsion," Washington recalled. It was the early autumn; the family—not only Washington, his brother and sisters, and his mother, but also grandparents, an aunt, plus another family called Kuntz—piled into four wagons. "The whole village turned out to see us off—many cried, all looked sad, because our departure was the death knell to the prosperity of the place." Later on, oil would be discovered in western Pennsylvania, and that would change its fortunes—but by then John Roebling had sold all his land there. This move marked a radical change in the life of the Roebling family, great and small. "We left the friends and companions of 20 years, most of whom were never to see each other again—I think my mother grieved more than anyone, the change in her life was to be so great—Ten years later I visited the place, but I was already a stranger."

The trip to Trenton took a week by wagon, canal boat, and railroad. It was not a luxurious journey—although on the boats there was rice pudding for dessert every day, a delight. One boat was so crowded that the family slept on floors and tables. Crossing over the Delaware River, as they neared their destination, twelve-and-a-half-year-old Washington saw a ship for the first time: "quite an event in a boy's life," he said. When they finally arrived at their new home—a two-story dwelling built of brick, next to the new wire mill that Charles Swan had been constructing according to John Roebling's detailed directions—Swan's wife was preparing dinner for the exhausted party. "How she opened her eyes when she saw the hungry army invade her premises," Washington wrote.

Trenton had already begun to thrive by the time the Roeblings settled there. Mahlon Stacy, the principal settler of early Trenton in the seventeenth century, had been a Quaker—one of the many fleeing persecution in Britain. Since those early days the place had been made the capital of the state, thanks to its central location—well situated between Philadelphia and New York—and its excellent access to transportation by water and over land. John Roebling's fellow industrialist, Peter Cooper, had urged the move; in 1847 he had been a cofounder of the Trenton Iron Works, and he likely thought that Roebling would be not only a good neighbor, but a good customer. The land John acquired—for one hundred dollars per acre, his son noted, also remarking that in 1894, the time of writing, the same land was worth $22,000 per acre—abutted both the Delaware and Raritan Canal and the Camden and Amboy Railroad, allowing shipping by every method and in all directions. One of the chief qualities recommending Trenton, too, was its flatness: for his wire mill and ropewalk John Roebling sought level ground at least 2,500 feet in length. The mill would be about a mile from the city center. At John Roebling's behest, Swan had three buildings constructed on the site—a small, two-story family home, a wire mill, and an engine house. The buildings and all the machinery for the mill were designed by John Roebling himself: "He knew exactly what he wanted," his son reported. In the first year of operation 250 tons of wire rope were manufactured—forty thousand dollars' worth of business.

Washington's schooling continued, at first under the tutelage of Dr. Francis A. Ewing, "a gentleman of the old school, resembling old Benj. Franklin with his silk knee breeches and embroidered vest," Washington recalls. Washington himself didn't think much of the education he received there; he was, he said, "no further ahead than I was at Henne's school, three years before. This simply proves that I should have been allowed to stay in Saxonburg a few years longer." But Washington's education took a practical turn when John Roebling suffered an accident that would curiously foreshadow the one that would later cost him his life. Just before Christmas 1849—the newest Roebling offspring, Charles, had been born just a few days before—John was in the wire mill, where the machinery had been started up. "While watching the operation he stood near the weight box of the counter twist machine"—Roebling's particular, ingenious device to twist the strands of wire together into rope, and one that operated under an extremely high tension—"unconsciously seizing hold of the wire rope which pulls it up, his left hand was drawn in, into the groove of the rope sheave," Washington recounts. Swan heard his agonized cry and reversed the machine to free John Roebling's mangled arm, at which point "he fell backwards . . . apparently lifeless." He would recover—but would lose most of the use of his left hand and wrist. "All flute playing & piano playing (except with the right hand) came to an end, while drawing was more difficult and many actions were seriously impaired." And yet, "it was after his accident that he accomplished his greatest engineering works."

But the shock of the accident was great—so great, it seems, that he couldn't write at all, although his right hand was absolutely fine. He insisted on his water cure as part of his recovery plan; although he had consented to initial treatment by Dr. James B. Coleman, Trenton's foremost surgeon, he was soon ignoring the physician's advice in favor of "sweat packs, douching, squatting in tubs and indulging in all the latest fads of Priessnitz." Despite that, he was up and about in a month's time—and ready to travel by February. But he still couldn't write, and needed help to dress—and so Washington, not quite thirteen years old, was drafted in to be nurse, secretary, and general aide-de-camp. Over the winter and into the spring of 1850, the boy would accompany his father as he toured the aqueducts being constructed at Lackawaxen, Delaware, Neversink, and Rondout Creek, a series of adventures the boy would never forget.

"The country at that time was very wild and unsettled," Washington recalled. "Wolves could be heard howling at night." Not only wolves were the cause of the boy's trepidation; his father's temper made him quake as well. "I served as a very poor amanuensis, writing from dictation. I was only 12½ years old, with an underdeveloped handwriting of my own, the cause of much parental complaint." The hardships were psychological and physical. They set off for the Neversink in the coldest weather, traveling by canal boat, by railroad—and finally, to get to the aqueduct, a three-mile ride in an open sleigh. This sounds delightful, but it was not. "A driving Northern sleet storm" had to be got through: "Having no overcoat or underclothes I nearly perished from cold—never shall I forget that ride—" That his eldest son—now effectively in his employ—should not recall being given any sort of warm clothing for such a journey is a dramatic indication of John Roebling's parsimony where his family was concerned. No wonder Washington noted his father's predilection for fine clothes with bitterness.

Over the rest of the winter and well into the spring, Washington's schooling was interrupted by his father's need for a companion on his travels. In May 1850 they headed off for High Falls, New York, and the Rondout Creek aqueduct; for part of that journey Washington found himself riding an ox when the boat meant to take them to the site was too crammed full of officials from the canal company. "No sooner was I hoisted on him, when he started to run away . . . full tilt. I stood it for 3 miles when I rolled off his round back and he turned and made back for home—the motion is like that of a dromedary producing sea sickness and spinal concussion."

Washington accompanied his father to Harrisburg to collect a final settlement of money owed John Roebling for the wire ropes manufactured for the portage railroad. At the hotel in which they had planned to stay, every room was taken—except one reserved for James Buchanan, at the time the most important politician in the state, and later the fifteenth president of the United States. But as Buchanan didn't arrive, the Roeblings were able to take the room. "The host informed us that we were receiving a double honor, that of occupying the best room in the house, and what was more, sleeping in Mr. Buchanan's bed." That was no honor to John Roebling: Buchanan was a Democrat, too soft on slavery in John Roebling's view, nothing more than a "sanctimonious hypocrite." John Roebling would be a generous supporter of the newly formed Republican party—the party of Lincoln. In Harrisburg, too, the boy was able to visit the state assembly chamber, which he expected to find a scene of gravitas. "As I was taking a front seat in the gallery I was hit in the face with a big paper ball and then saw the whole chamber was filled with a crowd of excited men all yelling and shouting at each other in boisterous merriment—each one had a supply of ammunition, paper wads, old shoes, hats, rag dolls which they were throwing at each other with great glee." This was, he discovered, the "time-honored" system by which the last day of the session was celebrated—"and it was just my luck to strike it."

Looking back at his father's life, Washington would come to think of these aqueducts as being among his father's finest achievements—for all they are, to the generalist's eye, far less lovely and less glamorous than his later grand bridges. The stiffness of the trunks holding the waterway worked, effectively, as a truss—so that the cables had only to carry the load of the water itself. When High Falls Aqueduct was dismantled in 1921, tests revealed that the tensile strength of the wire was 94,166 pounds per square inch—comfortably above Roebling's design requirement of 90,000. The wire was still in a good condition, despite having been exposed to the weather for seventy years. "It has often occurred to me," Washington wrote, "that if the Allegheny aquaduct had been built, or could have been built 10 or 12 years earlier, my father would have built at least 20 or 30 of them."*

THE EARLY YEARS of the 1850s, for John Roebling, were mostly taken up with building his wire rope business rather than planning engineering projects. "My father had realized quite often that professional engineering was an intermittent precarious occupation," Washington recalled. "He therefore never lost sight for one moment of the desirability of engaging in some kind of manufacturing business, which might go on without much special attention, leave him some time for engineering and yet bring in enough money to tide over a rainy day and gradually lay up a small competency." But laying the groundwork for such a business was not an easy matter; in essence,

---

*John Roebling's Delaware Aqueduct—now known as the Roebling Bridge—running over the Delaware River between Minisink Ford, New York, and Lackawaxen, Pennsylvania, is the oldest existing wire suspension bridge in the United States. Converted to carry road traffic, it was designated a National Historic Landmark in 1968.

John Roebling was starting an industry from scratch. Not only did the wire need to be drawn out, it needed to be annealed—treated with very high heat to prevent it becoming brittle, with acids to remove uneven scaling, and later with lime, both to lubricate it and to neutralize the acids. Roebling had no experience in any of these matters. That he became one of the greatest manufacturers of the age is further testament to his drive and industry.

Despite his scientific rigor, Roebling's passion for alternative medicine raged unabated. While single-minded in his determination when it came to his engineering projects or the wire mill, he was mercurial when it came to choosing apostles of his water-cure creed. Priessnitz, Washington says, "the original God of water cure," lost favor in the elder Roebling's eyes; instead, John Roebling turned to a succession of quacks, one of the most striking of whom was George Melksham Bourne, who twenty years earlier had been a prominent picture-dealer in New York. His artistic days behind him, he had become another of John Roebling's "fakirs," as his son had it; "he came with letters—lectured on Water, hygiene, Phrenology etc. and captured my father body & soul." Bourne described himself, in his publication *The Home Doctor*, as a "Thermal-Sudatorist and Pioneer Water-Cure Physician"; he believed in the curative power of steam baths and sweating over cold baths and freezing wet towels, hence *sudatorist* from Latin *sudare*, to sweat. Bourne's scorn for conventional medicine was as savage as Roebling's: "How long will the people permit these barnacles and leeches of the medical profession, to stand in the way of progress, to practice their iniquitous system, which is designed to keep their victims suffering and enslaved to their machinations?" Bourne was allowed to come to the house and dictate what the whole family ate and drank—a punitive regime. "On each childs [sic] plate was put a conical heap of Graham flour 3 inches high—this was chewed dry, for 1 minute per chew by the watch (half a teaspoonful) and at the word of command swallowed & so on until it was gone or secretly spat out," Washington wrote. "Whole mats of dates & figs were bought, full of worms which were considered a luxury"—since both meat and vegetables had been banned, along with coffee and sweets. "This lasted a week, we had fallen away to emaciated skeletons, sneaking around the neighborhood for scraps out of the swill barrel."

By an extraordinary turn of events, the Roebling children were saved from this nutritional nightmare by the tragic death of Bourne's own son, who was only a year or so younger than Washington. One Saturday the boy vanished—there was no trace of him, though he was hunted for high and low. "I suggested he might have gone swimming & been drowned—So I stripped and dived for half an hour actually finding the dead body in the bottom of the canal—all efforts at resuscitation were in vain." It is perhaps unsurprising that Washington, in recalling this incident, registered no sorrow for this boy he hardly knew—rather, relief that "the subsequent funeral broke up the starvation diet for good & our lives were saved." Bourne, Washington also notes, ran off the following week to California—in company with another man's wife.

Following his journeys with his father, Washington Roebling was enrolled in the Trenton Academy, one of the state's—and the country's—oldest schools, founded in 1782. Both Washington and his younger brother Ferdinand are listed in the register during 1851–52. An escape from home, however, was not an escape from industry; school texts of the era stress hard work, discipline, and submission to authority—qualities with which Washington was not unacquainted. The "Diligent Scholar . . . goes home as quickly as he can, he has so much to tell his parents and to do for them," admonished *Cobb's Juvenile Reader No. 2*. "The idle boy is almost invariably poor and miserable," said *Wilson's Third Reader*; "the industrious boy is happy and prosperous."

4

—

# "*I was not a chip off the old block*"

THE PATENT SYSTEM—BY which an inventor could see his name attached to an innovation, and could therefore profit from it—was enshrined, at least in embryo form, in the Constitution in 1787, though the system didn't truly become workable until 1836. Then a patent act was passed, providing for a patent office, which would award patents on the basis of originality, novelty, and utility—judged by a corps of examiners. The rate of technological advancement can be seen in the numbers of patents issued, doubling and quadrupling as the decades passed. In 1840, 477 patents were issued; in 1850 that number rose to 986; by 1860 the number had more than quadrupled: 4,588 patents were issued that one year. The American landscape, and Americans' way of life, were being irrevocably altered by steamboats, by canal systems, and by new methods of plowing the land, of threshing and milling grain, and of producing coal and iron.

But the railroad would work the greatest change on the country, its people, and its landscape. In 1850 Nathaniel Hawthorne reflected in a notebook what change the railroads had brought to the rural Massachusetts he knew so well. "The station at Newcastle is a small wooden building, with one railroad passing on one side, and another on another, and the two crossing each other at right angles," he wrote. "At a little distance stands a black, large, old, wooden church, with a square tower, and broken windows, and a great rift though the middle of the roof, all in a stage of dismal ruin and decay. A farm-house of the old style, with a long sloping roof, and as black as the church, stands on the opposite side of the road, with its barns; and these are all the buildings in sight of the railroad station . . . Anon you hear a low thunder running along these iron rails; it grows louder; an object is seen afar off; it approaches rapidly, and comes down upon you like fate, swift and inevitable. In a moment, it dashes along in front of the station-house, and comes to a pause, the locomotive hissing and fuming in its eagerness to go on. How much life has come at once into this lonely place! Four or five long cars, each, perhaps, with fifty people in it, reading newspapers, reading pamphlet novels, chattering, sleeping; all this vision of passing life! A moment passes, while the luggage-men are putting on the trunks and packages; then the bell strikes a few times, and away goes the train again, quickly out of sight of those who remain behind, while a solitude of hours again broods over the station-house, which, for an instant, has thus been put in communication with far-off cities, and then remains by itself, with the old, black, ruinous church, and the black old farm-house, both built years and years ago, before railroads were ever dreamed of." The railroads were the chief symbol of progress.

There could hardly be a more dramatic demonstration of the power of the railroads to erase distance and division than a railway bridge over Niagara Falls. In 1845 Charles B. Stuart was in charge of locating the Great Western Railway of Canada; he wrote to several prominent engineers soliciting bids for the crossing. Two engineers replied. One was John Roebling, the other the man who Roebling, and eventually his son, would come to see as his great rival: Charles Ellet. Ellet was a remarkable man, a few years younger than Roebling. Born in 1810 in Bucks County, Pennsylvania, he was raised to be a farmer but left home at seventeen after finding work surveying along the Susquehanna River. He would go on to work on the Chesapeake and Ohio Canal, all the while saving money so that, in 1831, he could enroll himself at L'École Nationale des Ponts et Chaussées, in Paris—a counterpart to the Bauakademie in Berlin where Roebling had been schooled. He was only twenty-two when he proposed to Congress a wire suspension bridge across the Potomac with a one-thousand-foot span; this "met with no encouragement in the National Legislature," since "the practicability of works of this nature had been demonstrated only in the mind of the Engineer." A decade later, he built the first wire suspension bridge in the United States, across the Schuylkill River at Fairmount in Philadelphia, Pennsylvania.

Three years after that bridge was completed, Ellet replied to Stuart's request for a Niagara proposal. "A bridge may be built across the Niagara below the Falls," he wrote, "which will be entirely secure, and in all respects fitted for railroad uses. It will be safe for the passage of locomotive engines and freight trains, and adapted to any purpose for which it is likely to be applied. But to be successful it must be judiciously designed, and properly put together; there are no safer bridges than those on the suspension principle, if built understandingly, and none more dangerous if constructed with an imperfect knowledge of their equilibrium."

John Roebling agreed that it would certainly be possible to stretch a wire suspension bridge across the drop of the falls, and one strong enough to carry the railroad, too: he was most particular about his knowledge of the equilibrium of such a construction. "It cannot be questioned that wire cables, when well made, offer the safest and most economical means for the support of heavy weights . . . The greater the weight to be supported, the stronger the cables must be, and as this is a matter of unerring calculation, there need be no difficulty on the score of strength. The only question which presents itself is: can a suspension bridge be made stiff enough, as not to yield and bend under the weight of a railroad train when unequally distributed over it; and can the great vibrations which result from the rapid motions of such trains which prove so destructive to common bridges, be avoided and counteracted?" Roebling had his reply ready. "I answer this question in the affirmative . . . The larger the span, the stiffer it can be made, on account of its great weight, which is necessary to insure stability . . . To counteract the pliability of a cable, stays must be applied, by which a number of points, which must necessarily correspond with the knots of vibration, are rendered stationary, and so that the stays and cables act in concert in supporting the bridge."

However, Ellet got the contract: it was awarded in 1847, the same year that, over in Britain, a cast-iron beam bridge over the River Dee collapsed while a four-car passenger train was passing over it. *The Illustrated London News* gave a dramatic report of the accident: "The train which leaves the Chester Station at half-past six o'clock had just arrived at the new iron bridge which crosses the river Dee, at the extremity of the race-course, when the furthest portion of the three iron arches or spans composing the bridge gave way with a tremendous crash, carrying the whole of the train (with the exception of the engine and tender, which reached the other side in safety) into the river below. Nine persons were taken out in a dead or dying state, and several others mutilated and injured in various ways. The stoker of the engine was thrown from his place upon the tender, and killed upon the spot." The bridge had been designed by no less an engineer than Robert Stephenson—son of George Stephenson, known as "the father of the railways," whose innovative locomotive, the Rocket, revolutionized the industry in 1829. Five people died in the disaster; some, in the aftermath, wished to sue Robert Stephenson for manslaughter.

The Dee Bridge gave way because of metal fatigue in a girder. It was, of course, of an entirely different design than what either Ellet or Roebling was proposing. All the same, the idea of a suspension bridge strong enough to carry a locomotive and all its cars was a striking one. Ellet's bridge was to cost $190,000—$10,000 more than Roebling's bid, but Ellet was then the better-known engineer. The first wire had been got across the gorge thanks to the five-dollar prize Ellet offered to the first boy to fly a kite across the chasm; fifteen-year-old Homan Walsh got the cash. By late summer of 1848, a temporary bridge had been erected, and opened to the public—which is when problems between Ellet and the bridge company arose. The bridge was a popular attraction, and before long five thousand dollars in tolls had been collected; money Ellet proposed to keep. Washington, in his account of Ellet's falling-out with his employers, told a dramatic tale: "By the time the footbridge was completed Ellet was in trouble with his directors . . . he took possession of the bridge, allowed no one to cross except on paying toll to him—fortified both ends, planted cannon, armed his men & defied the militia on both sides including the sheriffs—it took a regiment to dislodge him." Perhaps Washington heard this story from his father, the rival engineer; he admits it was not necessarily accurate. "The above is from hearsay which after 55 years becomes truth." That said, a broadside released at the time substantiates the story of a scuffle at the bridge, with Ellet setting up gates at either end to keep the directors of the bridge company off the structure.

Ellet was fired. Now John Roebling's plans were accepted: he was appointed engineer in 1852. Ellet had proposed that trains and horse-drawn carriages would run side by side, on the same level—which never would have worked anyway, Washington argued, the arrangement being "without an adequate stiffening truss—The main thing Ellet relied on was dead weight—stone—to stiffen his bridge—nonsense!" But his warnings must be taken with a healthy pinch of salt: his vehemence springs in part from a Roebling rivalry with Ellet that went beyond Niagara Falls. For Ellet's next project was a bridge over the Ohio River, at Wheeling, a job John Roebling also wanted. Ellet's bridge over the Ohio—then the longest clear span in the world at 1,010 feet—opened in 1849, and remains in use today, a national landmark. So Washington Roebling's remark that "when it came to practical details Ellet was a failure" must be put down to antagonism, rather than accuracy.

Whatever Washington Roebling's conflicts with his father, he constantly defended the Roebling name and reputation, and he admired the aspects of his father's personality that lay behind his great success. Recalling those days in Niagara Falls he wrote, "The leading feature of my fathers [sic] character was his intense activity and self reliance. I cannot recall the moment when I saw him idle—He commenced work right after the morning meal, drawing, planning, scheming, always devising something or perfecting an old idea—I am not out of the way when I say that for every bridge that he actually built he made about 50 plans that were never executed."

The towers of the bridge over Niagara Falls, built in the Egyptian Revival style, were made of Canadian limestone; for the most part the rest of the bridge was made of wood, as the manufacture of structural iron or steel had not yet begun at that time. Roebling's innovative design of the four main

*Reinforcing the anchorage of John A. Roebling's bridge*
*over Niagara Falls, 1857*

cables meant that no connecting arch was necessary above the bridge floor to stabilize the towers; the wire for the cables was brought over from Manchester, in England.* The lower deck was completed first, in 1853; the upper deck, which would carry the railway, was built last. The eight-hundred-foot-long span was stiffened with a wooden truss between the lower and upper deck; the great weight of the towers provided further stability. On March 18, the *New York Times* reported, the first train made a crossing of the bridge, drawn by the locomotive *London*, one of the largest engines of its day, weighing twenty-three tons alone.

"The great work of the age is completed . . . What has been pronounced by some of the most eminent engineers in the world both impossible and impracticable, has now proved both possible and highly practicable. It is no longer a chimera, but a reality, whose future promises benefits of inestimable value . . . At 10½ o'clock on the morning of the 18th the first train started from the British [Canadian] side, and, crossing the bridge at moderate speed, ran on the track of the New-York Central Railroad. Consisting of locomotive, tender, and twenty two heavily laden freight cars, it covered the entire bridge. The weight of the train was about *three hundred and fifty tons*, but neither this nor the furious snow squall, by which its passage was inaugurated, caused the slightest perceptible vibration in the structure." The drama of the setting did not escape the paper's correspondent. "Crossing the Niagara River at a height of two hundred and forty five feet from its surface, commanding on one side a full view of that mighty water-fall, the wonder of the world; and on the other a partial one of the renowned whirlpool, whose fearful eddies defy the approach of the boldest navigator and the most fearless adventurer, the Suspension Bridge presents one of the most imposing sights the eye ever rested on. All homage is involuntarily rendered to the mind that conceived, the skill that designed, and the enterprise that gave rise to this magnificent structure."**

---

*Among the early drawings in the RPI archive for the bridge is one showing a tower joined by a pointed, Gothic arch that looks remarkably like the arch of the later Brooklyn Bridge.

**John Roebling's breathtaking bridge is gone now, though the crossing is still in use and some of the stone abutment on the New York side remains in place. An arch bridge, known as the Whirlpool Rapids Bridge, designed by Leffert L. Buck, replaced the structure in 1896–97; "Sic Transit Gloria Mundi," as Washington wrote.

Not everyone was so taken with Roebling's vertiginous construction: Mark Twain, in "A Day at Niagara," admired the place as a "most enjoyable place of resort," but "then you drive over to Suspension Bridge, and divide your misery between the chances of smashing down two hundred feet into the river below, and the chances of having the railway train overhead smashing down on to you. Either possibility is discomforting taken by itself, but, mixed together, they amount in the aggregate to positive unhappiness." But another of the era's celebrities was drawn to the site for just the reasons that caused Twain to quake—Jean-François Gravelet, better known as Charles Blondin, the great nineteenth-century *funambule*. He made his first crossing of the falls on a tightrope in the summer of 1859. In the Rensselaer archive there is a stereoscope image of Blondin on his rope, John Roebling's bridge nearly parallel in the background. Washington's fine, neat script can be seen on the side of the card: "W. A. R. saw this performance in 1853"—an error, since his father's bridge was not even completed until 1854. But no matter: what his little note reveals is his sense of connection to his father's extraordinary bridge.

ON MARCH 13, 1854, Washington wrote to his father's right-hand man, Charles Swan, in Trenton. The tone of his letters to Swan is always friendly and fond—from Washington's boyhood into his manhood, there is always much more warmth in his correspondence with Swan than in that with his father. He asked if Swan could forward to him "a collection of logarithmic tables by Baron de La Vega, published in Vienna and written on the left hand in Latin and the other in French." The book was on his shelf at home; he told Swan exactly where to find it. A few weeks later he wrote "just after having swum across the Hudson"—and was therefore tired out. He went on: "There is no want of me having plenty to do. It comes of itself, if you only want it, and sometimes there is a little too much of it."

That year, at the age of seventeen, Washington decamped to Troy, New York, and the Rensselaer Institute to begin his formal instruction as an engineer. The city of Troy remains a beautiful place. Its original inhabitants were the Mohicans, an Eastern Algonquian Native American tribe. During the seventeenth and early eighteenth century the tribe—following disastrous wars with the neighboring Mohawks—sold most of their rich, well-situated land to the Dutch. "According to the deeds, land was sold for rugs, muskets, kettles, gunpowder, bars of lead, fur caps, shirts, strings of wampum, strings of tobacco, a child's coat and shirt, knives, hatchets, adzes, pouches, socks, duffel coats, bread, beer, a piece of cloth, a cutlass, axes, jugs of rum, blankets, guns, Madeira wine, pipes, and five shillings," a historian of the city notes. "Some would say today that it wasn't a very good trade." In 1789 the place's name was changed from Vanderheyden's Farm to the one it bears today—and Troy began its transformation into "the Silicon Valley of the 19th century." Located near the most northerly navigable reaches of the Hudson River, and the eastern end of the Erie Canal, Troy was in the perfect location for transportation and commerce when water was the best way to carry goods from place to place. Industrial mills could be powered by the falls that flowed over the city's high bluffs; thanks to this lucky combination of circumstances, Troy was the fourth wealthiest city in the nation by the time of the 1840 census. Samuel Wilson, slaughterhouse-man and meat-packer, made his fortune in Troy in the nineteenth century—and is purportedly the model for the United States's national symbol, Uncle Sam. The first detachable shirt-collars were made in Troy, along with the first machine-made railroad spikes and horseshoes. During the Civil War, the deck plates of the USS *Monitor* were rolled out from Trojan iron.

Twentieth-century industrial decline saw a falling off in the city's fortunes; not a few of the ornate brownstones that grace the city's center are boarded up and empty, though there is much talk of an economic revival. But perched on high ground, looking very much like a city on a hill, is a piece of Troy's history that continues to thrive: Rensselaer Polytechnic Institute, known to all as RPI.

The founder of what is now called Rensselaer Polytechnic Institute was also of Dutch origin. Stephen van Rensselaer was born in 1764 into a wealthy and well-established family. He was elected to New York's State Assembly in 1789, before rising to become a state senator and eventually lieutenant governor; he was a member of the commission tasked with exploring the route of the proposed Erie Canal in 1810 and was president of the state's first board of agriculture. In 1824—the year before the Erie Canal opened—he launched a novel school near his family's vast estate. "I have established a school at the north end of Troy, in Rensselaer county, in the building usually called the Old Bank Place, for the purpose of instructing persons, who may choose to apply themselves, in the application of science to the common purposes of life," he wrote. "My principal object is, to qualify teachers for instructing the sons and daughters of farmers and mechanics, by lectures or otherwise, in the application of experimental chemistry, philosophy and natural history to agriculture, domestic economy, the arts and manufactures . . . It seems to comport better with the habits of our citizens and the genius of our government to place the advantages of useful improvement equally within the reach of all."

In certain regards that desire to establish "the application of science to the common purposes of life" was not novel; the words are a conscious echo of the London prospectus of the Royal Institution, established in 1799. But the idea to place such an education "equally within the reach of all" was a peculiarly American idea. "Civil" engineering—as distinct from military engineering—was in itself a new concept. At the time Washington entered the profession, most college-trained engineers had learned their trade at West Point, as soldiers; most other American engineers had simply learned on the job. In France, L'École Nationale des Ponts et Chaussées had been established in 1747 and L'École Polytechnique in 1794; Polytechnisches Institut in Vienna was founded in 1815 and the Königliches Gewerbe-Institut (as it was eventually called) in Berlin six years later. The University of London began the teaching of engineering in 1840, in the same year a chair of civil engineering and mechanics was established at Glasgow. RPI remains the oldest continuously operating school of science and school of civil engineering established in any English-speaking country. Van Rensselaer worked closely with Amos Eaton, a geologist who he had met while surveying for the Erie Canal, and made him the first senior professor at the school, which had its official opening on Monday, January 3, 1825. Eaton, a prominent scientist, has been called "the father of American geology"; he was a botanist, a chemist, and a zoologist, too.

The "Rensselaerean" system of learning was unique—active, not passive, focused on the practical applications of education and with the students, not just the instructors, giving lectures to demonstrate what they had learned. The school was advanced in other ways: from the start some provision was made for the education of women. In a note dated October 29, 1828, Amos Eaton wrote "in his private capacity" that "at the urgent solicitations of several judicious friends, a lady, well qualified for the duty, will take charge of two experimental courses in chemistry and natural philosophy, in each year, for ladies: similar to the courses proposed for gentlemen in the annexed circular. They will be nine-week courses, at the same times and at the same charges . . ." In 1835 there is a record of "informal examinations" for women.

A code of bylaws described a curriculum that encompassed all the sciences. "Each student shall give five lectures each week on systematic botany, demonstrated with specimens, for the first three weeks, and shall either collect, analyze and preserve specimens of plants, or examine the operations of artists and manufacturers at the school workshops, under the direction of a professor or assistant . . . four hours on each of six days in every week, unless excused by a professor on account of the weather, ill health or other sufficient cause. For the remaining twelve weeks, each student shall give fifteen lectures on mineralogy and zoology, demonstrated with specimens; fifteen lectures on chemical powers and substances not metallic; fifteen lectures on natural philosophy, including astronomy; and fifteen lectures on metalloids, metals, soils, manures, mineral waters, and animal and vegetable matter—all to be fully illustrated with experiments performed by his own hands; and shall examine the operation of artists at the school workshops, under the direction of a professor or assistant, four hours on every Saturday, unless excused as aforesaid." In 1835 a department of "Mathematical Arts" was established specifically to teach engineering and technology. During the morning exercises students would give lectures on what they had learned; these were "closely criticized." Lectures were to be illustrated by a total of 1,200 experiments, performed by the student himself.

This demanding system began to pull in students from across the land; hard work in the nineteenth century was celebrated as a powerful virtue. In 1848 the school had twenty-two students and five instructors; by 1855 there were 114 students and eleven instructors. One of them—his father paying sixty dollars a year for the privilege—was Washington Roebling.

WASHINGTON DEPARTED FOR Troy under tumultuous family circumstances. In his memoir he recalled that he left for the school when he did in part because his father "suddenly returned from [Niagara] falls with Lucia Cooper"—whom he would marry in 1867. The daughter of a Dublin-born engineer, she had been born in 1820, and as an unmarried woman lived with her sister, a Mrs. Griffin, in Niagara Falls, where she met John Roebling. What exactly was the nature of their connection at this point—the brilliant, domineering engineer in his prime, and the thirty-four-year-old spinster—is not known, and Washington never referred to the relationship the pair had before their marriage, except once, when acknowledging that his beloved mother had never been the kind of helpmeet who could be John's social equal. Johanna Roebling was still very much alive in 1854; so no doubt John Roebling's arrival in Trenton with Miss Cooper caused some "wrangling," as Washington had it. "I was ordered . . . to pack my little kit, say good bye to the academy & go along to Troy N.Y." He was, he says bluntly, "dropped at the Troy depot to shift for myself, was allowed a dollar a day for all expenses . . . then commenced 3½ years of starvation and perpetual mental anguish— when I look back at it I often wonder that I survived." RPI, he acknowledged, was the best engineering school in the United States—but it was also "the most heart breaking, soul grinding, system crushing institution in the whole world." The director of Rensselaer in his days—Benjamin Franklin Greene, himself an early graduate of the place—was, he said, the hardest taskmaster that ever lived. The boys who managed to graduate "left the school as mental wrecks."

Hamilton Schuyler, the Roeblings' first biographer, would dismiss Washington's bitter recollections of his college days. He suggested that he might have "a little exaggerated in his memories the strenuous conditions of his student life, and his delight in forceful verbal utterances carried him beyond the boundaries of sober fact." Furthermore, Schuyler asserts, "he never appeared to suffer very much in later life from the evils which he depicts, for obviously he never became one of the 'mental wrecks' which he accused the Institute of producing." However, if his memories were "exaggerated"— enough to provoke intense and powerful recollections when he was in his seventies—one could argue his feelings were all the stronger for that.

Undeniably the technical and engineering curriculum was brutal. To begin: "Eight weeks in learning the use of Instruments; as Compass, Chain, Scale, Protractor, Dividers, Level, Quadrant, Sextant, Barometer, Hydrometer, Hygrometer, Pluviometer, Thermometer, Telescope, Microscope etc., with their applications to Surveying, Protracting, Leveling, calculating Excavations and Embankments, taking Heights and Distances, Specific

Gravity and Weight of Liquids, Degrees of Moisture, Storms, Temperature, Latitude and Longitude by lunar observations and eclipses." To continue: "Eight weeks, Mechanical Powers, Circles, Conic Sections, construction of Bridges, Arches, Piers, Rail-Roads, Canals, running Circles for Rail-Ways, correcting the errors of long Levels, caused by refection and the Earth's convexity, calculating the height of Atmosphere by twilight, and its whole weight on any given portion of the Earth, its pressure on Hills and in Valleys as affecting the height for fixing the lower valve of a Pump; in calculating the Moon's distance by its horizontal parallax, and in the distances of Planets by proportionals of cubes of times to squares of distances." This is a fragment of what the Rennselaer boys had stuffed into their heads in the days before Google, in the days when all mathematics, calculus, and algebra was done with pencil and paper in columns and columns of figures.

At seventeen, Washington Roebling was younger by at least two years than many of his classmates; most were at least nineteen and many quite a few years older. He was far too young, he wrote later, "to stand the racket"—and he also carried the burden of being the son of an eminent engineer, which came with its own cost. "Young fellows have to learn how to take care of themselves and not let themselves be imposed upon," he recalled. "As regards that I am just as helpless today as I was then—cheek is what I needed—not education—to this day I still let everyone ride over me—I was not a chip off the old block." Throughout his life Washington Roebling's determination, his drive, and his great courage were admired by all who met him, so the helplessness evinced in this passage may have been a passing fancy, a feeling written and then forgotten; but perhaps it lay under the whole of his life, a secret and a burden released upon the page. Without fear and doubt, it has been said, there can be no courage: if those difficult feelings are not mastered then what seems like brave action might simply be foolhardiness—and Washington Roebling was never foolhardy.

But his inability to stand up to his father's unreasonable demands weighed on him all his life; his sufferings at Rennselaer were simply further proof of his inability to resist. "I have long been of the opinion that a graduate should not work with his father," wrote Washington—a man whose whole life was overshadowed by that work. "At least such a father as I had—When I ventured to have a different opinion on some professional problem I was called a fool—my teacher was a damned fool . . . If I still dared to hold an opinion then a storm of vituperation followed by a hurricane of personal abuse and attacks on my life. In his frenzied rage, the spit would fly, his arms clove the air, he jumped & swore and cursed with horribly distorted features—any man with more spunk than I had would have killed him," Washington wrote with devastating plainness.

And yet Washington would send his own son, John, to RPI—expressing regret for doing so, and not just on his son's behalf. In one of this memoir's most poignant passages he writes: "Because an engineer has a son, it does not follow that the son would be an engineer—My own son should have been a doctor—My fathers [sic] father was a tobacconist and rolled cigars. To imagine my father rolling cigars would make a horse laugh . . . I sometimes think the above mentioned professor for studying the character of boys should be a woman—a woman's intuition is so unfailing that one glance enables her to arrive at conclusions which a poor man may arrive at in a year & then he is wrong."

Washington may have put on a brave face to Charles Swan, but in his later recollections he was more forthright. The brick-built house at 97 Ferry Street where Washington lived in Troy still stands; within its walls Washington was frantic, exhausted, hungry. "I did not have proper food to keep up my nervous strength, and nobody cared for me—The work never stopped day and night—Every night saw me past midnight, laboriously writing out the copious lecture notes of the day, notes that I cannot even read to day—and when I look at them I cannot realize that I wrote them—Before breakfast already busy cramming on the lectures of the day before."

Day after day, Washington had to stuff his head full of figures and facts. His surviving lecture books—thick, lined tomes in which Washington dutifully copied out the flood of information poured over him—are exhausting even to look at. In one there is a neat paragraph on "Rigidity of Suspension Bridges," in which his father's influence can clearly be seen. "In the older constructions, the bridge was very flexible; the mere walk of a man would make it shake to and fro. Flexibility was considered an essential condition, inseparable from the idea of a Susp. Bridge and accordingly several fell from oscillations caused by the measured tread of animals, such as oxen, horses or soldiers. This flexibility can be overcome by lattice-work parapets, by longitudinal chords and a stiff flooring—For protection against the effects of wind, extraordinary precautions must be observed. For probably one half of suspension bridges in this country have been blown down, showing a pal-pable neglect on [the] part of engineers in this respect . . . Much speculation has been entertained among European engineers as regards the adaptation of Suspension bridges to Rail Road purposes, principally owing to the fail-ure of one or two common Suspension bridges in England, which had been hastily fitted up for a Rail Road track, with scarcely any provisions for rigid-ity or stiffness—But we have at least one example in the Niagara bridge when this was successfully accomplished."

It was impossible to retain all this. "My estimate is that five times more was taught than could be properly assimilated," Washington concluded. "Memory was enormously stimulated—Now of all the faculties I find that memory is the lowest in order—Its cultivation stifles originality, checks the power of mental analysis, promotes servile imitation, besides weakening self-reliance." If horses and dogs, he noted, could be taught to remember, it could hardly be called an impressive or necessary skill.

In these, his later recollections of his time at Rensselaer Institute, there is only exhaustion and bitterness. He did not write of any sustaining friend-ships or encounters from those days; and most particularly he did not choose to memorialize one powerful friendship with a young man whose name is unknown—and who committed suicide the year before Roebling's class grad-uated. That class had begun with sixty-five students in it, and ended with only a dozen taking degrees: "proof," Washington wrote, "of the terrible grind." But this nameless young man also suffered, as two surviving notes from him show, from an affection for Washington Roebling that the latter could not reciprocate. In the autumn of 1856 the young man had made some sort of plea to Washington to deepen their friendship—and to "make allow-ances" for his feelings. He signed himself "your friend"—but his name, on the note he wrote, has been erased, either by the young man himself, or by Washington. "Dear Friend," the young man addressed Washington, telling him that "you are the only person in the world whose love and friendship I cherish, and upon whom all my affection is concentrated." He had asked Washington a question a few nights before—and upon Washington's answer, he wrote, "my life's happiness depends." From the young man's letter it seems as if Washington had been evasive in his reply; now, he begged for some kind of affirmation. "Something which appears only natural to me may perhaps appear incomprehensible or ridiculous to you," the young man wrote. He wanted a declaration of friendship to make their bond unique; he wanted a yes or no answer to this plaintive enquiry: "Am I, of all the people you know, with the exception of your parents, brothers and sisters, the one you love the most?"

If he could be sure of Washington's love, he said, "I shall be more content with my fate than I have ever been in my life. If I do not possess it, then God knows what I shall do." More than a month later the young man wrote to him again; the letter itself doesn't survive, but Washington copied it out from memory. His friend had apparently taken to using chloroform as a sedative; in case of an overdose, the young man wrote, Washington was to "pour cold water over my head, then breathe air from your mouth into my lungs and if there is no success get Dr. Bonetecon and tell him to cup me in the neck; as *ultima ratio* you may try Electro-Magnetism . . . If your efforts should prove fruitless do this: Keep of my things whatever you like, it is all yours!"

Washington annotated the letter carefully. "At this moment he suddenly staggered in, asking why I did not stay with him. Accordingly I went to his room—he took the letter afterwards so that I had no opportunity to copy it. The rest was merely an inventory of his property, together with some parting words of love." Those parting words of love—whatever they may have been—made a great impression on Washington's heart; perhaps it's little wonder that he kept them out of a memoir that was, in its way, an official document. He was willing to share his father's brutality, and what he saw as his own and his family's suffering at his father's hand; he would not share this. But it stayed with him—for many years later, one evening from a Union Army encampment in Virginia, he spoke of that loss, and his confusion, in a letter to Emily, whom he would marry in 1865. "My candle is certainly bewitched," he wrote. "Every five minutes it goes out, there must be something in the wick, unless it be the spirit of some just man made perfect, come to torment me while I am writing to my love. Are any of your old beaus dead? If I wasn't out of practice with spiritual writing I would soon find out.

"There is only one friend whose spirit I want to communicate with," he continued, "you have his picture with mine; he committed suicide because he loved me and I didn't sufficiently reciprocate his affection; I advised him to find someone like you for instance, but he always said no woman had sense enough to understand his love."

What photograph would Emily have had of her beau during the war? There is one taken while he was at Troy, and already he is a handsome man, with his wide-set eyes and shining hair swept over his broad clear forehead; he looks very youthful. She might have had that—or she might have had another, taken in 1861 when he joined the Union cause. Whichever image she kept on her dressing table, it seems remarkable that Washington would have given her not only his photograph—but a photograph of his dead friend, too.

Washington graduated in the summer of 1857. His thesis had been a design for a suspension aqueduct—just as his father had built. It is sixty-two pages in tiny, sepia-inked script. But he felt little qualified for much else. "One thing I have always regretted at Troy, that the boys did not have a few lectures on what might be called the conduct," he wrote so many years later, wishing they had been instructed in matters such as the importance of personal health, "the folly of thinking it noble to suffer." He also would have wanted some "explanation of stocks, bonds mortgages, other investments— interest, how to borrow money, take partners . . ." He knew about work; he didn't feel he knew much about life.

# "It is curious how persons lose their heads in times of excitement"

PERHAPS A LITTLE MORE of that kind of instruction would have done the whole country some good. The 1850s had been a boom time for the United States. The railroads expanded exponentially; gold was discovered in California. During the Crimean War Europeans began importing American grain since Russian grain was unavailable, which was a boon to the economy— but when that conflict ended in 1856 demand dipped steeply. As the country continued to import much more than it was exporting, gold reserves fell sharply, despite California's rich seams. The steamer *Central America*, carrying $1.6 million in Western gold and four hundred passengers, was lost at sea in a hurricane in 1857. Banks suspended gold payments and wouldn't offer credit; as a result, construction projects were halted, and unemployment rose dramatically. Thousands of businesses, including half of New York City's brokerage houses, failed: by December of that year, the loss from business failures in New York City alone was $120 million.

John Roebling, for all his hard-headed good sense, wasn't immune from the panic gripping the nation. That panic had really got rolling in Cincinnati, when the Ohio Life Insurance and Trust Company failed in August—the first big bank to do so. It was in Cincinnati that John Roebling was set to build a mighty bridge over the Ohio—but the crash would put an end to that plan, at least for the time being.

But in the boom that came before the bust, it appears that John Roebling recalled—however briefly—that it had been his intention, after all, to come to America in order to work the land. In 1856 Washington went home to Trenton for the summer before his final year at college—only to discover that he was to be ordered to accompany his father to Iowa. "He had the land fever!" Washington wrote, "like most Americans of the day." The country, settled first by Europeans on the Eastern Seaboard, was expanding ever westward, and the plains were rich country. As early as 1839, a correspondent in the *Buffalo (NY) Journal* had declared "that taking into consideration the soil, the timber, the water, and the climate, Iowa territory may be considered the best part of the Mississippi Valley. The Indians so consider it, as appears from the name which they gave it. For it is said that the Sioux, Sac and Fox Indians, on beholding the exceeding beauties of this region, held up their hands, and exclaimed in an ecstasy of delight and amazement, I-O-W-A, which in the Fox language means, 'this is the land.'" Later that year, it was reported that "one of the citizens of Cincinnati had just returned from a tour of Iowa and had stated that the prospects for an exceptional harvest were the best he had ever seen anywhere." In Cincinnati, nearly twenty years later, John Roebling heard similar claims. He bought 45,000 acres of land, selling his stock in the Delaware and Hudson Canal Company to raise the thirty thousand dollars needed: "The Del & Hudson got very mad at this and threatened to take his rope orders away!"*

Washington traveled out, by way of Cincinnati, to Grundy County, in the center of the state. It was, Washington wrote, "a limitless gently rolling prairie—with no habitation in sight." He noted too that the Indians—Sioux, Sauk, Mesquakie, and Winnebago—had "recently left," though it would be more correct to say that they had been driven out by the relentless tide of settlement, poverty, and war. By the time the Roeblings arrived, bison were long gone from this place, but there were deer and plenty of other game. The initial survey brought the usual battles with his father, made worse, no doubt, by Washington's new education. "Who could work with

---

*Some of the land was bought from former soldiers who had acquired it through the government's Bounty Land grants; in these cases, the president of the United States had to sign the deed through which the soldier passed on his property to a new buyer. By the deeds preserved, it is clear that John Roebling was still buying land in Iowa as late as 1861, since two of them bear the bold signature of Abraham Lincoln.

my father—He had learned Dutch surveying and mine was American—the two would not mix—Result, one perpetual volley of curses all day and all next day till everyone got tired."

While not idyllic, Washington's rural summer was still a welcome break from his studies in Troy. "I learned to plow & break the prairie sod," he wrote with pride. "My plow team comprised 4 big oxen, leaders were named Buck & Breck"—after the presidential candidate of that year, James Buchanan, and his running mate, John Breckenridge. The other two oxen were the biblical Samson and Goliath. "I learned how to plow a straight furrow a mile long and swear in Ox talk—Later on we ate up poor old Sampson [sic]—provisions had given out—we were reduced to one bag of dried apples and what prairie chickens I could shoot provided I hit them." But then powder and shot gave out, too. Rattlesnakes were a constant danger, "there was one every 10 feet," Washington recalled; one of the men on the farm died just by putting on a pair of boots—they'd belonged to a man who'd died of a rattlesnake bite. The fangs apparently had got stuck in the leather, and still had their venom when the next victim pulled them on. "There is no moral," Washington says of this tale. One morning, he claimed, he was nearly killed by a meteorite, too.

Despite such trials, his father found a renewed passion for a life of the soil. "He took a notion to live there," his son wrote, "and farm in the open free country of God." He built a house; a cousin came from Trenton to lend a hand. In September Washington headed back east, to Troy—and then only a day or so later his father received word that sufficient funds had been raised to build his Ohio bridge. He left the farm in the care of the Trenton cousin, who promptly made a hash of things. "He went to Kentucky & bought 25 prize boars @ $75 each—the next winter everyone froze to death in Iowa." As a result, his father "stopped the whole farce"—though the land was profitably sold later on.

Work began on the Covington–Cincinnati Bridge while Washington was back at Troy. Foundations were dug for the towers, first in Cincinnati and then on the Kentucky side; the building of the stonework commenced, up to the level of the bridge floor, eighty feet above low water. And then came the failure of the Ohio Life Insurance and Trust Company and panic spread across the land. John Roebling wasn't immune to the dreadful fear that had depositors all over the land withdrawing their hard currency from their bank accounts; as Washington notes evenly, "it is curious how persons loose [sic] their heads in time of excitement." With the wisdom of hindsight, Washington opined that his father—a wealthy man—might have made a killing buying stock in Chemical Bank when the price had plummeted, but "if I had ever suggested such a thing he would have killed me." Instead, Washington and Charles Swan were sent to the splendidly named Shoe and Leather Bank in New York City to withdraw nearly thirty thousand dollars in gold. "For a day the bank demurred, but finally gave it, with the remark that it was much safer in their vaults than in Trenton—It was so heavy that we could scarcely carry it." At a time when gold was worth a little more than twenty dollars a troy ounce, Washington and Swan would have had about one hundred pounds pounds of gold on their hands. "Arriving home we dug a hole in the cellar, put it in & sat on it for a long time until Mr. R. came home—After a while, things having quieted down, we were told to take it back to the same bank who received us with great derision."

Work stopped on the bridge over the Ohio, thanks to the Panic—and wouldn't begin again until the Civil War was nearly over, in 1864. For a while, after his graduation from Rensselaer, Washington wasn't "specially engaged in anything," so he went to the mill in Trenton to learn "the mysteries of rope making" under the supervision of Charles Swan. In 1858 he was sent to Pittsburgh, where John Roebling had won the contract to design and construct a bridge over the Allegheny River, at Sixth Street. Pittsburgh was well established as the "gateway to the west," first as a hub of transportation and eventually as a powerhouse of manufacturing: by 1860 there were 20,500 laborers in the 1,191 workshops of Allegheny County producing $26,563,000 worth of goods. A contemporary guide to the city, *Pittsburgh As It Is,* noted that the metropolis "will be the most progressive and accumulative city of the Union" thanks to its advantages of geographical position and natural resources, including coal and natural gas. "For it seems as if Nature had gathered all manufacturing facilities and hurled them in one mass into position there; in its vicinity being all the requisites for a permanent domination," a later historian noted; indeed, in decades to come Andrew Carnegie and Henry Clay Frick would create their industrial empires here. The old Sixth Street Bridge was wooden, and had stood since the early years of the century; now it was far too narrow; John Roebling's suspension bridge, which would be completed in 1859, would more than double the width of the carriageway.*

It was on the Allegheny bridge in Pittsburgh that Washington first saw wire cables being made in place; he wrote to Swan that he was half in the office and half at the work on the bridge. "I always manage to find enough to do so far. Business is quite brisk in town, all the rolling mills are going and any quantity of steamboats coming and going . . ." Pittsburgh, Washington wrote, was an exciting place. "Pittsburgh is getting to be a great place for murders, stabbings, etc. Something of that kind occurs every day." An opera troupe came to town; he saw *The Barber of Seville*, *La Bohème*, and *Don Pasquale*. John Roebling, too, was "passionately fond" of the opera— something they had in common. Some visitors complained of Pittsburgh's relentless industry, but not Washington. "Clark Fisher passed through the other day," he wrote of a Trenton friend. "I prevailed upon him to stop off a day and take a look at Pittsburgh. He was thoroughly disgusted with it, I can assure you. You see it happened to be one of those dark, cloudy, smoky afternoons, when the sun doesn't shine and the gas is lit at 4½; he can't imagine how anybody can live here and enjoy life. For my part I feel quite sorry that I will have to leave it." Toward the end of 1858 he recorded: "I am perfectly at home in this place and have ten times as many acquaintances as in Trenton."

Decades later his memories were much darker. His work on the bridge was simply to serve as "talking post" for his father. "All his outlandish notions on Religion, chemistry, politics and everything else were first thrashed out on me and if I could not or would not understand I was dubbed an ass and a scoundrel and was glad to escape with my life." His father's medical beliefs continued to take their toll on Washington. "This year he ate charcoal in quantity," he wrote of 1858, "of course I had to follow suit—I still live." In a little notebook he kept at the time, containing a "list of persons I have been introduced to in Pittsburgh, Pa.," there are also "Notes and remarks about the acute abdominal attacks I am afflicted with." The worst pains

---

*The bridge would stand until 1892, when "a huge Trussed structure [replaced] the graceful little wire bridge that the Pittsburgers [sic] were so proud of," as Washington noted. The present crossing, called the Roberto Clemente Bridge, was built in the 1920s.

John A. Roebling's bridge over the Allegheny River

came in the winter, "brought on by over-exertion in making a big snowball and afterwards catching cold." The pains were, he wrote, "excruciating"— but the remedy seems little better. It is noted in detail.

> In case of any future attack I shall try 1st very strong mustard plasters & drinking as much hot water or hot tea as I can bear, provided it stays on my stomach, if not, I must first vomit all the bile. Then to give the wind a chance to pass off I must use the rectum tube & fill the colon to its utmost extent, not with warm water alone but with some of the following recipes taken from Dr. Mattison's little book:
> 1. Take a tablespoonful of Turpentine & yolk of two eggs, beat them well together and add nearly a pint of warm water—
> 2. Take a tablespoonful of Turpentine—two of sweet or lard oil, beat them together and add about 1 pint of warm starch mucilage, made by boiling a tablespoon of starch first mixed with cold water . . .

His father had once pronounced that "there is no chronic disease of any part of the body without some phase of the same condition in the nervous & other tissues of the stomach. This is a fixed rule."

Once the bridge was finished, father and son returned to Trenton. John pressed for the completion of the Covington–Cincinnati Bridge, its half-finished towers looming over the Ohio; meanwhile Washington's younger sister Laura got engaged to one Anton Methfessel, a "pedagogue from Mühl-hausen" who had started a school on Staten Island—going strong today as the Staten Island Academy. Laura was, Washington says, a brilliant pianist; and he himself took violin lessons from a well-known local musician around that time. He joined an amateur orchestra, but it was "too much for my nerves." The strain of "dodging his father" clearly still told.

The country was under strain, too. "The winter of 1859 found us both at home," Washington writes. "The troublesome political condition which preceded the rebellion excited him very much, as he had but little sympathy with the South." In October of that year an abolitionist named John Brown—who, three years earlier, had massacred five proslavery men in Kansas territory—launched an attack on the federal armory and arsenal at Harpers Ferry, Virginia (later West Virginia), hoping to incite a slave rebellion. The assault was a failure: it was put down by a band of U.S. Marines, led by Colonel Robert E. Lee. Brown was captured and hanged for the crime of treason on December 2. His actions—and his death—were, said Henry David Thoreau, like a meteor across the sky. "I never hear of any particularly brave and earnest man, but my first thought is of John Brown, and what relation he may be to him. I meet him at every turn. He is more alive than he ever was. He has earned immortality . . . He is no longer working in secret. He works in public, and in the clearest light that shines on this land."

John Brown's doomed action made him a martyr and polarized political sentiment across the states, both North and South. Shortly after the raid, the Roeblings' state of New Jersey elected a governor, Charles Smith Olden, from the newly formed Republican Party. Not all New Jerseyans, however, were in favor of the party, or in favor of its presidential candidate, Abraham Lincoln—the lawyer from Illinois who, in a speech at the Cooper Union in New York in February 1860, pledged at least to prevent the spread of slavery into the western territories. The *Jersey City Standard* called him a "fifth rate man, with no public experience, with coarse and uncouth manners, whose oratory is a vile compound of clumsy jests and clownish grimace." In the election of 1860, New Jersey was the only state in the North not to award every single electoral vote to Lincoln. He was elected nonetheless; still, the Democratic-leaning *Daily True American* of Trenton rejoiced: "It is with no small amount of pride and satisfaction that we record the facts to be found in our table of returns of the electoral vote,* which, although not complete, show conclusively that the Rail-Splitter has been defeated in the State by a majority of about five thousand." The piece continues: "Whatever disasters may result to the country from the election of LINCOLN, which seems to be conceded on all hands, it will be a great consolation for the Democracy and Union men of this State to know, they are not responsible."

Despite the state's presumed antipathy toward him, Lincoln passed through Trenton on his way to his inauguration in February 1861, arriving around noon to find a crowd of some twenty thousand people massing along the route from the train station to the statehouse. No doubt Washington was kept too busy at the wire works to be among them; nevertheless, he wrote, "as the excitement in public affairs increased my fathers [sic] excitement at home increased, so that it was almost impossible to live at home." In December, South Carolina had seceded from the Union, urging the other southern states to follow suit; two weeks before the president elect arrived in New Jersey, Mississippi, Florida, Alabama, Georgia, Louisiana, and Texas left the Union and the Confederate States of America declared its independence in Montgomery, Alabama, naming the Mississippi senator Jefferson Davis its president. Nevertheless, Lincoln spoke that day of his desire to avoid conflict: "The man does not live who is more devoted to peace than I am," he said to his audience in the state assembly chamber. The assemblymen cheered. "None would do more to preserve it; but it may be necessary to put the foot down firmly," he warned.

At dawn on April 12, 1861, Confederate batteries fired on Fort Sumter, in Charleston Harbor—a precarious Federal redoubt defended by Major Robert Anderson and a garrison of fewer than ninety men. The *Charleston Mercury*, passionately secessionist, called the attack a "Splendid Pyrotechnic Exhibition"; by Sunday, April 14, Anderson—a Kentuckian, though strongly

---

*At this time, this meant the popular vote.

loyal to the Union—handed over control of the fort to the Confederates, and the American Civil War had begun.


JOHN ROEBLING WAS intensely agitated by the coming of war. "He constantly lamented that his health and age would prevent him from taking an active part in the conflict," Washington recalled. "Had he been a little younger he would have entered the army he said and become its commander in chief in a year!" The elder Roebling would eventually advance a loan of one hundred thousand dollars to the government to aid the cause of the war—an enormous sum at the time, and even now. And his son served in his stead. It was a story Washington would tell often, over the years. "One day, about Fort Sumpter [sic] time, we were eating dinner when my father suddenly remarked to me, Washington! You have kicked your legs under my table long enough, now you clear out this minute!* A potato that I was guiding to my mouth fell to the plate," Washington drily remarks. He got up, put on his hat, and walked out the door to join the cause that day: his name can be found as enlisting on April 16, 1861, with Company A of the National Guard of Trenton, for a term of three months; in June he transferred to the Ninth New York State Regiment.

Writing years later, he recalled a peculiar connection to the outbreak of the war. During one of his last vacations at Troy he had gone with his father to Washington, D.C., where General Montgomery Meigs was in charge of constructing the new Capitol building—at the time the largest project of its kind in the United States, Washington says. Meigs took John Roebling's advice as to the best method for securing the great hoisting derricks with guy wires. The meeting with Meigs would have "quite a bearing" on Washington's career in the Army, for he would serve under General Meigs's orders for a year and a half. But while in Washington they also visited the office of the colonel in charge of all the post office buildings in the country to discuss how such buildings might be fireproofed by the use of newfangled iron beams, made by the Trenton firm of Cooper & Hewitt. So the government

---

*I am reliably informed that, to this day, German fathers will say this to their sons.

beam inspector from the Trenton mill went along with them—Major Robert Anderson, of Fort Sumter fame. Anderson lost his hat on the journey, Washington recalls. "He was such a kind, obliging, mild, gentle, inoffensive harmless unaggressive man that I took him for a Sunday school teacher— Little did I know what the future had in store for him. He did not buy a new hat."

Many expected that the Civil War, like the First World War a mere five decades later, was going to be over quickly, as indicated by Washington's three-month enlistment term—"unless sooner discharged." "Things progress very slowly here," Washington wrote to Charles Swan. His company was stationed in the nation's capital; at the outset, by Washington's account, preparation was desultory. "Loafing in the camp seems to be the principal occupation not only of ourselves but everybody else. During the cool of the morning and evening we practice sabre exercise; probably we will know something about it six months from now." Soldiers were issued uniforms; but Washington was in charge of his own footwear. "Please tell 'Ferdy' to go to Grammetts' and order a pair of shoes for me, medium high around the ankles and with projecting soles, thus"—a little drawing in the letter showed exactly what he wanted—"soles double heavy. He has my measure from my last pair of shoes. Tell him that the last pair was a little tight around the instep and to make them a little wider; put on lacing leather strings and send to me express (prepaid) care of E. K. Knorr, 514 K St., Washington." Reading any soldier's letters from a war is a reminder that clothing, bedding, and food are often the things that matter most. Thanks to its supply lines the Union Army was always better provided for than the Confederacy—a contrast that became stark as the war wore on.

However, not everything was easy or pleasant. "I have a few moment's leisure time to write a few lines before dinner which will consist of one hard cracker and a piece of fat pork," Washington wrote to his sister Elvira a month or so later. Brother and sister were close; she would become an important confidant. "The crackers have to be broken with a hammer; here they go under the name of Uncle Sam's pies." He hadn't taken his clothes off since he left Washington, he told her; the men slept on the bare ground with only a blanket to cover them. "Such a life is very healthy provided you have plenty to eat; but such is not the case here . . . Provisions . . . are scarce in the whole army, there being a scarcity of wagons and horses to haul them." He was writing to her from Charles Town, Virginia (later West Virginia), "a mean little town," which had the distinction of being the place where John Brown had been executed. Brown, Washington told his sister, "was hung in a corn-field next to us. The site of the gallows is marked by a cornstalk and pieces of the gallows sell at $1 per pound."

The first real fighting he saw was at the Battle of Ball's Bluff, part of Major General George B. McClellan's operations in northern Virginia. McClellan, "Little Mac" or "Young Napoleon" as he was sometimes known, was a soldier beloved by much of the army but not, in retrospect at least, by Washington Roebling. McClellan would take command of the Army of the Potomac in the autumn of 1861, replacing the aging General Winfield Scott, under whom he had served during the Mexican-American War. But his military approach was too much marked by caution to please his commander in chief; he would be relieved of command not long after the bloody Battle of Antietam, in 1862. At Ball's Bluff, in October 1861, Union soldiers first mistook a line of trees across the Potomac River for a Confederate encampment; once across the river they were attacked by Confederate forces and no retreat across the river was possible, with not enough boats available for the men. As Washington recalled in his memoir: "McClellan, the ass, lost nothing and we lost thousands. At the river, we had no pontoons, not a boat. Our army, numbering 26,000, sat on the bank, the enemy looking on from the heights opposite." Eventually a rowboat was found, but it was only twenty feet long; hundreds had to be ferried over the river in it. When the enemy began to wake up "they thought it was some kind of circus performance, not dreaming that McClellan had ordered a small army to be taken over in one small boat . . . Most of us were dancing around . . . without shelter or protection, trying to dodge the hail of bullets from the bluff above. Many were killed here. I had a narrow escape; a bullet pierced the pommel of the saddle, but was stopped by a little piece of brass, from entering the body."

This would be the first of Washington's many narrow escapes, which are all the more striking when the scale of death and destruction all around him is taken into account. Though by his own accounts plagued by ill health before the war, and then for many years afterward when the building of the Brooklyn Bridge took its toll on him, Washington had the devil's own luck on the dreadful battlefields of the Civil War—a conflict in which three quarters of a million men, 2.5 percent of the country's whole population, would die.* A few months before Ball's Bluff, and fewer than fifty miles away, sixty thousand men had met on a field outside the town of Manassas, Virginia, and in the space of a dozen hours nine hundred men were killed and 2,700 wounded in this, the first major battle of the war; during the entire Mexican-American War of 1846–48, just over 1,700 men had been killed over the course of two years. With little adequate medical care available, poor sanitation, and few systems in place at the beginning of the war to deal with the dead and wounded, the horror can hardly be imagined. The calm observational quality of Washington's writing makes even more real the terrible sights he saw, the dreadful conflict he experienced.

Later in the war Washington would serve under General George Meade; Colonel Theodore Lyman, an officer also on Meade's staff, described the junior officer succinctly. "Roebling is a character, a major, aide-de-camp and engineer. His is a son of the German engineer Roebling, who built the celebrated suspension bridge over the Niagara River. He is a light-haired, blue-eyed man, with a countenance as if all the world were an empty show. He stoops a good deal, and when riding has the stirrups so long that the tips of his toes can just touch them; and, as he wears no boots, the bottoms of his pantaloons are always torn and ragged. He goes poking about in the most dangerous places, looking for the position of the enemy, and always with an air of entire indifference. His conversation is curt and not garnished with polite turnings. 'What's that redoubt doing there?' cries General Meade. 'Don't know; didn't put it there,' replies the laconic one." His was the look of a young man who had already seen a great deal of violence, not least in his own home. He had suffered there and in his schooling, and seen too the machinations of men in the conniving, striving days of nineteenth century industry and technology. Little wonder he spoke little and kept his thoughts, at the time, to himself. But his character was—and is—apparent on the page. The war killed hundreds of thousands, and broke many hundreds of thousands more; but in many ways it would be the making of Washington Roebling. In four years he would rise from private to colonel, and he would be known as Colonel Roebling for the rest of his long life.

---

*If that percentage of Americans were to die in a war today, the death toll would be seven million people. *American Experience*, "The Civil War By the Numbers," www.pbs.org.

# "The urgency of the moment overpowers everything"

BY THE TIME OF Ball's Bluff, Washington had been promoted to the rank of sergeant; autumn gave way to winter. "Sleeping on the cold ground with the thermometer at 10° is no fun; neither does it kill you," Washington wrote. The cold, at any rate, never dimmed Washington's sharp eye, and one story reveals how much expertise he had already developed in his mid-twenties. One night, Washington recalled, while setting up camp on a piece of frozen ground, the young soldier found himself tangled in a wire fence. "[I] found my blanket had caught in a splice—The splice looked familiar and yet it was different from our regular cable wire splice & after a while it occurred to me that I had seen that splice on the Wheeling bridge during a visit in 1858— This being so, the question was who would be apt to build such a fence unless it were Ellet himself—and so it proved." Incredibly, they had camped by chance on the land of John Roebling's great rival; perhaps no other man alive in the country would have recognized that splice. When Washington knocked on the farmhouse door the next morning he discovered that the engineer had "gone to the wars." Ellet had developed ram boats to be used against Confederate shipping on the Mississippi; he died of wounds suffered at the Battle of Memphis in June 1862.

Washington's company moved into southern Maryland, to be stationed opposite the Confederate battery at Shipping Point, at the mouth of Quantico Creek. The Confederates had blockaded the capital: the Potomac River was essentially closed off to Union shipping, while Confederate supplies could be taken through Charles County in Maryland. Washington made his own attempt at breaking the blockade when a couple of cannon arrived to his care, sent from the "loyal Americans of London," as he wrote later. They were

English guns: Whitworth cannon, unusual and advanced because—unlike almost every cannon of the day—they were breech-loaded rather than muzzle-loaded. "A trial was made one day under my charge from some high ground to see whether we could hit the Shipping Point battery, two miles off. I took aim and hit it every time, with what damage was then unknown, but we could see an ox-team topple over. Later on when that battery was evacuated we went over and found in the surgeon's office on the mantelpiece a Whitworth shot with the label 'This shot was fired by the Yankees on such a day and hit my house.' " The exercise, however, seemed mostly desultory according to Washington's later recollection: "Constant firing took place back and forth without result; a large round ball can be seen coming and is easily dodged." But his writing from the front at the time offers a rather less laconic description: "We had lots of sport yesterday firing into the rebel batteries at Shipping Point opposite," Washington wrote to his younger brother Ferdinand. "Usually the rebels do not think it worth while to answer to our shots . . . but this time they got mad and opened fire on us with 9 inch shell—the moment the guard sees the flash from their gun he cries out 'here she comes' and all hands tumble head over heels for the hole, where they lie in a heap until the shell has passed over or burst . . . the report reaches us about 4 seconds before the shell. Then the shell comes whizzing along slowly—with the fuze burning—if they should happen to burst over the pit we would all go to the devil."

The blockade ended in the spring of 1862—not because of any aggressive action on the Union's part, but because the Confederates, under Joseph E. Johnston, withdrew to better defend Richmond. The news came as a surprise to Washington, who observed the destruction of the abandoned batteries. "This Sunday has been the most eventful with us since we have been here," he wrote to his brother. "We were all sitting quietly in the sun making after dinner reflection, when the report was brought into camp that large quantities of smoke were visible on the other side. Immediately all available tree tops were manned by an eager crowd and the report was found to be true. The smoke and fire seemed to be increasing every moment, until, in a short time, all their camps, to the number of six or more, appeared one mass of smoke extending far up the Quantico. Presently flames began to shoot forth from the steamer *Page*"—a ship that had been instrumental in the Confederate blockade—"and also from 3 schooners which she had captured in the beginning of their career . . . By this time the flames had reached some of the magazines, resulting of course in some tremendous explosions." The batteries were then claimed for the Union.

A few days later Washington went over to the captured Confederate batteries. The Civil War was one in which the two sides were never equally matched: the industrialized North, with its railroads and factories, would always be the greater power than the largely agrarian South. Even as early as 1862 Washington observed, in his visit to the now-deserted Confederate stronghold, a premonition among its soldiers about the way things were headed. "I have just returned from a visit to the sacred soil of Secession," he wrote to Ferdinand. "Their camps are very filthy concerns, most of their huts being burrowed into the hillside to the rear of the battery," protecting them from shell fire. There were, he said, "plenty of provisions lying about, comprising beef, both raw and roasted, rice, potatoes, flour and cornmeal, dried apples and peaches, crackers, soap, coffee mills but no coffee; in fact all the necessities of life were present in abundance but no luxuries . . . A large number of letters have been found; their uniform tenor is hard times, and harder times coming."

But Union troops had no certainty of victory. In the summer of 1861, the Confederate navy had begun to reconfigure the steam frigate *Merrimack* as an ironclad; the famous Tredegar Iron Works in Richmond made two-inch-thick iron plate to protect the superstructure above the waterline; one-inch plate sheathed her 264-foot hull to three feet below. Bristling with guns, her prow had been made into an iron ram, so that the ship herself could be employed as a deadly weapon. Her weak engines, however, struggled against her great weight: a half-circle turn took a half hour. News of her construction prompted the Union to play catch-up; President Lincoln's hastily established Ironclad Board awarded its contract to a Swedish-born engineer, John Ericsson, who promised he could deliver his radical design in fewer than one hundred days.

The USS *Monitor* was like no ship anyone had ever seen. Assembled at the Continental Iron Works at Greenpoint, in Brooklyn, the *Monitor* was much smaller than the *Merrimack*—rechristened the *Virginia*, though the name has never stuck—and had only two guns to the *Merrimack*'s dozen. But the *Monitor*'s guns were mounted in an extraordinary revolving turret encased in eight inches of armor. Her flat deck was protected by iron plates nearly five inches thick: all the ship's vital machinery was protected by this shell; she floated barely two feet out of the water, making hardly any target for the enemy to hit. On March 8, 1862, the *Merrimack* turned her fire on the five Union ships guarding the James River, boasting more than two hundred guns among them. The twenty-four-gun *Cumberland* and the fifty-gun *Congress* were sunk by the Confederate ironclad in the space of a couple of hours, a feat no other enemy would accomplish until Japan's attack on Pearl Harbor in 1941. All the while the *Monitor* had been sailing south for several days; on Sunday morning, March 9, the two ships finally engaged. Neither could gain the advantage and the battle was reckoned a draw: but naval warfare would never be the same. "One conclusion . . . admits of no doubt whatever," observed the London *Times*. "There is an end of wooden ships for ever."

Washington wrote to Charles Swan that his first glimpse of the *Monitor* had startled him: "I had no idea that she was so very small as she really is . . ." In his memoir he recalled how, while they were loading up their equipment into boats at Fort Monroe, "news was received of the first coming out of the rebel 'Merrimac,' [sic] accompanied by great destruction, filling the whole North with consternation. We were too busily employed in twisting the tails of mules to make them go into the boat, and had not time to get much excited. I have always noticed that honest employment is an antidote to nervous apprehension." They sailed on to Hampton Roads and dropped anchor. "Next morning after the mist had lifted a curious sight presented itself; every vessel that had sails or steam seemed to be running away. The big boats went out to sea, the smaller ones up Chesapeake Bay. Presently we noticed in the direction of Norfolk the black roof of a huge barn floating towards us. Nearby we discovered through the haze a little cheesebox on a raft, arrived the night before." The huge barn was the *Merrimack*; the cheese box on a raft, the turret of the *Monitor*. "We were helpless and could not run away . . . We trained a 6-pounder on her but refrained from sinking her!" "Refrained" is surely ironic—were not able to sink her, he means. The *Monitor*, he writes, had to wait to turn her turret, "which made it slow"; the *Merrimack* "had to move bodily to train her guns on the small object." Once again, Washington had nothing but bitter words for McClellan's command. "The moral influence" of the battle "was appalling. It paralyzed the whole of McClellan's campaign, he did not know whether to go ahead, sit still, or what to do. So he did nothing as usual and nearly a month was wasted."

McClellan may have been twiddling his thumbs: but that spring was momentous for both North and South. In February, the Legal Tender Act created, for the first time, a national paper currency for the Union—the "greenback." The huge costs of the war meant that reserves of gold and silver (which up until then had been the only legal tender) were very much depleted; now paper could be used to pay anything except for interest on public debt, and import duties. Later in the year the Revenue Act would be passed— the first tax paid on individual incomes by residents of the United States. On April 16 of that year the Confederacy, already short of men, enacted the first conscription law in American history. And in the west, on April 6, the battle began for Shiloh, Tennessee; by the time it was finished, with a victory for General Ulysses S. Grant and the Union, it would prove to be the most destructive battle of the war thus far, with twenty thousand casualties. It was after Shiloh, Grant wrote later, that he "gave up all idea of saving the Union except by complete conquest."

Washington, back in Virginia, wrote to his father about the plans for the attack on the Confederate capital. McClellan's intention was to advance on Richmond by what was known as "the Peninsula," formed by the York and James Rivers in Virginia. In preparation for his attack, the Confederates, under Major General John Bankhead Magruder, began building their lines of defense, improving the fortifications that had served George Washington well in the Revolutionary War and constructing new earthworks and dams to flood the lowlands. Writing to his father on April 16, Washington described "half a dozen dams" running across the Warwick River, entirely flooding the country for miles around; "each dam is protected by a fort on the other side of the river—here will occur the hardest fighting." When the fighting began, he wrote, "if you want to get deaf for the rest of your life you had better get a pass from Stanton and come down and listen. The firing is incessant from both sides, continuing night and day, the enemy firing only heavy guns." The resulting loss to Union artillery was "severe," Washington said. Heroics didn't come with any guaranteed reward. "Our sharpshooters are showing themselves to advantage: some of them have killed over 30 already; one Californian . . . was so flushed with his success that he refused to be relieved. The consequence was that the next day some rebel marksman put a bullet through his brain." A couple of days later he wrote of seeing a squad of rebel prisoners being marched by: "they look horrid," he wrote, "but they can handle cannon well. Two of the topographical engineers had their left arms taken off yesterday." His letters to his father—practical, astonishingly observant, and richly descriptive of the terrain—reveal that very little passed him by.

McClellan's forces outnumbered those of the Confederacy by as much as ten to one. Washington's account in his memoir—"Another example of characteristic delay, 140,000 men against 5,000"—is exaggerated but gives a sense of McClellan's hesitancy. Famously Magruder deceived McClellan by marching small numbers of men past the same position over and over, thereby making them appear to be a much greater force. Lincoln despaired. "It is indispensable to you that you strike a blow," he wrote to his general. "I have never written you . . . in greater kindness of feeling than now, nor with a fuller purpose to sustain you . . . But you must act." McClellan, ever scornful of Lincoln, paid little heed: he wrote to his wife that if Lincoln wanted to break the rebel lines "he had better come and do it himself." Eventually the Confederates would withdraw; but the delay afforded them valuable time to gather the troops that would beat McClellan back from the Peninsula.

Washington had well and truly "seen the elephant"—slang at the time for experiencing combat. Now, however, his military life would take a different turn. As he was traveling with the army to Williamsburg, he received some startling news. "I was amazed by receiving an order from the Secretary of War to report at once to General Meigs, the Quarter-master General at Washington. I did not know what it meant; so I went, and got there with great difficulty." The summons to Meigs came because, as he told it years later, the attack of the *Merrimack* had put the frighteners on the Union commanders. Meigs had summoned John Roebling to discuss a blockade of the Potomac with wire rope. The plan was eventually abandoned, but "while consulting with Gen. Meigs, the latter thought that light military suspension bridges would be a good thing for the army to have. Mr. Roebling thought so too, mentioning at the same time that he had a son in the army with sufficient technical experience to put them up."

The diversion was significant on all counts. The battle of Williamsburg, on May 5, was a deadly one; thanks to Meigs's order, Washington escaped it. Of the order to the capital he wrote: "Perhaps it saved my life, as many in the whole battery were killed." But more than that: "It changed my whole career in the army; it was the cause of a year or two of hard, disagreeable, thankless, impossible work, because Meigs' ideas were all illusions. Suspension bridges can not be put up in a day across a stream, like pontoon bridges, especially as I had no force of men, no facilities, no tools, no nothing, except my honorable self. As these bridges came only occasionally I was put on staff duty in the meantime, which was most agreeable, and enabled me to take part in a great many battles, at least twenty, besides giving me a certain freedom of life which I prized highly."

The first thing Meigs asked for was a book about how to build military suspension bridges. Washington wrote it in three weeks. General Meigs, however, took all the credit—according to Washington's account. The manuscript titled "Manual on the Construction of Military Suspension Bridges," dated 1862, was illustrated in detail, and came complete with a list of equipment necessary, from pliers to ladders to wire and rope. "Plan and manuscript sent to Meigs" is scrawled on the top, in pencil.

Then—ropes having been made at Trenton—he was ordered to join General McDowell's command at Fredericksburg on the Rappahannock and "astonish the world with military suspension bridges." Washington had no choice but to ignore his own astonishment at the scale of the task that had been set him, and at how little official help he would get. Throughout his life, he would distrust the machinations of governments, of the public and private bodies whose demands he would have to answer: the stoicism, the ironic resilience he exhibited in years to come surely had its foundation in the work he did during the war. In June, it looked as if a bridge would be built at Front Royal, Virginia—but no sooner had the work started, than the Union forces were withdrawn and the work was abandoned. In a letter to his brother Ferdinand, Washington sounds not a little relieved: "It is rather a hard matter to build bridges here, the nearest place to get a plank is Alexandria, the same with every thing else. I had two double track road bridges to build, one over the north fork & the other over the south fork of the Shenandoah, each about half a mile above the junction. One had 3 spans of 125' & the other 2 spans of 125'. The water runs like a mill race; nothing in the name of a boat was to be had, and to get the length of the spans I had to swim across with the tape in my mouth."

He was sent to build a bridge over the Rappahannock, but was not encouraged at the prospect. "Here I stood in an utterly false position, without a man or mechanic or a tool or material to accomplish anything with," he

recalled. Drawings, surveys, estimates—all had to be drawn up, and he had
to send to Washington, D.C., for the equipment he needed, all of which was
granted "most grudgingly and after an infinite amount of official red tape."
Aside from equipment, of course, he needed men to build his bridge. "A
requisition was made on Gen. King for men, which he refused to grant; fortunately while I was talking to him he fell down half dead in an epileptic fit,
so that the Adjutant General let me have the men just to get rid of me." One
hundred runaway slaves were made available to him; some were more useful
than the Union soldiers, "none of whom were mechanics or had any skill
whatsoever." Black carpenters dressed the wood for the towers; he found two
white Southern carpenters, who had not yet joined the Rebel army: "they
were more useful than all the rest put together." Runaway slaves and white
Southerners together worked on Washington's bridge.

Lieutenant Roebling—he had been promoted in February of that year—
was chief engineer of the work; he was quartermaster, too, and in charge of the
commissary, organizing food and lodging for the men under his command.
He himself slept in the old stone jail, and his first night there gave him a
fright. "In the jail stood a huge long box; no one knew its contents; curiosity
drove me one day to pry open a board." Under his hand he was startled to
feel the contours of a still, cold face. "I was," he wrote laconically, "a little
startled" by the discovery—which proved to be a statue of Mary Washington, mother of George, which had been stored in the jail for safekeeping.

When the cables arrived, oxen had to be found to drag them over the
shallow river to be hauled up over the towers. When it was time to put up
the bridge's floor beams—gathered from local houses torn down for the
purpose—and hang the suspenders from the main cables, "Most of the men
got dizzy and ran away." The work was dangerous, too: to Charles Swan he
wrote that two men fell into the river while working on the bridge; another
nearly broke his back on the land span.

"Finally the little bridge, over a thousand feet long, was opened to travel
by marching a brigade of cavalry over it," Washington wrote in his memoir
of the war. "I question whether it was worth the trouble to put it up, certainly not as far as I was concerned personally." The bridge was in use for
just over a month. When the place was evacuated, General Burnside blew up
the anchorage, causing the bridge to collapse.

Major General John Pope had now been put in command of the Army of Virginia. This appointment was, according to Washington, "One of the great mistakes made by the people of Washington; [he was] a bombastic, vainglorious self-boaster, with no ability or experience . . . he lacked foresight, caution, and had no initiative; waited until it was too late and under-rated [General 'Stonewall'] Jackson." But for the moment, Washington's own life was a little more pleasant. "I now had staff duty of a most agreeable kind. I was very expert at reconnoitring, [sic] in making sketches and maps of the country as I went along, of finding the enemy's position, and getting a knowledge of the terrain. I had carte blanche to go anywhere I wanted in the whole army or surrounding country. One dangerous trip to Madison court house was made within the enemy's line. I was almost a spy." Once, surveying roads in Virginia, he was in the saddle for more than thirty-six hours, "riding my horse to death which I bought only a week before," he wrote to his father. But little was seen of the enemy during the trip—"and not more than 8 or 10 stragglers captured; on the way out we took by-roads but returned on the main road, and here it was at 5 o'clock in the morning that we surprised the rebel Maj. General Stuart and staff at breakfast . . ." Stuart got away because the horses of Washington and his fellow soldiers were too tired to give chase. In a "claim for indemnity" for the poor horse, he noted he had traveled nearly eighty miles in that time, with no forage for the animal at all.

To his father, he expressed dismay at some of the actions of the army in which he served. "While at the river I had occasion to see the ruin of the buildings at the famous White Sulphur Springs," he wrote. "Our men destroyed them the day before, an act only in keeping with all their other acts of vandalism. I have seen numerous instances myself where men with their families who considered themselves rich the day before, were homeless beggars the day after, not knowing where to get their next meal; most of the destruction is perfectly wanton, and not necessary, and only calculated to make the inhabitants your bitterest enemies."

That letter to his father was begun on August 24 and concluded in the first days of September after what became known as Second Bull Run, a Union defeat—at a cost of more than 22,000 casualties to both armies—that would bring the career of General Pope to a close. Just a week after the battle ended,

Michigan general Alpheus S. Williams wrote of "a splendid army almost demoralized, millions of public property given up or destroyed, thousands of lives of our best men sacrificed for no purpose." To Williams, as to Washington Roebling, the cause of the defeat was clear: General John Pope. "I dare not trust myself to speak of this commander as I feel and believe. Suffice it to say . . . that more insolence, superciliousness, ignorance, and pretentiousness were never combined in one man." If some of Washington's later writings about the war display the benefit of hindsight, he was always a clear-eyed observer, even in the thick of battle. Of the last day of the fight he wrote to John Roebling of the poor leadership that contributed to the rout: "During the night no disposition was made on our part to properly meet any charge in attack . . . our generals seemed perfectly content to sit still and allow the enemy to do what he pleased . . . The running of our men"— by which he meant, their desperate flight away from the fighting—"had already commenced, at least 10,000 were on the full go. Many had not even heard the whistle of a shot before they ran. This was a most humiliating spectacle, showing the utterly demoralized condition of the men." He wrote scornfully of Pope's ordering a retreat, despite the fact that "we fired 20 cannon shots to their one and in place of lead they fired stones from their muskets."

His own reserve of will was running dry, but he could not fail to admire the will of the enemy the Union Army faced. He was, he wrote to his father, "completely tired out and used up; I have not had one meal a day for the last 3 weeks, have slept on the ground every night generally without blankets and been in the saddle constantly; I have also been lucky in not getting shot . . . As for the future I have no hopes whatever; I assure you on Saturday night last I felt utterly sick disgusted and tired of the war . . . Our men are sick of the war, they fight without an aim and without enthusiasm, they have no confidence in their leaders except one or two . . . In the next place one Rebel is equal to 5 Union men in bravery that is about the proportion." One year later he returned to the battlefield of Second Manassas in the company of his new commanding officer, General G. K. Warren. Hogs were rooting up corpses from shallow graves, and he saw the skeleton of a Confederate soldier. "The ants had polished the 6-foot skeleton to a brilliant whiteness; a colony of wasps had established their nest within the skull and were flying in and out of the eye holes—a sad commentary on the glories of the martial field, but hardened as I was I could not resist shedding a tear."

At the time, there was hardly time for tears. The Army of the Potomac mounted a series of assaults against Lee's forces in Maryland, one of which would end in the bloodiest single day in American military history. After the Battle of Antietam in mid-September 1862, nearly six thousand men lay dead; another 17,000 were wounded. Washington—who had already remarked to his father at his luck on not getting shot—was in the thick of it. His corps, under Major General Joseph "Fighting Joe" Hooker, struck out over Antietam Creek: "McClellan then ordered Burnside and Hooker's corps to cross the creek that evening and get into position," Washington recalled. "We had no bridge to cross on, but built one over the deep creek in two hours so that our corps was over before sunset." It was, he wrote with some restraint, a "sharp fight"—"while speaking to Gen. Ricketts I was nearly killed by a cannon ball sucking the wind out of my mouth."

Washington, among Hooker's men, fought near the Dunker church that still stands on the battlefield: a plain, whitewashed structure built by a pacifist German sect who believed in full-immersion baptism. Burnside wasted time defending the bridge across Antietam Creek when his men could have waded the fords nearby: "He was practically behind the enemy—any other general would have won the battle in his position," Washington wrote. In the heat of late summer, the ghastliness of the fighting's result was hardly to be imagined: "The appearance of the battlefield was horrible; the hot . . . sun changed a corpse into a swollen mass of putridity in a few hours—too rotten to be moved. Long trenches were dug, wide and deep, into which bodies, thousands of them, were tumbled pell-mell, carried on fence rails or yanked with ropes, unknown, unnamed, unrecognized. This is the kind of glory most people get who go to war." Washington Roebling's remarkable map of the battleground, made on yellow paper just a day after the fighting ended, measured almost twenty by twenty-five inches, each field delineated differently; trees and buildings were carefully drawn in perfect detail. On the right of the map was marked the ford by which Hooker's men crossed Antietam Creek, and the line of his advance around to the right (over the top of the map). In the middle of the map, just south of a little cornfield was marked in Washington's tiny writing "place where Hooker was shot in the foot." The map has the purity of an abstraction, for all its precision; yet Washington would have drawn it when those bodies he described so dreadfully—and which can be seen in Alexander Gardner's famous photographs of the battlefield—would still have been lying where they fell.

In terms of strategy, the battle was a success for the Union—though McClellan allowed Lee's army to escape back into Virginia. "Feeble pursuit was instituted, but [Lee] recrossed the Potomac unmolested," Washington wrote. But the Confederate defeat frustrated the Southern states' hopes for British recognition, and precipitated one of the most significant events in American history. Five days after the battle, Lincoln told his cabinet of the covenant he had made with God: if the army drove the rebels out of Maryland, he would issue his Emancipation Proclamation. Unless the states in rebellion returned to the Union by the first of the year, the slaves held in those states "shall be then, thenceforward, and forever free."

Not long after the battle, Washington fell ill. "The excitement which had kept me up now passed away and I collapsed completely," he recalled. "A fever possessed me, made me delirious for several days; I crawled to a field hospital just established, and laid for ten days." Whatever caused his illness, his father's belief that all sickness could be resisted if the correct mental attitude was brought to bear never left him, no matter how much he protested against the idea. "I could have avoided it all by more self-control!" Washington insisted. "All diseases are unnecessary." After he recovered he and his corps was stationed at Harpers Ferry, Virginia, where a bridge over the Shenandoah River had been destroyed. "It seemed advisable to replace it." With a personal order from General McClellan, Washington set about a task that took him four months, to the end of the year. He sent for old David Rhule, one of his father's foremen, to help him out as "a civilian expert"; Rhule "was anxious to do something for his country." The soldiers seconded to work on the bridge—there seemed to be less trouble about that this time— "looked on it as a picnic and were glad to come." Trouble came, as ever, from the generals in charge, notably the Baltimore-born General John Reese Kenly. "This general has very queer ideas about bridges," Washington wrote to his father in the last days of 1862. "[He considers] that they are merely avenues by which the Rebels might get after him and are, therefore, very dangerous structures." When Kenly appeared, three quarters of the floor beams had been put up on the bridge; the general believed that putting planks or gangways up to reach the flooring might allow the enemy ingress. Washington was stumped. "As none of my men could fly, I was compelled to lay down a track of two planks wide up on which to jack out the stringers. This fact came to the ears of the General, some days since; I was sent for in great haste, received a great lecture upon the enormity of the crime committed, etc etc and was ordered to remove every plank that night." Any planks put up during a day's work had to be removed each and every night. "Well, to work in this style was perfect murder and I had determined to leave the bridge and go to Washington for new orders," an exasperated Washington wrote. But then Kenly was relieved by General John Henry Kelly—"The latter is a man, and consequently affairs here assumed a new aspect directly." Looking back at the work at the beginning of the new year, Washington, in a letter to his exacting father, allowed that he was not displeased—though a little note of self-criticism displays the characteristic Roebling trait, a desire for constant improvement. "The bridge has turned out more solid and substantial than I at first anticipated; it is very stiff, even without a truss railing, and has been pretty severely tested by cavalry and by heavy winds; the weather was so bad of late that the men could scarcely work half the time . . . I am sorry I did not make my cables a little stronger . . . so as to be more in keeping with the rest of the bridge."

President Lincoln was after improvement, too. In November 1862 he finally got rid of George McClellan, replacing him with Ambrose Burnside as commander of the Army of the Potomac. But Burnside wouldn't last long: in December 1862, the Federals suffered a catastrophic defeat at Fredericksburg, Virginia: over the course of four days' fighting Union losses ran to more than thirteen thousand men against the Confederacy's five thousand casualties. In January 1863, "Fighting Joe" Hooker was put in command; and in March 1863 the Conscription Act was passed, the first draft in the Union, to feed the seemingly insatiable appetites of the cannons' mouths. That, along with the enactment of the Emancipation Proclamation at the beginning of the year—which led many white workers to believe that newly freed slaves would drive down wages—gave rise to what became known as the Draft Riots in New York City that summer: more than one hundred people were killed in the riots, the worst in American history.

In the spring of 1863 Washington had applied to General Meigs for a new assignment of duty. "I think he was tired of military suspension bridges by this time," he remarked drily in his memoir. He was sent to report to General

Woodward, and then to General Cyrus Comstock, the army's chief engineer; but after the Battle of Chancellorsville, in May 1863, Comstock was transferred to the Army of the Tennessee, in the western theater of the war. And so Washington was turned over to General G. K. Warren, chief topographical engineer of the Army of the Potomac—"with whom I contracted a lasting friendship and with whom I served during the remainder of the war." Warren's influence on Washington Roebling's life would be, in more ways than one, much greater than the young lieutenant could ever have imagined at the moment of their meeting.

"The army was recovering from the failure of Burnside's attack on the Heights back of Fredericksburg," Washington recalled in his memoir, in reference to the wave after wave of Union troops that had fallen assaulting the Confederate position at Marye's Heights. But morale was improving. Hooker was popular and was infusing "a new spirit" into the men. Now, Washington's engineering skills were put to use in building "corduroy roads"—cut logs laid down side by side on the ground—as it was "mud season" and any travel would otherwise be impossible. "All real Americans know how to swing an axe." No doubt his rural Pennsylvania childhood came in handy, here. Around the same time, he discovered that "the suspension bridge I had built the year before over the Rappahannock had disappeared. So much useless work is done in a war!"

With Hooker in ultimate command, Washington crossed the Rappahannock as the army headed for Chancellorsville: a battle that would bring victory to Robert E. Lee, but at the cost of General Thomas Jonathan "Stonewall" Jackson's life, a loss the rebel army could ill afford. Washington recalled crossing the Rappahannock at night and catching sight of his friend, a Captain Cross, with whom he had been at Troy—only to see Cross "killed by a bullet through the forehead." At then, at noon on the first of May—the second day of the battle—Washington received a summons from General Hooker himself. As Washington recalled it, the general spoke to him in no uncertain terms. " 'I have determined to receive the enemy on my bayonets here at Chancellorsville,' Hooker said. 'I want you to ride ahead to Gen. Slocum and tell him to stop the advance and return here with his command.' To hear was to obey," Washington recalled. General Henry Slocum—who didn't like Hooker one bit, and who was already advancing his division—reacted with fury. "When I reached Slocum the steeple of Zion Church, Salem, was already in sight. When I gave my orders from Hooker, Slocum turned on me with fury, saying 'Roebling you are a god-damned liar, nobody but a crazy man would give such an order, when we have victory in sight. I shall go and see Gen. Hooker myself, and if I find that you have spoken falsely, you shall be shot on my return.'" Slocum couldn't believe that his commanding officer was effectively ordering a retreat, or at least a holding action rather than an advance. "Off he went, the advance was stopped. The battle of Chancellorsville was lost right here." Admittedly, Washington's account of Slocum's behavior may well have been colored by their postwar encounters; the real conflict between them was yet to come.

The high ground, Washington recalled, had already been gained; and Chancellorsville was low and indefensible. "Hooker remained obdurate. He had that ugly streak in him, which a man gets when he is only half drunk." Hooker's drinking was the subject of much bitter debate; some argued he was much more than half-drunk at Chancellorsville. In correspondence with Gamaliel Bradford, a historian of the Civil War, Washington later recalled that perhaps by the standards of the time, Hooker's consumption was not necessarily notable—but still, it had its effect. "Fifty or a hundred years ago it was a common custom for people of all ranks to take a toddy several times a day, be it whiskey, brandy or a bitters. Such was Hooker's habit, so long continued that he could not break himself of it . . . I never saw him what is called real drunk, nor did anyone else. But the consequence of such indulgence is that when an emergency arises, requiring a strong bracer, the system cannot respond and a collapse ensues." Hooker, he insisted, was "a used-up man."

Once again, at Chancellorsville, Washington had a narrow escape—as did his commanding officer, thanks to his subordinate's quick thinking. The tale told in recollection perhaps has a bit of the drama of an old soldier conjuring past glories, yet still bears repeating. "The view of the battlefield was a harrowing sight. Swept by cannonshot and bullet in all directions, encumbered by the dead and dying, affrighted fugitives running in all directions, not knowing which way to turn . . . While riding over to the first corps a confederate came out of the woods close by and fired at me with deliberate aim—an excellent line shot, but too low, as the bullet passed through the sole of my boot and belly of the horse." But that was not the end of it: "On the porch of the Chancellor House a bullet struck me in the forehead. A very stiff visor on my cap deflected it into the cheek of a bystander, who cursed me with this other cheek." And as he leaned against a column of the Chancellor House—which gave the little Virginia hamlet its name, being home to Frances Chancellor, a widow, and her family—watching the battle, Hooker came out of the house and leaned against the same column. "At that moment I saw a ball coming which bid fair to hit him. I yelled 'Get back, General!' He moved less than a foot, just enough not to be killed." He was however knocked down—and out. Washington was convinced he was dead, and rushed into the house crying "Hooker is killed!" His staff dragged the general off into a tent. "Oh, would that I had kept silent and allowed him to be killed. The battle of Chancellorsville might still have been saved!"

The battle, as far as Washington was concerned, ended ignominiously. He was sent off into the woods to lay out an entrenched line as a further safeguard—but got lost among the trees, and then it started to pour. "That night we retreated. Every corps for itself. No one to give orders . . . The astounding farce had come to an end." Hooker, he opined, was "a moral fraud." He reckoned the general might well have wished to be lost in the woods, too. "He would have liked to hide in the bushes, but with thousands of eyes upon you the commander has not the privilege of the private." To Bradford he later wrote: "I sometimes wonder why Lee and other confederates take so much credit for winning this battle, when this poor half demented man [Hooker] was doing his utmost to help him win."

OUT IN THE west, Union fortunes improved. Ulysses S. Grant's forces converged on Vicksburg, an important stronghold on the Mississippi, which surrendered on July 4. Back east, Washington began some of the most unusual work of the war.

As early as 1794, the French army had used aerial reconnaissance: Captain Charles Coutelle and General Antoine Morlot rose above the battlefield of Fleurus, in the Austrian Netherlands, in the balloon *L'Entreprenant*, as the French Republican Army fought the Austrians and the Dutch. Their balloon was filled with hydrogen, and tethered to the ground: visibility could be as much as 18 miles. During the course of the day-long engagement on June 26, Coutelle and Morlot remained aloft, communicating with the ground via the tethering cables, questions and orders traveling up and down in a bag. This was, essentially, the first battle ever to be won thanks to control of the air.

Across the Atlantic, by the middle of the nineteenth century, New Hampshire native Thaddeus Sobieski Constantine Lowe was coming to public attention. Born in 1832, Lowe was an early convert to the nascent science of flight: at the age of eighteen he had attended a lecture on lighter-than-air gases, and his future was set. By 1859 he was planning to fly across the Atlantic in his airship, the *City of New York*. *Harper's Weekly* reported that the ship was constructed "of twilled cloth, oiled and covered with three coats of a peculiar kind of varnish, manufactured by Mr. Lowe, and calculated to render the balloon envelope air-tight. Six thousand yards of cloth have been consumed in the work, and seventeen sewing machines and have been unceasingly occupied in stitching the pieces together." The airship—later rechristened the *Great Western*—never made it across the ocean, but an attempted flight from Cincinnati, Ohio, to Washington, D.C., caught the attention of the federal government. Lowe had taken off just a week after the fall of Fort Sumter, in April 1861; blown off course from Cincinnati, he landed in South Carolina and was promptly arrested as a Northern spy. (The local rural people were astonished both by his tall silk hat—and by the fact that water he carried in rubber bottles had been frozen by the cold of the upper atmosphere: "One man asked how anyone but a devil put so large a piece of ice through the nozzle." He was set free thanks to his connections to the local intelligentsia.) Eventually he presented himself, and his ingenious balloon, to President Lincoln, who was especially impressed by Lowe's dispatch of the very first telegram from the air to the ground. Lowe was made chief of the Corps of Aeronautics of the United States Army—which is how Washington Roebling came to spend his twenty-sixth birthday in the basket of a balloon.

Lowe's tenure as chief aeronaut was short-lived. He became ill with malaria, contracted in the mosquito-infested swamps near the Chickahominy River, and he chafed under military command; he resigned his commission at the beginning of May 1863. But his balloons—made practical by his invention of a portable hydrogen gas generator, allowing the balloons to accompany the army anywhere in the field—still served; and Washington, as he tells it, "was elected to make the observations" as the army moved north. "A captive balloon was allowed to go up 12 to 1500 feet," he recalled. By "captive" he means that it was tethered to the ground, rather than floating free. "The ascents were made before sunrise, being the only time when you can observe the morning campfires, otherwise the thick woods hide the view." Washington's physical courage was challenged, at first, by flight. "The first ascent makes one a little nervous," he wrote. "The balloon turns around constantly, mixing up the points of compass. In a high wind it slants so that you are almost thrown out. To use glasses and maps required dexterity and still more was needed to know what you were looking at." But after that first ascent, "the rest are easy." And it was, he knew, in some ways at least an enviable task. "In clear weather the view was grand. The Blue Ridge slowly rose above the horizon in the west. The Potomac appeared as a silver thread, even the Chesapeake rose to sight. The slightest whisper is heard from below, but from above down almost nothing is heard."

AT A STRATEGY conference with President Jefferson Davis in Richmond, in the middle of May 1863, Robert E. Lee revealed his plan to move his army into Pennsylvania, taking the war into the Union's home territory. Washington Roebling, up in a balloon at dawn, saw which way Lee was heading. "One morning the campfires were missing, a fact which I communicated as probably indicating Lee's westward march," he recalled. Two days later, he wrote, two Confederate deserters confirmed the fact—and so, in the middle of June 1863, "we commenced to follow Lee." When it turned out that there were no good maps of Pennsylvania available, Washington was sent off to Philadelphia to acquire some; having secured them, he was able to make a flying visit home to Trenton where he found everything "in despair and confusion. They were so prostrated with gloomy forebodings . . . I took the next train back to rejoin our army which was the only safe place I knew of." His father, Washington recalled, was quite desperate at the thought of what might transpire. "He was sure that Lee's army would beat ours, that he would capture Philadelphia, that he would come up and take Trenton, burn up his wire mill, his rope shop, his fine new house, and make him a beggar on the face of the earth. Down in the army we did not realize how the people at home were scared to death. I returned to Gettysburg with an indifference which my father could not comprehend." His mother wept as he left; and it was a significant parting. "The next time I saw her she was lying on her death bed, never to recover."

Washington made his way back south, hoping to meet with General G. K. Warren in Baltimore, but Warren had gone. And so he headed for Frederick, Maryland, the church bells ringing to summon the people to defend the earthworks around the town, as Lee was reported to be near. "This was dubious news for me," Washington wrote. "It was a glorious June day, clear, bright and warm. I rode on the great pike for 30 miles without meeting a human being—most incomprehensible for me." The eerie solitude, he later learned, was on account of the Confederate cavalry having swept past the day before; the local population had run away or gone into hiding. He slept one night under an apple tree; still, all was deserted, although "I knew whole armies were near by." He kept riding and riding; eventually he thought he heard the sound of cannon; when he put his ear to the ground he could hear it plainly, thundering in the distance. Having realized he was in Pennsylvania, he slept that night in an old iron mine—and the next day, at last, met up with the Fifth Corps at Littlestown, fewer than a dozen miles from their destination: Gettysburg.

Despite the might and men of the Union, in the summer of 1863 the war still seemed to hang in the balance. Lee—who had been offered the command of the Union Army but refused it at the start of the war, saying he could not take up arms against his native state, Virginia—was a courageous and visionary soldier; many Southerners fought, as they perceived it, against the invasion of their homeland.* When Washington met up with the marching line of Union troops, he discovered that there had been another change in the Union command: after Chancellorsville, Hooker had been replaced by General George Gordon Meade. "How different was Meade [from Hooker] at Gettysburg! The qualities of a great commander showed in his every act!"

---

*Historian and novelist Shelby Foote expressed this sentiment in his account of a ragged Confederate soldier captured by Union troops, and asked—since it was clear this man could own no slaves—what he was fighting for. "I'm fighting because you're down here," came the reply.

Confederate general A. P. Hill had learned of a supply of shoes in Gettysburg—and authorized his division to acquire them. There they encountered Union forces, who had expected a fight in the vicinity: the town was at a crossroads, surrounded by ridges and hills, good vantage points for defense. The battle commenced in earnest on July 1; the next day Lee ordered General James Longstreet to attack Union forces at Cemetery Hill, just south of the town itself. Longstreet resisted, thinking that was just what the enemy would expect. But his commander left him no choice; Meade rushed reinforcements into the fray, toward a wheat field, a peach orchard, a maze of boulders called Devil's Den, and a hill called Little Round Top. Had the rebels got their artillery up on that hill, they could well have gained the advantage over the field—a brigade of troops from Alabama moved to seize it, and there was nothing but a Union signal station in their way.

General Warren recognized that Little Round Top was undefended. Roebling's brigade—commanded by a former student of Warren's at West Point, Captain Patrick O'Rorke—had been marching along the road beside the hill when Warren commandeered them for its defense, calling to O'Rorke to bring his men up the hill double-quick. Suddenly, as Warren's biographer said of him, "the methodical engineer [was] transformed into a demon of blazing energy. If Warren, the experienced, cool Regular, declared that an emergency threatened, the danger must be there." Up the hill the company stormed—guided by Lieutenant Roebling.

In his memoir, Washington stated that not until he revisited the battlefield in 1907 did he really understand what happened that day, such was the confusion all around him. He saw the wheat field and the peach orchard that had been "the theater of bloody conflicts"; and he recalled "the dreadful carnage" he had witnessed when General Meade had sent them in the direction of Little Round Top because he had heard "'a lot of peppering going on over yonder' and had said, 'Suppose you ride over there and see what is going on and if there is anything serious see to it that it is met properly.'" Peppering—Meade's exact, understated word, Washington assured his readers—was indeed what they had found. Washington rushed up the hill, "finding a signal officer crouching behind the rocks—as soon as I showed my head above the stones, bullets began to whistle about my ears. I could see confederates in the woods in front—I rushed back to Warren and we both came up. In these few minutes the enemy had approached in force . . . Not a moment was to be lost. The summit of Little Round Top must be held at all hazards . . . The way was so steep that the horses could not pull the guns. We had to help pushing them up by hand." Sharpshooters picked off the men who got to the summit: Washington himself—once again—had a narrow escape. "Had Warren arrived five minutes later it would have been too late and we would probably have lost the field. In the desperate fighting that raged around this spot a person forgets himself. The urgency of the moment overpowers everything. After it is all over then your nerves think about breaking down."

After they felt holding the hill was certain, Washington rode over to Meade with the news and "gave him at least one ray of comfort" and then returned. The sharpshooters continued to ply "their murderous trade far into the night with deadly effect." In desperation, thirty Union men were detailed to fire at each Confederate flash—"hoping that one ball at least might find its billet. Next morning one corpse was found in the cleft of a great rock without a mark on it. The concussion of a number of shots must have killed him."

Many years later Washington wrote to General James F. Rusling of Trenton in reply to a query about his service during the war. Of Little Round Top he said drily, "There is no credit attached to running up that little hill; but there was some in staying there without getting killed." To the left of the brigade that now held the high ground, Joshua Lawrence Chamberlain—who, only a year before, had been a professor of rhetoric and languages at Bowdoin College—wheeled the Twentieth Maine around and charged the Confederates with bayonets fixed, their ammunition having run out: it was one of the most famous assaults of the war. The holding of Little Round Top was a turning point in a battle that is widely seen as the turning point of the war. The next day, July 3, Lee would order a charge, led by Major General George Pickett, against the center of the Union line. "Pickett's Charge" has been said by many to be the high-water mark of the Confederacy, an expression of the Southern states' supposed gallant ideals. Rebel soldiers "fell like leaves in an autumn wind," the historian James McPherson has written. Fourteen thousand Confederate soldiers surged forward; only half returned. "The results of this victory are priceless," wrote the diarist George Templeton Strong in New York. "The charm of Robert E. Lee's invincibility is broken."

Before the charge, a cannonade of Rebel gunfire erupted in an attempt to break the Union lines. Washington, down from Little Round Top, had got himself back to the "absurd little two-room log house, misnamed headquarters" when the Confederates opened fire. Washington described 138 cannon discharged as one volley, the firing being kept up for an hour: "Nearly forty-five years have elapsed since that day," he wrote in his memoir. "The impressions left by that are as vivid now as then. I fail to find words which would portray the absolute paralyzing horror of the situation. It was of no use to try and run away; all you could do was to stand still and commend your soul to God . . . A fragment of shell flying in through the door knocked down the table at which Col. Paine and myself were looking at some maps; we promptly ceased that occupation." A surveyor and fellow soldier, William Paine would become one of Washington's most trusted aides in Brooklyn. A group of horses were tied up outside the house; "a dozen were quickly killed." He described "the sickening thud of a ball passing through a horse . . . neither was it an infrequent sight to see a poor beast fly to pieces from a shell exploding inside of it." As a rule, however, the enemy's guns overshot their mark: "Hence the line of our troops lying flat on the ground behind a slight rise were reasonably protected."

The cannonade stopped in advance of the charge. "Every available gun of ours, over a hundred in number, had orders to open on Picket's [sic] line as soon as it appeared." Washington was near the firing batteries of cannon. "All the enemy who actually penetrated our front were captured or killed. The field they had advanced on was strewn with dead, the wounded, and those who had thrown themselves on the ground to escape annihilation. No troops could have withstood our terrible fire." Perhaps with the benefit of hindsight, Washington wrote that he was certain this was a momentous day. "We had a feeling that this was the last effort on the part of Lee, not because there was not plenty of fight left in his men, but because they had nothing to eat. The country had been ransacked of all available food supply, and as he had no railroad leading south to his own magazines, there was no alternative left but to starve or retreat."

That night he lay down to sleep in the open furrow of a plowed field. Despite a downpour so heavy that many of the wounded were drowned where they lay, he slept soundly, he recalled; and the next day rode across the wreckage of the battlefield. "You could almost step from corpse to corpse. Many bodies still moved spasmodically, showing they were not quite dead. Our little headquarters had become a surgeons' shambles, with a pile of legs and arms six feet high back of the little window . . . After you have seen a certain amount of misery and suffering one becomes hardened to it. There is room for only so much pity in the human breast." He reflected in his memoir upon all the books that had been written about what had taken place at Gettysburg, "mostly by people who were not there, and based upon facts that were slowly gathered during months and years afterwards." But so it must be: "The eye witness can only describe the great events as they pass before his vision." As for credit and reputation, "That is given by newspapers to the first man who strikes their fancy, and there it stays."

More than 165,000 soldiers were engaged at Gettysburg, at a cost of more than fifty thousand casualties. And yet General Meade did not pursue Robert E. Lee and his army, even in the wake of victory. "Great God!" Lincoln cried when he heard of Meade's reluctance. "What does it mean? . . . There is bad faith somewhere . . . Our Army held the war in the hollow of their hand and they would not close it." Washington recalled the outcry at the time: "The papers raised a fearful clamor . . . Dispatches came hourly, Why don't you attack! Why don't you reap the profits of your victory! None of these people knew or cared about the real situation . . . Nothing is so disheartening as the attack of a vile, irresponsible, rascally newspaper, whose ignorance is only exceeded by its animosity." It was decided to attack "the next day"—July 4—but Lee had already fled. "The bird had flown that night," Washington recorded. But according to his account, an attack against the Confederates' line would have been fatal, for in falling back they had made a strong entrenched line, "built of logs and earth, well chosen, ditched in some places with strong abatis [branches of felled trees placed to entangle attackers] in front. We could not have taken it. That we did not make the attempt was a great disappointment to Gen. Lee."

The Army moved instead to Harpers Ferry, where Washington was able to observe the fate of the work he had done some months before. "When we arrived within a mile or so I examined with a field glass the suspension bridge which I had built the winter before across the Shenandoah. I could see that the enemy had cut the suspenders and let the floor fall into the river, but the cables hung there apparently undamaged . . . I immediately got a detail of men and a number of wagons and commenced to load up timber and lumber and boards, obtained by tearing down houses and robbing the Baltimore & Ohio railroad. When the troops arrived I was already busy rebuilding it. In two days I had it finished and part of our army crossed on it." A year later the Confederate general Jubal Early would destroy the bridge entirely. Washington always knew that a bridge a friendly army can cross is also one an enemy can make use of; on at least one occasion he had left instructions for destroying his own handiwork, should that prove necessary in the face of an advancing army: "Cut the Wire Ropes where they emerge from the ground at either abutment, by means of a cold chisel and hammer. If no cold chisel or hammer should be available, let a man knock out the pin from one of the shackles connecting the wire ropes; Or, cut down a wooden trestle pier with axes."

7

_7_

# "I am very much of the opinion that she has captured your brother Washy's heart at last"

WASHINGTON'S WAR NARRATIVE, INSERTED almost involuntarily, it seems, into what was meant to be solely a biography, ends just after the Battle of Gettysburg. The war was not the subject he was meant to be addressing; and yet he found it impossible to keep away from the subject. "When I start to write about the war I never know when to stop," he remarked to General Rusling in 1916. But Washington was an assiduous correspondent during the war, writing to his father, to his brother Ferdinand, and to Charles Swan. Letter writing was a way to while away the time when the army was doing, well, nothing. "We are at our old work again, sitting still with all our might and main," is a typical refrain. That remark was made in August 1863; General Warren had been put in command of the Second Corps—filling in for General Winfield Scott Hancock, who had been badly wounded at Pickett's Charge—and Washington was his principal aide-de-camp. Washington recognized that the army's presence in the country alone was costing the Confederacy dear: "I think that in a few days our cavalry pickets will be along the banks of the Rapidan," he wrote to Swan late in the summer. "In one respect this occupation of the country is good because our army devastates the land like a cloud of locusts, and compels the inhabitants to move farther into Virginia and then become a burthen to the people there . . . One thing is certain that this army will now take Richmond no matter who the generals are."

But all the same, every yard of ground had to be fought for, and Washington's war ground on. In very late November and early December 1863, in

Orange County, Virginia, General Meade made an attempt to strike at the right flank of the Confederate Army; but the field fortifications that Lee had prepared in the little valley at Mine Run proved a match for the Union Army—as Washington himself, along with Warren, discovered personally. "At break of dawn by light of the moon Warren & I crawled on our knees close up to Lee's works, found them fully manned—high & strong, built the year before—no assault could have succeeded—Ten thousand men would have been slaughtered . . . Mine Run was lost the day before when the works were unoccupied and we could have walked in, but waited for nothing."

Despite the mud and slaughter some diversion was to be had. On February 22, 1864, Washington Roebling found himself invited to a ball. "You know the 3rd corps had a ball some six weeks since and the 2nd was determined to put that ball into the shade entirely," he wrote four days later to his sister Elvira, his closest confidante in such matters. The evening was, as far as Washington was concerned, a spectacular success. "Our supper cost 1,500 dollars and was furnished by parties in Washington. The most prominent ladies of Washington were present from Miss Hamlin"—daughter of the vice president—"Kate Chase"—the striking, politically powerful daughter of Salmon P. Chase, Lincoln's treasury secretary—"and the Misses Hale down." The Hale girls were the daughters of the senator from New Hampshire. But these women were not the reason for Washington's letter to his sister. "Last but not least was Miss Emily Warren, sister of the General, who came specially from West Point to attend the ball; it was the first time I ever saw her and I am very much of the opinion that she has captured your brother Washy's heart at last. It was a real attack in force, it came without any warning or any previous realization on my part of such an occurrence taking place and it was therefore all the more successful and I assure that it gives me the greatest pleasure to say that I have succumbed."

What they said to each other that night, the way they danced, what she was wearing, the gleam of the candlelight on the buttons of his officer's tunic—all this is lost, but Washington's lines to his sister are true and clear. Still, he wasn't quite ready for the news to get out. "Now don't go like a great big goose and show this letter to everyone, will you dear?" He admonished his younger sister. "No don't! You are my favorite sister you know just as she is the General's favorite sister and can therefore appreciate my feelings. I can just appreciate your feelings at reading this letter and therefore await your

.

speedy answer with impatience." He added a postscript, just the kind of detail a young man notes about his beloved: "She gets a sore throat once in a while and is additionally charming therefore." He signed himself off, as ever, "Your affectionate brother Wash."


IN THE EARLY autumn of 1620 the *Mayflower* set sail from Plymouth, England, to the New World. Among its passengers was Richard Warren, a merchant of London, and one of the young colony's most valued members. Upon his death in 1628, it was remarked that his loss was a sore one, for he had been "an useful instrument and during his life bore a deep share in the difficulties and troubles of the first settlement of the Plantation of New Plymouth." He was among those who first encountered the land's native inhabitants, as William Bradford recorded. His wife, Elizabeth, survived him— as did their seven children, two sons and five daughters, ensuring that Warren's line would be among the most prolific of the original settlers. Down the generations in what came to be called the United States of America, descent from these came to be seen as a great mark of honor—and it was to these that Emily Warren's family traced their line. Toward the end of her life she would write a little book of her own, *Richard Warren of the Mayflower and Some of His Descendants*, placing herself and her family firmly in that illustrious genealogy. She notes that Joseph Warren fought and died at Bunker Hill; and General James Warren was "a friend and correspondent of the immortal Washington" and later served as speaker of the House of Representatives.

By the early years of the nineteenth century the family was well established. Emily's paternal grandfather was John Warren: born in Dutchess County, New York, in 1765, he died the year of Washington Roebling's birth, 1837. Warren owned a three-hundred-acre farm, a blacksmith shop, a gristmill. He too was the father of seven children, including Emily's father, Sylvanus, and Cornelius, who was a judge in the Court of Common Pleas and later a member of the U.S. House of Representatives. Sylvanus married Phebe Lickley, and was a New York State assemblyman and supervisor of Philipstown, which included the village of Cold Spring, where Emily was born in 1843, the fifth of the family's surviving children. Her brother, G. K.—named for his father's close friend, Gouverneur Kemble, a prominent diplomat and industrialist who had established an important cannon foundry at West Point—was thirteen years her senior.

Emily's older brother always took a close interest in his very much younger sibling. At twenty-two, when she was nine, he wrote to their mother: "I was very sorry to hear of the illness of Robert and Emily, and am somewhat alarmed about Emily yet. They are so young . . . that they need a good deal of care. You must dress them warmly and give them the best of shoes to keep their feet dry." And in a time when the education of women was not often viewed as a priority, G. K. offered to give up his own salary to pay for Emily's schooling. "There are times in every man's affairs when he must invest not for immediate but for prospective benefit. That is especially true with regard to the interest of young ladies," he wrote. "They should be as well educated as can be afforded, and the opportunity given, by taking them in a proper society, to enable them to choose a suitable husband." In 1859, when she was sixteen, Emily went off to the Georgetown Academy of the Visitation in Washington, D.C., then in its sixtieth year, where she studied "Profane History, Ancient and Modern; Geography, Mythology, Prose Composition, Rhetoric and Grammar, French, Algebra, Geometry, Bookkeeping, Astronomy, Botany, Meteorology, Chemistry and Geology"—in addition to courses in housekeeping and domestic economy, needlework, painting, and music.

Her intelligence, liveliness, and charm were always apparent to those around her. Not long after they met, Washington started up the process of introducing Emily to his family. "I have sent two of your pictures home. I bet they will say that you are awfully fat; your pictures give no idea of the peculiar grace of carriage that you possess; few people know how to move gracefully especially when they are not so very small."

The teasing tone of this letter gives a fine flavor of their correspondence—or at least, of Washington's. Only half of their long-distance courtship remains, for Washington kept no personal letters at all from the time of his service in the war. To Emily he wrote in the summer of 1864: "Has it not often struck you as perfectly wonderful how two people can write each other day after day about almost nothing. I have burnt almost all of your letters after receiving them, but I know that is the way I would feel if [I] were to attempt to look over the huge pile at once. I think a bundle of old yellow stained love letters is a very melancholy object of interest to contemplate."

After that first encounter on the dance floor he wrote her constantly. It was, as he had told his sister, a *coup de foudre*. He described his beloved to Elvira; she was, he wrote, "slightly pug-nosed, [with a] lovely mouth and teeth . . . and a most entertaining talker, which is a mighty good thing you know, I myself being so stupid. She is a little above medium size and has a most lovely complexion." He admitted, as he did to Emily herself, that he didn't really like any photograph of his true love. "Some people's beauty lies not in the features but in the varied expression that the countenance will assume under various emotions." And it's true that the pictures taken throughout her life never seem to do Emily justice, even when exposure times became shorter as the nineteenth century drew to a close; she looks solid and stolid, and seems to be, as Washington recognized, a person whose appeal couldn't be captured in stillness.

That he adored her there can be no doubt. Although he kept that "list of persons I have been introduced to in Pittsburgh, Pa.," there is not much evidence for earlier romances; rather, evidence to the contrary in a letter from his mother, written to him while he worked on the Sixth Street Bridge. "The rumor is certain that Annie Train is married . . . He is said to be a teacher from the Normal School. You probably know him. She is said to have declared that she could have had Washington Roebling any time she wanted to. Did you propose to her so urgently? I think she would have become an old maid if she had waited for you." Emily was worth the wait. Clearly a powerful physical attraction existed between them, even given the constraints of mid-nineteenth-century courtship. "Pray tell me what is love," he wrote to her about six weeks after they met. "Is it kissing each other, is it tickling, hugging etc one another, is it writing billy duxes,* kicking each others' shins— that must be it I think—the shins —" Just a week earlier he had written from army headquarters, asking if she would send him a cat to clear out the rats that plagued soldiers and officers alike: "Send me no Tom," he wrote archly, for "he would be too lazy & would be captured by some old maid at any rate before long—Can you give me some explanation of an old maid's love for a Tom cat, it surpasses my understanding—of course no personal allusion is

---

*A "billy dux" is the delightful American approximation of the French billet-doux.

meant, you goosy . . ." He called her "my darling, mine, mine, mine" and left a little mark at the very bottom of the page: "This round spot is for you to kiss, send me one too—we must do it by proxy you know." And in an undated letter, written after their formal engagement, he asked her to send him rather more: "Tell you what let's do. You draw a picture of our Wedding night as you think it will be, and then I will send you one in return. Will you Pet or don't you think yourself equal to the task. Take a drink before you commence. I wish I had the knack of drawing as you do . . ." He closed this letter ardently. "Good night my precious broad hipped beauty—Your adoring Washy."

Luckily Washington would be well informed of just what would occur on their wedding night; toward the end of 1864 he told her that he and some comrades had captured a book in enemy territory, "Advice to Married People." They were all reading it "with great gusto." Learning that she was reading what was, apparently, a rather explicit book for the time, he wrote: "I see your taste is perverted past all redemption; a person soon gets tired of reading all piquant and racy books of that kind. I shall, however, constitute myself your purveyor in that line and get you anything you want; I want you to know as much in that line as I do." He told her that he had "a good deal of the hugging nature in me" and teased her that once they were married there would be no escape: "Why, I always pretended it was such a bore to be kissed by you, so that you made it a point to give me twice as many, which was just what I was fishing for. A little trouble in getting something always adds to the zest of it. You must not apply that rule though after we are married. I shall claim everything then and bite you if you don't obey willingly."

He wrote to his father and informed him of his plans with no little trepidation. But fierce John Roebling's thoughts on marriage, and on his son's future, are compassionate to a surprising degree. "My dear Washington," John wrote to his son on March 30, 1864, "Your communication of the 25th came to hand last night, and I hasten to reply. The news of your engagement has not taken me by surprise, because I had previously received a hint from Elvira in that direction. I take it for granted, that love is the motive, which actuates you, because a matrimonial union without love is no better than suicide. I also take it for granted, that the lady of your choice is deserving of your attachment. These two points being settled, there stands nothing more in your way except the rebellion and the chances of war. These contingencies having all passed away, you and your young bride . . . will be welcome at the paternal house in Trenton. Our house will always be open to you and yours, and if there is not room enough a new one can be built on adjoining ground . . . As to your future support, you are fully aware, that the business in Trenton is now suffering for want of superintendence, and that no increase or enlargement can be thought of without additional help. Of course I do not want to engage strangers, and it is you therefore who is expected to step in & help forwarding the interests of the family as well as yours individually . . . Should you be in want of money at any time, let me know. I conclude with the request, that you will assure your young bride of my most affectionate regards before hand, and before I shall have the pleasure of making her personal acquaintance."

Washington was clearly startled by the quick warmth of this reply. He sent what he called "the portentous letter" from his father on to Emily with a note, addressing her as "Ma chère fiancée." But now there was no turning back. "I am beginning to get scared now; everything has gone so quickly that I hardly know what to do now; the idea of the chief difficulty being surmounted so easily has knocked me all in a heap as they say; can't you let me back out (say no), the thought of being tied to the side of a loving wife, isn't it horrible for a young man to contemplate." The little joke displayed his reticence. One can hardly doubt that his early childhood, his observation of his parents' relationship, were not the best preparation for a marriage of true minds. But he was wholly relieved and delighted that his father accepted his engagement so readily: "I am so pleased at the kind way in which he writes; I supposed I would probably have to enter into a long correspondence with him, but he settles the whole matter at once in one short dear letter." He wondered, to Emily, what his father might really know about love: "From the manner in which he speaks about love I am inclined to think that he has had some experience in his younger days, although I have never heard him make the slightest allusion about his personal affairs before he was married."

Now, Washington instructed, she needed to secure the approval of "that fraternal relative of yours known familiarly as GeKay; set your woman's wits to work and devise some ways or means of how to do it. Shall I simply say, 'General, I want to marry your sister' whereupon he probably remarks 'the hell you do.' " She assured him she would take on this alarming task, which delighted him. "We are all right now; let no doubt or distrust of each other ever arise in our hearts after this; let us have perfect confidence in each other; that alone can be the basis of true & lasting love; where love exists, mutual respect and esteem naturally follow." That Warren approved of his beloved sister's choice is evidenced by the gift of a fine sword to the junior officer: "It is just like his own, long may he wave."

In Culpeper County, Virginia, the rain poured down day after day, cats and dogs, he wrote to Emily from his tent; it swept away roads and bridges, so that there were no trains, and no mail. In these hours he told Emily of the friend who had killed himself for love of him when they were young men at Troy, evidence of how secure he must have felt in her affections—secure enough, too, to share his own anxieties about their future together. "Nothing influences my sense & judgment in regard to you so much as your blind confidence in my faith and constancy," he wrote to her at two o'clock in the morning. "You have no idea how powerful a hold upon me that state of your feelings imposes; it is much stronger than anything you could impose by your mere presence or influence when with me. When you are with me there is always an involuntary tendency on my part to flee from your influence, to emancipate myself as it were from your presence; that entirely ceases when I am away from you and have nothing to think of except your goodness and the perfect self-sacrifice you are ready to undergo on my behalf. You are a dear good girl, that's all I can say, and you love your Wash and I hope you will continue to do so because I assure you he loves you . . . If God spares me I will turn up all right some of these days . . . To this day I wonder why you love me; so pray dearest, satisfy my anxious soul once more on that vital point! Perhaps you could not help it dearest; is that it dearest Em? To day the 9th N.Y. went home, the regiment I came out with; my heart felt a little heavy when I thought of you; but I trust to Providence and the prayers of that dear Emmy for her loving Wash."

Neither could have imagined just exactly what Emily's self-sacrifice would eventually entail; but it's hardly surprising that the idea of sacrifice weighed heavily on Washington's mind as the Army of the Potomac ground toward Richmond. In early May, in Spotsylvania County at what became known as the Battle of the Wilderness, General Warren initially resisted an order to attack into the dense underbrush, fearing, Washington wrote in later years, "that the troops would be lost and out of sight as soon as they entered, with nothing to be gained." Ulysses S. Grant, then the commander of all Union troops, wasn't having any of it: he sent "a peremptory order to Warren that unless he instantly attacked . . . he would cashier him on the spot!"

Once again, at the Wilderness, Washington escaped unscathed, but a few days later he wrote of the horror of the aftermath of those bloody days: nearly nineteen thousand Union casualties had fallen in the battle. He found himself on a steamer on the Potomac, the *Daniel Webster*, which had been pressed into service as a hospital ship, and Washington himself had been pressed into the service of those who had fallen. He was raised to the rank of major, but higher rank provided no respite from horror. "We have about 400 wounded men on board," he wrote, on a night so stormy they could not get into the capital but had to lie at anchor. "Dr. Corson is the only doctor on board . . . We carried every man on board ourselves and it was the hardest work to do it that I have done for many years . . . we had no single convenience, no stretchers, stores, medicine, food or bedding. I got some bandages, whiskey, etc., from a surgeon here and Dr. Corson and myself have been busy all the evening dressing and bathing the wounds of the poor fellows on board. Some of them have had nothing to eat for four days, nor have they had their wounds dressed since they were dressed on the battle-field. We managed to make a few bucketsfull of gruel and gave our own crackers to pass around among the wounded. They devoured it like wolves and seemed to be contented after that to lie in the filth and waste straw we have them lying on. The stench of their wounds is horrible, and the whole thing disgusting in the extreme. We have probably not seen anything compared to the battlefield itself."

Over the course of the Wilderness Campaign, General Warren and his corps would lose twelve thousand of the twenty-eight thousand men under his command, all within a forty-three-day period at the battles of the Wilderness, Spotsylvania, and Cold Harbor—the last a dreadful, lopsided Union defeat in which Federal troops used the bodies of their dead comrades to shore up their defensive earthworks. Washington wrote to Emily of a personal loss: "Another one of my best friends in the army has been killed—Colonel [Frank A.] Haskell who used to be on our staff in the 2nd

Corps.* One goes after the other with perfect regularity. It reminds me of the work of the guillotine in the days of the French Revolution." But it was clear to him that the Confederates had fewer and fewer men to send to the front: he had already seen one of the effects of the shortage. "The next thing we will see are the female guards," he wrote, among the Confederates. "They have sufficient patriotism to come out if that is all that is required. We did capture a full-fledged artillery woman who was working regularly at the piece. She was very independent and saucy, as most Southern ladies are." A month later, from Petersburg, where the Union Army would lay a siege against the heavily defended city, Washington wrote of seeing how "old men with silver locks lay dead in the trenches side by side with boys of thirteen or fourteen. It almost makes one sorry to have to fight against people who show such devotion for their homes and their country."

Washington and the men around him fell victim to forces beyond their control. He expressed little sense of a belief in a great cause, rather the resignation of a young man who knew, and who had always known, that he must do his duty, and that his fellow soldiers would, as well. "They must put fresh steam on the man factories up North, the demand down here for killing purposes is far ahead of the supply," he wrote to Emily bluntly on June 23, 1864. Just a week before he had been part of the initial assault on Petersburg; now Warren's men were settling in for the long siege. "Thank God, however, for the consolation that when the last man is killed the war will be over. This war, you know, differs from all previous wars in having no object to fight for; it can't be finished until all the men on either the one or the other are killed; both sides are trying to do that as fast as they can . . . The biggest heroes in this war are the privates in the line—the man with the musket. When I think sometimes what those men all do and endure day after day, with their lives constantly in danger, I can't but wonder that there should be men who are such fools. I can't call them anything else. And that is just the

---

*Franklin Aretas Haskell was remembered not only by Washington Roebling, but by history, too. A native of Vermont and a graduate of Dartmouth, like Roebling he was in the thick of the fighting at Gettysburg; and his account of the battle, written in a letter to his brother, is now recognized as one of the most important primary sources for the conflict. It has been published and republished many times.

trouble we are laboring under now—the fools have all been killed and the rest think it is about played out to stand up and get shot."

Just before the Wilderness, Washington caught his first real sight of the commander in chief of the "man factories," as Washington had it. He had glimpsed Lincoln when he had arrived in Washington, D.C., in 1861; but in 1864 the president traveled down to Culpeper County to review the army before the Battle of the Wilderness. In an account written in a letter decades later, he described Lincoln's participation in the "cavalcade." "The President was mounted on a hard-mouthed, fractious horse, and was evidently not a skilled horseman. Soon after the march began his stove-pipe hat fell off; next his pantaloons, which were not fastened on the bottom, slipped up to his knees, showing his white home-made drawers, secured below with some strings of white tape, which presently unraveled and slipped up also, revealing a long hairy leg.

"While we were inclined to smile, we were at the same time very much chagrined to see our poor President compelled to endure such unmerited and humiliating torture. After repairs were made the review continued, but was shortened on his account. I never saw him again and was in Covington, Kentucky, when I heard of his assassination."

The day before he wrote of his admiration for those privates in the line, he missed having his photograph taken by none other than Mathew Brady, whose glass-plate images (and even more those of his assistants) would revolutionize not only the art of photography but also the world's perceptions of war. Brady had come to take pictures "of G. K. & staff, luckily I was absent on business and I am spared the pain of seeing myself perpetuated in my present appearance," which was, he wrote Emily, "dirty, ragged and I might say lousy—there isn't a brass button on my coat, I haven't had a shirt collar on since we left Culpeper and when my only shirt is washed I lie abed." When she was anxious for his safety, he tried his best to reassure her: "Bad news you know always travels much quicker than good news, so when you hear nothing just hope for the best." Emily, it seems, was planning a visit to Trenton around this time: "You must make yourself especially interesting to Mr. Swan; I know beforehand that he will be tickled to death at it." Certainly she did visit Washington's sister Laura out on Staten Island—and it seems was not much encouraged by what she found among "all the Dutch uncles & cousins etc." as Washington wrote. Again, he strove to reassure her;

she had been imagining for herself, most likely, a rather more sophisticated life than what she saw there. Her family were not immigrants come to the New World to seek a better life; they were the Mayflower pilgrims, the settlers of that world. "The tone of your letter is one of sad resignation and even your Wash seems of scarcely sufficient weight to counterbalance the scale. And well might it be so if your life were doomed to [be spent] among that Dutch crowd on Staten Island . . . However you must take heart my dear, all of our family is as much American as you could wish with the exception of Mother and she never had the opportunity."

The very German family life she saw in Staten Island—and the fate of Washington's mother, Johanna—continued to trouble Emily as she considered what future lay in store for her as the wife of an engineer, and her place in his family. For while Washington was laying siege to Petersburg, out to the west, in Cincinnati, John Roebling's bridge over the Ohio had been revived, and was a sure-enough bet to increase land values. "Valuable corner property on 2nd and Greenup Streets," ran an ad in the *Cincinnati Enquirer*. "The property from its location, being nearly opposite the New Suspension Bridge landing, offers extraordinary inducement for investment," its seller assured. Even during the war, the city had prospered; in 1861, 336 new buildings had been erected, of which "but 27 were of wood," wrote Charles Theodore Greve in a centennial history of the city. By 1865 Cincinnati's population had increased to two hundred thousand souls, and by the following year another eleven thousand would be added, even taking into account a bad bout of cholera that cost more than two thousand lives. But that year, too, a telegraph system for the police and the fire department was installed; "the fire department was regarded as vastly improved thereby."

For Washington, dug in at Petersburg, his father's bridge over the Ohio was still an abstraction, and the fate of the Union still hung in the balance. General William Tecumseh Sherman was moving toward Atlanta, but slowly; Confederate general Jubal Early marched toward Washington: the London *Times* remarked that "the Confederacy is more formidable than ever." Washington revealed some annoyance when he replied to Emily, who was, understandably, anxious about what exactly awaited them at the construction site on the Ohio River. "You ask about Cincinatti," he wrote on July 27. He never could get the hang of how to spell the place. "Well, I hardly know; In the first place I must get out of the Army—then after a little while we get married;

then we will have a royal good time in each others [sic] society alone for a little while—then we will probably stay quietly in Trenton for some months, which will bring us towards the latter part of April, by that time you will want to go to Cold Spring to show yourself to Mamma & sister in your new capacity of Mrs. Roebling; then in case the work at Cincinatti is sufficiently advanced I shall go there, leaving you at home for the time being; when I once go there I shall have to stay until Christmas before I can go home again; you can come out & see me for a month or two if you like, I shall want you to come anyhow, if I don't go to Cincinatti next year why we will stay in Trenton, live virtuously together as man & wife fear God & walk in his ways . . ." Her letter in reply plainly expressed her unhappiness with this plan—for a variety of reasons. Her fiancé was equally plain in answer.

"I have before me a long letter of yours—how to answer it I scarcely know as yet," he wrote on August 2. "I would be false to myself not to acknowledge that most of its contents deeply pain and grieve me, and that its repeated perusal gives me the conviction, faint but nevertheless existing, that the girl whom I love, the only one for whom I have ever felt a true, lasting and sincere affection—no longer esteems me as a woman ought to esteem her lover . . . God knows that if I ever had an earnest desire, it is to live together with you always. I will not be so uncharitable as to say that I don't care whether you ever see your old home again, you will never receive a hint to that effect from me—my proposition to that effect in my letter was merely to give myself time to find a suitable place to live before sending for you . . . When you cite my mother's case you do so with but a partial understanding of the facts; my father has enjoyed the best education that could be given in his days—wherever he is he always moves in the first society of the place and although he acquired the English language after coming to this country, he is as seldom taken for a foreigner as I am; my mother on the other hand never enjoyed the advantages of a good education, and it is only within the last 10 years that she has acquired sufficient english [sic] to get along in ordinary society and daily life." From what she had observed of his family, she clearly feared she would be whisked away from her relatives at Cold Spring, never to see them again—and made to live a solitary existence of housewifely drudgery while Washington swanned across the country, keeping the best society. "I feel too lowspirited to write any more," he closed. "If you knew how wretched your letter has made me feel you would not write me another of the kind."

But Washington's low spirits came from another source, too, after the unsuccessful attempt to break Lee's lines at Petersburg that became known as the Battle of the Crater. A plan had been hatched—by Pennsylvania soldiers who were miners in civilian life—to dig under the Confederate lines: eventually they would burrow a tunnel more than five hundred feet long beneath the earth. Four tons of gunpowder exploded underground burst a great hole up through the surface: "an immense mass of dull red earth was thrown high in the air," ran the account in the *New York Times*. "Those near the spot say that clods of earth weighing at least a ton, and cannon, and human forms, and gun-carriages, and small-arms were all distinctly seen shooting upward in that fountain of horror, and fell again in shapeless and pulverized atoms. The explosion fully accomplished what was intended. It demolished the six-gun battery and its garrison of one regiment of South Carolina troops, and acted as the wedge which opened the way to the assault." But there was just as much confusion in the Federal lines, and whatever advantage had been gained was quickly lost. For Washington it was "a sad, sad day," as he wrote to Emily just after the battle on July 30th. "The first temporary success had elated everyone so much that we already imagined ourselves in Petersburg, but fifteen minutes changed it all and plunged everyone into a feeling of despair of ever accomplishing anything; few officers can be found this evening who have not drowned their sorrows in the flowing bowl." Even though her brother was a general, it must have been impossible for Emily to imagine the carnage Washington witnessed day after day after day. "This business of getting killed is a mere question of time," he wrote. "It will happen to us all sooner or later if the war keeps on . . . The most inspiring sight is the flock of buzzards constantly hovering over us and waiting for their feast. Those birds are at least impartial because they eat both sides alike; the same I suppose is true of worms."

And so the war rolled on, with a presidential election in the offing. George McClellan, whose obstinacy and lack of action had so troubled Abraham Lincoln, was put forward as the Democratic candidate for the presidency; the Democrats planned to sue for peace should the party win the office. But in September Atlanta was taken, and all its supply chains by road and rail destroyed; General Philip Sheridan's men carried out Ulysses S. Grant's instructions to turn the Shenandoah Valley into "a barren waste." On the Union side, diarist George Templeton Strong was ebullient. "Glorious news this morning—Atlanta taken at last!!!" he wrote. "It is (coming at this political crisis) the greatest event of the war." Mary Chestnut, wife of former South Carolina senator James Chestnut, conveyed the opposite sentiment in her journal, on behalf of those in the Confederacy. "These stories of our defeats in the valley fall like blows upon a dead body," she wrote after learning of Grant's campaign. "Since Atlanta I have felt as if all were dead within me." Those who opposed President Lincoln allowed ugly sentiment full voice. The "beastly doctrine of the intermarriage of black men with white women" was now "openly and publicly avowed and indorsed and encouraged by the President of the United States," ran a piece in the *New York Freeman's Journal & Catholic Register*. "The Filthy black niggers, greasy, sweaty, and disgusting, now jostle white people and even ladies everywhere, even at the President's levees."

Washington's letters from that autumn pick up on the better spirit victory brought to the Union forces; and no doubt—even given some premarital wrangling—his engagement cheered him, too. "Well Sir, how do you flourish now-a-days?" he wrote to Charles Swan that fall. "Who are you going to vote for, 'Old Abe' or 'Little Mac'? . . . In the next place what do you think of my Emily? Don't you think she is some? She writes to me that she has made 'Ferdie' feel very bad; poor fellow, how I pity him. The only thing I fear is that she is so strong that she will be able to lick me in case we ever have a fight." And when Lincoln was reelected—the first president since Andrew Jackson to win two terms—he wrote to Emily in fine spirits on November 8. "Hurrah for old Abe; we have already heard that he's elected through the telegraph. The vote of New Jersey has not come in yet, but that don't matter because I was not there to vote."

However, not all the news on the home front was cheerful. Washington's mother had not been well for some time. Her life had worn her out. Seven children had survived an arduous childhood, a frontier life made harder still for children and mother, thanks both to John Roebling's extreme frugality and his eccentric medical beliefs. And now, as she weakened, John Roebling was in charge of her medical care, which proceeded under his direction. "Mother is improving slowly, but I think surely, under the hydropathic treatment and the diet connected therewith," John Roebling wrote to his son on November 17, 1864. "I left her in the care of Dr. Brinkman . . . a Physician of much experience and intelligence. Dr. Coleman is not competent to manage her case and knows too little about water at any rate. All physics will be poisons to her and must be avoided. Mother is so low and so much reduced, that her entire recovery will take a year." But no recovery would come. Just a few days later Washington received a telegram—not from his father, but from Charles Swan, addressed to his commanding officer and brother-in-law to be, Major General Warren. "Inform Maj. Roebling his Mother died this P.M," it simply said. "The greatest giver of us all [is] gone," he wrote to Emily at Christmastime that year. Her simple memorial is in Riverview Cemetery in Trenton—beside her husband's much grander and more ornate monument. As it went in life, so it did in death, and Roebling never forgave his father his treatment of his mother. "The poor woman was glad to die, even at 48," he wrote in his memoir. "She had had a hard life—worn out with hard incessant work, many children, her nerves racked by the never ending, everlasting, continuous senseless useless scolding on the part of her husband, she gave up in despair—he would allow her no doctor—she literally died from being stuck in cold water all the time." Dr. Coleman—not competent by his father's reckoning—was "smuggled in so as to be able to give a death certificate."

Her only joy came when John Roebling was on his travels, her eldest son recalled. "The few rays of sunshine fell on her path only when her husband was away on a long engineering trip. I could write pages on our unhappy family life, but what's the use—every family has a skeleton in its closet—In a few years we will all be dead and it will all be forgotten."

Except that it would not be forgotten; not least because he chose to leave his account of her suffering. An engineer and a geologist, he had been schooled by both inclination and instruction to take the long view; and yet that perspective would never erase his emotions.

In the letter Washington wrote many years later to General Rusling, he said that he left the army "shortly before" the victory at Five Forks, in Virginia—a victory that nevertheless led to General Warren being relieved of his command by General Philip Sheridan, a failure that would haunt Warren, and his family, to the end of his life. Warren's Fifth Corps had provided vital support for Sheridan in attacking the road junction at Five Forks—but not vital enough, by Sheridan's lights. "Just imagine Sheridan sitting on a fence, sending a staff officer every five minutes to Warren to hurry and save him and his Cavalry from being captured by Lee's troops—and when Warren does come (after wading through an icy creek up to their middle), saves Sheridan and wins the battle—then Sheridan turns on him and cashiers him—Grant himself has often acknowledged in private that it was all wrong—but never had the manhood to right the wrong."

That battle took place on April 1, 1865; Washington, having been brevetted lieutenant-colonel on December 6, 1864, served until January 21, 1865. He clearly regretted not being able to support Warren both in battle and then, later on, in defending his reputation; for years after the war had ended, Warren would persist in his efforts to clear his name. In the spring of that year, just weeks before Lee would surrender at Appomattox Court House, Washington was promoted to colonel, "for gallant and meritorious service during the war." But first, he and Emily would be married.

# "All beginnings are difficult, but don't give up"

IN THE *ALBANY LAW JOURNAL* for Saturday, April 15, 1899, a short piece appeared about halfway through the issue entitled "A Wife's Disabilities." Just two weeks earlier the author, Emily Warren Roebling, had graduated with high honors from the Woman's Law Class of New York University in a ceremony at Madison Square Garden. As the *Trenton Evening Times* reported, "There were forty-eight women dressed in college caps and gowns in the class, and an imposing row of professors and alumnae in more elaborate robes, making the exercises very impressive. Added to these were the masses of flowers, the solemn pile of parchment rolls and the hundreds of friends who had come to be witnesses of this proud moment in the lives of the young women in the front seats"—only the ninth class to graduate from the women's program. The wife of Colonel Washington A. Roebling, as the Trenton *Times* was careful to describe her, was fifty-five—very much a mature student, and one whose maturity had served her well. Her essay, a few paragraphs of which the paper reproduced, had won her the prize for the best essay. It is a striking argument for equality, written at a time when the legacy of the Thirteenth Amendment, freeing all the slaves held by the confederacy, and the Fourteenth Amendment, giving all persons in the United States equal rights under the law, were still being hotly debated. "Under the old common law a married woman was classed with parties incompetent, infants, lunatics, spendthrifts, drunkards, outlaws, aliens, slaves and seamen. As a single woman this incompetency did not exist, but the sacred rite of marriage conferred upon her the honor of ranking in legal responsibility with idiots and slaves," she wrote.

Things had improved somewhat, she allowed, since William Blackstone, in the England of the eighteenth century, assured his readers that "the disabilities which the wife lies under are for the most part intended for her protection and benefit"—but not enough, despite the passage of laws that at least allowed married women to own property in their own right. Emily Roebling stated that a woman's entitlement, in a marriage, was still not given its due. "Does she not contribute largely to his success or failure in life? Must she not bear poverty and reverse of fortune with her husband when they come, and shall she not lawfully share in all the profits of his success and prosperity?"

IN JANUARY 1865—in a country still at war—all this lay far off in the unimaginable future; but Emily Warren would, undoubtedly, contribute largely to her husband's success in life. She would participate fully in its struggles, bearing reverses of fortune with a fortitude this essay must, however guardedly, reflect. But for now, Washington and Emily had to build a household, and practical matters as much as romance were the order of the day. "If any body wants to make you a present and don't know what to give tell them a dozen silver Desert [sic] Knives would be acceptable," Washington wrote briskly to his bride-to-be from Trenton. "I have a place in a box to put them in but have none of my own to put in. I have the wedding ring already— properly engraved . . . Are you going to hold your boquet [sic] in your hand or in a boquet holder. Have you got one?" (Even today, one might remark, the bouquet-holder issue remains a subject for debate in matrimonial situations.)

They were married on January 18, 1865, in a double ceremony in the red-brick church on Cold Spring's Main Street in which her brother, Major Edgar Washburn Warren, married Cornelia Barrows. Washington had bought the wedding ring in Philadelphia, and had it engraved *E. W. & W. A. R. Jan 18 1865*; it had cost seventy dollars to have the wedding cards sent to Cold Spring by express. In a little pocket notebook—which looks almost exactly like the notebooks he would carry while working on the Brooklyn Bridge just a few years later—he poignantly recorded his wedding preparations. He wrote a careful list of what he would need, revealing how close he still was to his sister Elvira:

Stovepipe hat
Nailbrush

12 linen standup collars
White satin cravat
6 pair of 10 ½ stockings
3 pair of drawers
undershirts I have
6 handkerchiefs
Elvira makes the shirts
White Kids [gloves]

He noted further that from his tailor he would require:

Overcoat
Suit of clothes, swallowtail?
Dressing gown
Pantoufels*

He further reminded himself of the need for "Watch (gold) & chain." On the last page of the notebook he drew a little puppet or doll, with four jointed limbs and a string to pull for motion; the expression on the doll's face is somewhat agonized. Knowing Washington was pulled between his responsibilities as the war drew to a close and the work he knew would await him in Covington and Cincinnati, and his new responsibilities as a husband and head of his own household, a century and a half later it's hard not to smile at his sketch. Their wedding photograph, taken a few weeks after the ceremony, found him in trousers that look a size too large—seated, with Emily standing behind him, her hand on the back of his chair. The couple gaze across each other, out toward the middle distance, their expressions unreadable. But this was the beginning of a remarkable partnership that—despite her later thoughts on the legal status of married women—built a legacy not only in love, but in steel and stone.

"EVERYTHING HERE IS black, black, no one wears anything else; it is of no use; your purple silk checked with white would be as black as your black silk

*A contemporary word for house slippers; one may only hope the term is revived.

in a week. No amount of washing keeps the hands clean. I am wearing my flannel shirts now as the wash bill comes too high with white shirts," he wrote to Emily from the smoky Ohio. They had stayed in the Roebling house in Trenton for a few weeks; then he had gone on ahead to find them a place to live. The scale of what he saw when he arrived at the bridge site in Cincinnati astonished him. "Our bridge here is an immense thing, it far surpasses my expectations, and any idea I had formed of it previously. In a short time I shall be accustomed to the enlarged dimensions of everything and then it will be looked upon like anything else." To Charles Swan he noted that the work "is the highest thing in the country; the towers are so high a person's neck aches looking up at them."

One of his jobs was to help drum up business for the wire rope mill back in Trenton. In March he wrote to Ferdy, back at the mill, checking whether the business cards he had asked to have printed were ready yet. "I should like to have a lot to send around here. This is a great manufacturing town, with any quantity of hoisting machines, etc." By early the following year he would be telling his brother that he had got hold of most of the business gazetteers—the telephone books of the day—for the western states; he wanted one thousand circulars to mail out. Their father, he noted drily, didn't like to advertise—"because 15 years ago he advertised in Boston and did not get a single order." No matter: "I have the opportunity of getting the business Gazeteers [sic] of most of the northern states where it will pay to send out circulars, and I mean to do up the business thoroughly; it will probably take a year, and involve considerable expend [sic] for addressing envelopes & stamps and gazetteers to be paid by the Trenton office. The first lot of 1000 will make for a beginning."

In the spring of 1865 Emily arrived from Trenton. Washington would later write that "my father was violently opposed to the idea of an engineer taking his wife around with him—he never did, so why should I." It was a bit of rare resistance against his father. The couple had found a place at what Washington described as a "Sesesh house"—by Washington's account there were plenty of folks who remained sympathetic to the Confederate cause, especially on the Covington side, where Washington and Emily eventually lodged. For twenty dollars a week he and Emily took room and board with a family called Ball. "I hope he will improve the grub as the season progresses," he wrote to Ferdinand. "Nobody else boards there, so that we form part of the family. They are great rebels you know, and on that account people think we are rebels too. All the respectable people in town with but few exceptions are sympathizers as they are called." Thanks to the postwar business boom, they struggled to find a place of their own: "they are very scarce; you need to have looked a year in advance."

But the new prosperity allowed the bridge work, which had been abandoned before the war, to be revived. In the 1850s, John Roebling had reckoned the bridge would cost six hundred thousand dollars; now he knew it would run to nine hundred thousand—and indeed, in the final report to the stockholders of the bridge company, the figure would come in at just under one and a half million dollars.* There had been a 200 percent increase in the cost of labor and materials in the intervening years; and now the bridge's girders would be made of wrought iron, rather than wood.

Even during the war Cincinnati had continued to thrive, transforming from a riverside settlement into the largest commercial and manufacturing center of the West—thanks largely to the slaughterhouse and meatpacking industry. As early as 1815, pork, bacon, and lard were the city's second largest exports; by 1825, forty thousand hogs were packed in the city; the "Queen City" would come to be known as "Porkopolis." In 1863 the city's population was recorded as being 186,329; by 1865 it had risen to 200,000 souls. But despite the city's prosperity, the future of the abandoned bridge hadn't always been certain. "Left without the moral and financial support of the proud Queen of the West," John Roebling wrote once the work was done, "the Covington enterprise was allowed to sleep, and that sleep came very near terminating in its final dissolution by the threatened sale, at public auction, of the splendid masonry of the Cincinnati tower, carried up 45 feet above the foundation, in order to satisfy the proprietor of the ground, whose claims had not been finally settled." A narrow escape. "The great exigencies of the war, by the movements of troops and materials across the river, made the want of a permanent bridge all the more felt." During the war, a local architect, Wesley Cameron, had built a pontoon bridge out of old coal barges over the Ohio to move men and matériel from Covington to

---

*Not bad for a bridge that still serves its community today, nearly a century and a half after its opening.

Cincinnati, but the city's paper, in reporting official pique at civilian petitions to cross by pontoon, showed just how much a permanent bridge was wanted. "It should be understood that the pontoon bridge across the river is only eligible to the military," ran a notice in the *Cincinnati Enquirer* in September 1862. "The guards are continually annoyed by the application of civilians and ladies who are anxious to cross merely, we should imagine, for the novelty of the thing."

THAT "SESESH HOUSE" belonging to the Ball family is still in Covington today—one of a pretty row of houses with a river view, at Fourth and Garrard Streets on the northwest corner. John Roebling was nearby, either at the home of Jesse Wilcox, a businessman who was president of the board of trustees for the bridge company, or at a boarding house owned by Wilcox; the bridge offices were just around the corner.

Washington agreed with his father that the bridgework should have been pursued even when economic prospects seemed darker; "You have to strike while the iron is hot," he wrote. And by June 1865 he could convey to his brother Ferdy, back at the mill, some of the excitement of the work, the rhythm of his days, musical interludes included. "A stone fell from the Cov. Tower the other day, and broke off the end of the little bridge causing a delay of a day. Fortunately no one was hurt; it occurred through the breaking of a socket, there being a bad weld in the handle . . . how it held so long is more than I can understand . . . The big wire rope created as much sensation as the elephant would; we had a hard time hauling it; the timber wheel man gave up at first, and we had to load it for him . . . To night we are going to see *Don Giovanni*. I hope it will be good. End of next week we will commence the cornice of the towers, which is a sign that they are approaching their completion . . . Charley writes to me and wants money. It reminds me of one beggar begging off another." Their younger brother Charles was at school in Staten Island—the school started by their brother-in-law, Anton Methfessel; but Washington, clearly, was kept on as tight a financial rein as his brother the schoolboy.

Edmond F. "Frank" Farrington, able master mechanic for the Roeblings, when the bridge was completed published *A Full and Complete Description*

*of the Covington and Cincinnati Suspension Bridge with Dimensions and
Details of Construction*, a clear and concise account of the work. The towers under construction were "two in number, standing a little more than
1,000 feet apart," when completed, "on either bank of the Ohio River." Their
foundations were composed of "immense square oak timbers, bolted together
transversely, and all the open spaces filled with cement." The stonework
rose to a height of two hundred feet; the towers were eighty feet long and
fifty-one feet wide.*

The iron wire that arrived hadn't been made at the Roebling mill, but
rather in Manchester, England—by Johnson & Nephew, founded in the
eighteenth century, the firm that had also supplied wire for Niagara. But it
wouldn't be carried across the river until the autumn; there was much work
to do before then. "Our work is getting along a little slowly," Washington
wrote to his brother at the end of June, "we are so near the top that we can't
work fast. The other work is far ahead of time: I mean the wire shops, cable
making machinery, etc. The wire from England is now arriving; it will all be
here in time.

"Climbing the towers is perfectly awful. The ladders are almost straight
and about 200' high now, and when you go up 5 and 6 times a day your
back begins to ache. I used to go up on the rope until father positively forbid it.

"Can you recommend me any way by which I can make more money than
I do now? That is the question agitating my mind." Asking his father for a
raise didn't appear to be an option. That summer it was, he wrote to Ferdy,
"as hot as you may imagine." "It was the fashion then to wear shirts that
buttoned in the back," he would recall years later. "They always stood open,
giving the sun a full chance on your hide." Toward the end of the month the
towers—which, when their elaborate brick turrets were finally added, would
rise to 240 feet—were ready to receive their saddles, the grooved plates over
which the great cables would ride. These were hoisted into position: nine-ton
"bed plates," nine by eleven feet, were embedded in the masonry at the top of

---

*These days it looks as if the foundations of the towers are, like the towers of the Brooklyn
Bridge, built under water, for the waters of the Ohio flow around them; but this is not the
case. At the time of the bridge's construction the river had not yet been widened by the series
of dams that gave it its modern breadth.

the towers; the saddles themselves weighed thirteen thousand pounds each
and had curved and grooved backs to receive the cables. And as the bridge
towers grew, the city, and the country, tried to settle itself after the trauma of
the war, but it was clear that wounds were still fresh. "No one can form any
conception of the utter ruin and exhaustion of the whole South," ran a report
in the *Cincinnati Enquirer.* "Railroads are worn out, the rolling stock either
destroyed or useless; fences and houses many of them burned and what are
left are badly shattered; horses and mules carried off by the armies; stock of
all kinds is very scarce."

The making of the new bridge's cables, Farrington wrote, was "an interesting process, rather difficult to describe intelligibly to those who never
witnessed it." The Manchester wire was oiled with linseed before being
spliced together to form a continuous wire, straightened, and "drummed up"
on enormous wooden reels kept on the Cincinnati anchorage. The drums
were seven feet in diameter and sixteen inches wide; each would hold a ton
of wire. Straightening was done by passing the wire through wooden blocks;
splicing was accomplished by filing the end of each wire flat and tapering,
for about four inches at the end; the edges were then nicked with a series of
notches so the wires would catch against each other, and then the whole join
was wound with fine wire and painted. "So strong are splices made in this
manner, that experiments prove that they will bear a greater strain without
breaking, than the sound wire."

In September the first wire rope was passed over the towers and the river—
at last connecting the two shores. This would not be part of the finished
bridge, but would allow the construction of a footbridge for the workers.
One end of the rope was passed over the floor of the anchor pier on the Cincinnati side and then up to the top of the tower, just as the stones had been
hoisted there; it was then coiled down into a flatboat, which was towed to
Covington—the wire being paid out as it went and sunk to the bottom of the
river. Hoisted out and passed over the Covington tower, it was pulled out of
the river when the water was clear with a steam-powered engine. "We are getting along tolerably," Washington wrote to his brother on October 5. "The
footbridge was finished yesterday; it is a mighty long timber structure and
apt to blow away in a storm; I hope we will not have any serious ones until a
couple of strands of the cable are run. By beginning of next month we may
begin Cable making; most likely it will be middle of November." The oak

bridge—three feet wide, made from slats a little more than six inches wide—
would in fact be broken up three times by high winds: the last time the whole
section that ran through the Covington tower was thrown completely into
the river.

As Farrington wrote—in his always colorful style—after the footbridge
was built, "communication [was] established FOREVER!" The rickety-
seeming structure became quite the attraction. "Several ladies crossed at different times, and a few ventured even to climb the ladders to the top of the
towers," Farrington reported. "The writer was once coaxed into gallanting
two young ladies over the bridge, and up on to the Covington tower, one
starry night. The damsels behaved bravely, and beyond a little desperate hugging and squealing in the most giddy places, no demonstration of nervous-
ness were made." A "fair and fat" Methodist preacher tried to get over, but
lost his nerve; a "brave fellow from Cincinnati" who "eluded the vigilance of
the watchman" one foggy morning "came to his knees when part way over."
The watchmen on duty, it was clear, had to keep a sharp eye out, for as the
paper reported, "hundreds of persons make applications daily to be allowed
to cross the foot-bridge . . . but all are refused, as the crossing would seriously interfere with the operation of the workmen."

What Farrington didn't recount was that—by Washington's account—
John Roebling nearly had them both killed as the first rope was towed across
the river. "John A. took charge of this personally," Washington recalled.
"He thought he could make a hemp rope fastening at the end—this gave
way when the strain came on, the cable flew over the tower into the river, and
narrowly escaped killing us both." Washington didn't hesitate to remark the
failure of his father's plan. He had no love for the footbridge, and it was his
own skin, not his father's, that was most at risk all this time. When the footbridge was damaged by the winter weather, it was Washington's job to see
that it was fixed—"always at the risk of my life. On one occasion I could
only return by walking on the main cable back to the top of the tower—
Before reaching it my strength gave out—below me was sure death—How
I managed to cover the last 100 feet is still a hideous nightmare to me."

But there was nothing for it but to continue. "We are getting along well,"
he wrote to his brother back in Trenton in mid-October. "Commenced
splicing. Got cradles up and working ropes over; will run out wires in a
week, and expect to get a strand made & finished before we come home."

For Christmas, he meant—and now the work of making the cables could really begin. Today the bridge looks rather different than it did when it first opened; an additional cable was added over each tower in 1895 to strengthen the bridge for the increased traffic it now had to carry. It looks slightly less graceful than it did in its earliest days, but it remains an elegant construction, and a foreshadowing of what the Brooklyn Bridge would be, with its diagonal "stays" sweeping down from the main cables. The stays and vertical suspenders are made from wire rope twisted just as hemp rope would be twisted, and these ropes came from the Roebling mill in Trenton. The main cables, however, which were made of the English wire, are strands of wire "laid up" parallel to each other, the tension in each wire carefully regulated. The best way to envision what's going on inside a bridge cable is to look at the elegant drawing done by Washington in 1899, when the Williamsburg Bridge was being built. That bridge was designed by another engineer, Leffert L. Buck, but the cables were made by the Roebling firm: the image shows the pattern of the individual strands. A layer of wire is then wrapped tightly

*Cable-making for the Covington–Cincinnati bridge*

round the whole cable, clamped securely at intervals, compressing the cable into a solid, strong mass.

He had thought the cable-making would take six months—but it took nine, in the end. His and Emily's lodging improved before the holidays when he left Ball's boarding house for a place with a Mrs. Dodge: "the change is vastly for the better." By the end of January the following year he wrote to Ferdinand about getting ready to tie off strands, lower them into their correct position in the saddles, and regulate their tension. His practical tone in letters at the time doesn't reveal the continuing danger of the work—cable-making is hard, breaking in new hands is "an ugly job." Of the work as a whole, "few men have the nerve to do it," he said simply, decades later. The most dangerous part of the work was letting off the strands as they were finished, and connecting them to the anchor chains that would hold them in place. The tension on the wire was up to forty tons; if a wire got away it could easily kill any man within its long reach—as would happen in Brooklyn near the end of the strand-making, years later. "The management of this dangerous business fell to my lot as usual and caused me many gray hairs."

That *as usual* reveals again the conflict with his overbearing father. Writing his memoir, the work long in the past, he expressed his dissatisfaction not only with his father's manner but also his engineering decisions: the stay system was carried "to excess" in the Cincinnati bridge, Washington opined, though it was his own "hard steady grind in cold weather from 7 A.M. to 6 P.M." to put them up; the floor of the bridge, too, he thought was too light for the load it would have to carry, even at the time of its construction (and indeed, a new steel truss and floor beam system would also be added in the very last years of the nineteenth century). John Roebling insisted that the ends of the cables, where they were fastened to the anchor chains, be encased in cement: he was "a great believer in the preserving quality of cement on iron imbedded in it"—twenty-five years later, Washington noted in his memoir, it was discovered that the cement had, in fact, badly damaged the iron.

But there was no arguing with him, even by a soldier, a colonel, who had seen bloody battle during all the awful years of the war. "To discuss such matters with Mr. Roebling was an impossibility—His violent temperament instantly asserted itself at the slightest difference of opinion, and as I stood toward him in the position of son to a father, who claimed powers of life and death over his children, my only refuge was to save my life by flight. If men endowed with such temperaments would only confine themselves to crushing their rivals, controlling mobs, overcoming great physical obstacles, etc. it would be all right but they do not—the innocent, the harmless the deserving are all crushed alike."

Work progressed through the spring, despite the gales blowing down the Ohio, and Washington—perhaps hoping that the growth of the family business would improve his financial situation—worked just as hard at finding more business for the mill as he did on the bridge. One important invention of the nineteenth century was about to transform the wire-rope industry, and bring John A. Roebling's Sons the kind of financial success its founder might hardly have dared to dream of: the elevator. In 1857 the Haughwout and Company department store opened on Broadway and Broome in Manhattan—the building, with its innovative cast-iron facade, still stands. "The elegant and varied assortment of new goods with which we shall open has probably NEVER BEEN EQUALED IN THIS COUNTRY" the owner announced on March 17 of that year in the *New York Times*. But the piece didn't mention the store's true distinction: the first real passenger elevator ran between its floors, installed by Mr. Elisha Otis. While Otis is widely credited with the "invention" of the elevator itself, his patent was for the safety device that would eventually make high-rise buildings a real possibility; this new device had its first outing at Haughwout's. In the beginning elevator ropes—just as in the days of the portage railway—were made of hemp; but before long the work of Mr. Otis and the work of Mr. Roebling would be brought together. Washington would recall that in the early days Charles Swan thought elevator ropes "a nuisance"; "And yet forty years later we are making millions upon millions of feet per annum and their manufacturing comprises nearly one half of our business—All beginnings are difficult, but don't give up."

By the end of April 1866, the cables were progressing nicely—and he was mailing thousands of flyers off to Michigan and Indiana. In June the cables were nearly finished, despite the "young hurricane" that damaged the footbridge on the Covington side; the cables would be wrapped by the end of July. By the autumn, the end was in sight—but as the winter drew on, Washington's dedication to his work affected his health—a premonition of worse to come. He suffered a severe attack of pleurisy "brought on by climbing the steep ladders of the towers, then, having the perspiration suddenly checked by the cold wind on top—for some days I was delerious [sic]—(no medical attendance of course)—The pleuritic pains attending the adhesions that formed followed me most of my life." And every day was "hard work on the bridge," made worse by John Roebling's unsatisfactory truss system: the truss underneath the bridge floor should further stiffen the bridge, but this was "a truss in name only," in Washington's opinion. But almost all of the responsibility for this bridge lay with Washington, not his father; during the last six months of the construction John Roebling had been on site for only a few weeks.

On New Year's Day an ad was placed in the *Cincinnati Enquirer.* OPENING OF SUSPENSION BRIDGE, ran the banner. "The suspension bridge will be opened for the crossing of Wheeled Vehicles of every description, TO-DAY, JANUARY 1, 1867." It was signed "A. Shinkle, President"—president of the Covington and Cincinnati Bridge Company. Shinkle, whose lovely, imposing mansion still stands just on the Covington bank of the Ohio River, was a great entrepreneur of his age, a man who started life as a cook on the flatboats that ran on the river, and rose to be a coal magnate and much more. "Amos was in the water works, gas works, street cars, glass works, coal, tobacco, whiskey, stone, hardware, bridge, real estate," Washington wrote. "Such men deserve to succeed." And he was, too, a tireless promoter of the bridge—an enterprise out of which he would do very well indeed.

On the day itself the river was so choked with ice that the ferry couldn't cross: "but for the bridge there would have been a great detention in the trade of the cities," the *Enquirer* noted sagely. The Adams Express Company was given the honor of driving over the first teams—their first wagon drawn by eight horses, the second by six, the third by four, all of them decorated with "flags, plumes, &c." as the paper reported. "As the procession entered the bridge the cheers of the people made the welkin ring, which, with the stirring music of the band, created no ordinary enthusiasm." It seemed as if all of Cincinnati—and Covington—had come to join the festivities; it was estimated that fifty thousand foot passengers made their way across the bridge, never mind the stream of wagons and carts. "If ever any person entertained doubt of the strength of the bridge, the extraordinary test applied to it yesterday should satisfy the most skeptical. The fact is that the monster structure did not sway in the least, but under the immense weight appeared much stronger than when only a few workmen were engaged upon it. The bridge is eminently a success and too much praise can not be awarded to the projectors and proprietors of this great enterprise." The bridge was 2,200 feet long from Front Street in Cincinnati to Second Street in Covington; the main span over the river was 1,020 feet long, and ran 100 feet above the river's low-water mark. "Viewed as a whole it is immense—GRAND," Frank Farrington wrote in his report. The *Cincinnati Enquirer*'s opinion echoed the master mechanic's. "Viewed in detail, it excites wonder and admiration at the genius which invented and the skill and daring which executed the design." To cross the bridge cost three cents, for a foot passenger; a horse and carriage was fifteen cents. To take a sheep across cost the shepherd a penny.

The new bridge was a great success—but Washington was already thinking ahead. In April 1867, he wrote to Ferdinand on the subject of how to keep a good supply of the Roebling product easily to hand. Now was the time, he said, to open a wire rope store in New York City. "It has become a matter of absolute necessity to do so and now is the time," he wrote. "If I am to have a decent store it is necessary to be able to do all the little jobs that the business may bring along. It is not to be a mere office but a regular store with 20 or 30 coils where people can buy on the spot," he said. And it seemed as if there would be more than "little jobs" to be had in New York: for that same month the Albany legislature had voted through the act that authorized a private company to build and operate a bridge over the East River—a bridge to be completed by January 1, 1870. A month later, the new New York Bridge Company—headed by Henry Cruse Murphy, a lawyer, former editor of the *Brooklyn Daily Eagle*, and former mayor of Brooklyn, now a senator for the state of New York and a man who would be a driving force behind the bridge—appointed John Roebling chief engineer of the as yet undefined construction. "Confidence on the part of the public and of those whose money was to be invested in the undertaking would best be insured by employing an Engineer who had achieved the most successful results, and who was thus most likely to accomplish this great enterprise," the company's record would state.

Pittsburgh, Niagara, and now Covington–Cincinnati had proved that John Roebling had no peer in American engineering, and he would now begin to focus his energies on the extraordinary project brewing between the separate cities of New York and Brooklyn. His eldest son would be—as he had been on the Ohio—invaluable to his father; but he would fall under his father's sway in more than the matter of business. For while John Roebling had been in Cincinnati, the year before, they had ventured out one evening for a very particular meeting. "Last evening, Tuesday Feb 27 . . . Father and Emily & I went to Mr. Colchester the celebrated medium," Washington wrote to Ferdinand. "We are all confirmed Spiritualists now; father & I are good mediums. From Willie we had several communications, also from mother, and numerous other dead relatives and friends. Little Willie seemed highly delighted that he had a chance to talk to us and promised to communicate often with father. Among other questions I asked what you are doing that particular moment; the answer was that you were quite indisposed that evening, and had already retired. This was about 9 o'clock in the evening. I wish you would write to me in regard to yourself whether that was in fact."

THE BORDERS BETWEEN what could be proved and what must be believed were thinly drawn in the middle of the nineteenth century because they were in the process of being established. Photography and the telegraph enabled a human presence to be transmitted or preserved at a distance: many thought that the distance between this earthly world and the spirit world might also be bridged. In the wake of the deadly cataclysm that was the Civil War, a great thirst developed to communicate across the final divide. Photographers such as William H. Mumler, who began operating in Boston just before the war broke out, produced *cartes-de-visite* showing the sitter alongside his or her deceased loved ones; a few years after her husband was assassinated, Mary Todd Lincoln sat for Mumler, who produced an image of a robust Mary shadowed by the spectral presence of her late husband, his hand on her shoulder. Spirit photography was a natural outgrowth of a national passion for spiritualism that had been burgeoning since the 1840s. America was still a new place; its inhabitants were living on a frontier that was physical, technological, spiritual; it was an atmosphere in which men like Mumler, and the spiritualist Andrew Jackson Davis, could arise and flourish—and persuade even men of science like John Roebling into believing their trickery.

The summer of 1867 Washington and Emily—now pregnant—sailed for Europe. Things had not been easy at home. His father had now married Lucia Cooper, the woman he'd first met when working on the Niagara bridge, by Washington's account; she was, he said, "a harmless creature who ought to have been married to a different type of man, twenty years sooner," he wrote in his memoir. But there was more to Washington's disquiet than the fact of his father's new union. "My four years service in the Civil war had left me with broken down nerves," Washington recalled. "The incessant excitement, risk of life, and hardships leave their mark." And the Covington–Cincinnati bridge, following hard on the war, had been challenging, to say the least. "The building of a large Suspension Bridge is nearly as bad . . . In recognition of these conditions it was planned that I should make a short trip to Europe." It was his first trip abroad, and the couple would, in the end, be gone for nine months. But it was not really a trip for his health—it was one in which he would visit engineering sites and manufacturing operations across the continent with the aim of getting the best understanding possible of which firms could best serve the burgeoning project in Brooklyn. This trip, he would recall, "took me through many countries and brought me in contact with many eminent people—I was then comparatively young and strong and enjoyed everything." His "special investigations," however, "were directed towards deep foundations, more particularly to the use of compressed air in Caissons, all of which had a bearing on a possible Brooklyn Br of the future."

They sailed on the first of July, and would go from London to the north of England, to Paris, and across Germany—quite a trip not just for Washington, but for Emily, too: she was some months pregnant, by then. Her determination to stay by her husband's side seems evidence of her powerful desire to escape the fate of her late mother-in-law, a woman left behind. As they traveled, back home the spirits were revealing themselves.

John Roebling may have remarried—and yet Johanna was still a real presence in the Roeblings' Trenton household. Washington's father never really recovered from her loss, and he regretted his ill treatment of her. "In a higher sphere of life I hope I meet you again my Dear Johanna!" he had written in the family bible after her death. "And I also hope that my own love and devotion will then be more deserving of yours." As if in answer, Ferdy wrote to Washington in November 1867, "The latest sensation we have had here are spiritual communications from Mother . . . Ed. Riedel is the medium." Riedel was their cousin and foreman of the rope shop in Trenton. "Last Saturday night as he and father's draughtsman were sitting in their room, they heard three knocks right under Ed. He did not know what to make of it so they

examined the room and porch and all around, the knocks still followed . . .
They told father of it the next morning and in the evening they formed a circle
in father's office but got no communication, but as soon as Ed. went to his
room & pulled his boots off he heard the knocks, they then called father &
all went to the kitchen, where there is no carpet and the knocks were louder.
They then used the Alphabet and found out whose spirit it was. No answer
would be given to anyone but Ed. She said she was in the 4th Sphere and Willie, Mary, Grandmother & Grandfather were with her there. Various questions were asked, father suggesting them, but none of any account and finally
she said she would come back again in two weeks."

Even before Johanna's death John Roebling's inclinations were toward the
unification of matter and spirit. In his lengthy treatise "Truth of Nature," a
rolling inquiry into just about everything imaginable, his handwriting slopes
strongly across the pages, right to the very edge of the paper. Under "Reality
and Appearance" he wrote: "Hegel has taught, that things must be thought by
man, before their inner truth can be known. Man's mind must become identified with the inner truth of things, before Nature can be understood . . .
When things change their form, their appearance, when ice is converted into
water, and water into steam, they are negations, self-contradictions. Hence
in Nature there is no affirmation without negation . . . existence is a process,
a process of change, affirmation and negation at the same time. A thing is
and is not." He wondered where our physical and spiritual boundaries might
be said to lie: "Now if mind can be formed out of matter, cannot matter be
formed out of mind? Surely the mystery is no greater in the one case than in
the other! Human reason is forced to conclude, that there is but one Absolute Existence, one Absolute Being which is the Universal Spiritual Being, all
else proceeds from it and all else refers to it."

In John Roebling's distinctive hand are the sequence of questions that
was put to the spiritual presence of his late wife as she began to appear,
dated just a few days after Ferdinand wrote to his brother.

1. Are you here in company of other Spirits?
2. Is your mother with you?
3. Your father?
4. Your sister?
5. Little Willy?

Mary Herting, Johanna's sister, had died in 1859 at the age of thirty-three; she had contracted scarlet fever at the age of two, which had left her deaf. Willie Roebling was the little boy who had been lost to diphtheria in the first year of the Civil War. His death had affected John Roebling power-fully. The loss "has produced a great vacancy in our family, which will long be felt by mother & myself," John Roebling had written to Washington in October 1861. "Such a sweet little boy, with such a loving disposition and so intelligent a nature is not easily forgotten. I had no idea, that any death could affect me as much, as this has done."

The dogged questioning carried on, with occasional answers noted, in pencil, at the side. "Is my friend Overman there?" John wanted to know. Frederick Overman had been, Washington would later recount, his father's "bosom friend," a man who could run a newspaper or a blast furnace but whose interest in metaphysics had been what had drawn him to the elder Roebling, and John to him. "My mother was never enchanted with the Over-man friendship," Washington recalled. "She had no use for metaphysics." No wonder, with all she had to do. "Is your son Washington a medium?" John asked—and a penciled "yeas" revealed that he was. Washington's youngest surviving brother, Edmund, just thirteen, was a medium too. John Roebling asked if he himself was a medium; no, he was not. "Could your earthly life have been prolonged by better medical treatment?" he asked his late wife. "No," came the answer—conveniently proving John Roebling's own theories. The line of questioning, as it carried on, was clearly designed to offer evi-dence of what John Roebling already believed. He wasn't just correct: he now had official confirmation from the other side.

"Do you remember, my dearest, the conversations I have had with you about the Spiritland & Spheres?" *Yeas*, she answered. "You remember the opinions and views taught by Andrew Jackson Davis on this subject?" That too she recalled; and how could she not? John Roebling had always approved of the "Poughkeepsie Seer"; perhaps a factor was Davis's apparent belief in reason. "Reason is a principle belonging to man alone. The office of the mind is to investigate, search, and explore the principles of Nature, and trace physical manifestations in their many and varied ramifications," Davis wrote. He believed that the goal of matter was the production of spirit; death, he believed, didn't really exist at all. "The process of death is simply one whereby the existing relationship of body and spirit is dissolved"—that was all. There were six "spheres" of existence for the spirit to pass through after death; that spirit moved through those spheres in a progression before reaching the final, "Supercelestial" sphere—Paradise, effectively.

The son of poor, uneducated parents, Davis preached spiritual regeneration and social reconstruction; he believed that, after death, spirits gravitate to a level reflecting the degree to which they had lived, in life, in harmony with Divine Law, traveling up through six spheres before reaching "Summerland," the Spiritualist paradise. The *New York Times* had grave doubts about the seer's work. "Whenever an uneducated and densely ignorant person undertakes to write a particularly stupid and trivial book, he delights to entitle it the 'philosophy' of something," began the paper's review of one of Davis's later volumes. "When Mr. Andrew Jackson Davis, who has gained considerable notoriety as a ungrammatical prophet, and an unnecessarily tedious writer of twaddle, gives his views upon the best way of entering into intercourse with the spirits, and describes the manners and customs of the dead, he naturally calls his book *The Philosophy of Spiritual Intercourse.* The amount of philosophy contained in Mr. Davis's experiment upon the English grammar is about on a par with the amount of trustworthy information which he conveys."

But Johanna's spirit would confirm that Andrew Jackson Davis's theories were both correct "on the whole" and also "in detail." When he asked her how she occupied her time in the spirit world, "doing good" was given as the reply; John Roebling noted: "The same as on Earth, you never thought of yourself, you always thought of others." Yes, again, is written beside his words. "You are very happy in your Spirit home, are you not? Do you acquire a fuller and truer knowledge of the Creator by the study and comprehension of his works? Are the facts of spiritual intercourse between men in the spheres and men on this earth, discussed in your circles?" "Are you tired of answering questions?" John Roebling enquired of his late wife's toiling spirit; not surprisingly, she answered *Yeas.*

Even death didn't prevent John Roebling from wishing to settle an argument. He chased after his friend Overman again in the spring of 1868—though Overman had been dead for sixteen years by this point. An amateur chemist, he had died inhaling the arsenical fumes of some lime ores he was analyzing, according to Washington's account. After confirming the cause of his death, John Roebling drilled his late friend on his beliefs—and how they must, surely, have changed after his own death.

"You did not believe in spiritualism in this life? 'No.'

"I remember discussions I had with you on this subject, & on Andrew Jackson's Revelations. You had little faith in those Revelations. You remember this? 'No.'

"You remember your old friend von Reichenbach, and his discovery of Odic Forces? 'Yeas.'"

Baron Doctor Carl von Reichenbach was an eminent German scientist who became interested in what he called the "Odic Force," an energy he believed pervaded all living things. It appears that Overman, while still alive, was not convinced by this notion.

"You appeared to have little faith in them, yourself, when alive, but you looked upon Reichenbach as a man, honestly engaged in the search after truth?" Roebling asked, at least giving his late friend the benefit of the doubt as to his thoughts on von Reichenbach's character. Overman's spirit, needless to say, replied in the affirmative.

"Have you been reconsidering this subject in your Spirit home?" came the next question. Yes, is the answer, as it is to the next: "Is Reichenbach correct in his views?" Indeed, Overman confirmed from beyond the grave.

John Roebling would not allow even death to separate him from the triumph of his own beliefs. And yet, laced throughout these extraordinary records are hints of a father's ordinary sorrow—a side of John Roebling very rarely seen. GOOD EVENING ALL OF YOU, came little Willy's message one evening; his father asked him plaintively: "Willy are you grown any?"

Just before he asked after little Willie, John Roebling asked his late wife again whether Washington—an ocean away as his father sat rapt around a wooden table—was a medium. Yes, he is, came the answer, and a "writing medium," too: meaning that he, like Riedel, could have taken this spiritual dictation from his mother on the other side. But Washington himself remained somewhat closed-mouthed about the whole affair, for in his memoir he hardly discusses his father's spiritual life. "After water came philosophy as an amusement," he wrote. "He was a great follower of Oken now"—Lorenz Oken was a German physician and natural philosopher with strong Transcendental leanings—"of Andrew Jackson Davis . . . of Swedenborg and later on of Emerson—"His mind was ever investigating and never satisfied." He didn't appear to pass judgment on his father's beliefs; but they arose, largely, after Washington had left home. He was never the victim of those beliefs, as he had been the victim of his father's brutality, or his passion for the water cure.

Washington Roebling and his pregnant young wife moved through Europe at his father's behest, reporting on everything he saw, from the building of the Blackfriars Bridge across the Thames in London, to the Universal Exposition in Paris, to the silver mines of Bohemia. And on this trip—even as momentum was building in New York and Brooklyn for that bridge across the East River—Washington saw a chance of how he might escape from under his father's thumb, and he determined to take it.

9

# "I will have to go to work at something"

"IN THE WINTER OF 1866–67 Mr. Roebling received a letter from a certain boat builder in N.Y. City whose name I forget—He had a little shop for building small boats on South street—This man asked Mr. R to make an estimate for an East River Bridge to Brooklyn," Washington recalled in his memoir as he set out to tell the story of the beginnings of the Brooklyn Bridge. It's a story about money—because the unnamed boatbuilder claimed that he had "influential backing" for the building of a bridge; a claim that was, apparently, nonsense. But a week of John Roebling's time was given over to the estimate: such a bridge would cost four and a quarter million dollars, John Roebling said. It was a calculation that would foreshadow some of the trouble that would come in the following years, for it demonstrated John Roebling's tendency to underestimate the true costs of his work in his initial proposals. "The estimate made over a year later," his son noted, "accompanying the first report, the result of long and careful calculations, after an actual survey, amounted to 7¼ millions of dollars." That discrepancy caused "no end of trouble," Washington acknowledged; for it would later be claimed that the three-million-dollar difference was the sum that the city's corrupt politicians hoped to steal from the project. For just as the bridge project was getting under way, William M. Tweed, Democratic "boss" of Tammany Hall, the organization that effectively ran the city's politics, was about to be elected to the state senate, further consolidating his grip on the city's administration and finances. Before the bridge was finished the "Tweed Ring" would collapse and the florid career of its founder would end in ignominious disgrace—but not before tens of millions, perhaps hundreds of millions, of dollars had been stolen from the city; and the Roeblings' bridge would not escape the ring's taint.

But New York City was a place to make money in the middle of the nineteenth century. In those years, "the metropolis became the principal facilitator of America's own industrialization and imperial (westward) expansion. Capital flowed through and from its great banking houses and stock exchanges to western rails, mines, land and factories; it became the pre-eminent portal for immigrant laborers; and it exported the country's industrial commodities as well as its traditional agricultural ones." Even before the war 598 licensed hacks, 4,500 carts, and 190 express wagons clattered along the city's streets; omnibuses carried one hundred thousand passengers a day. In the 1850s, fifteen thousand vehicles passed Saint Paul's Chapel on the corner of Broadway and Fulton every day. A visiting journalist from London found himself abashed by the hustle and bustle. "The throng and rush of traffic in the business part of New York is astonishing even for London," the *Times* remarked. "There is a perpetual jam and lock of vehicles for nearly two miles along the chief thoroughfare."

But in the summer of 1867, not long after the charter was made that brought the New York Bridge Company into being, Washington and Emily were far from home—and able to read the *Times* in London itself. From the Royal Hotel beside the Thames—just at Blackfriars, where the brand-new bridge was being constructed—Washington began a sequence of detailed letters, to his father, to Charles Swan, to his brother Ferdinand, that would not only supply crucial information to John A. Roebling's Sons regarding the state of engineering, iron- and steel-making, and construction in Europe at the time, but also, as ever, display Washington's sharp eye for character and detail.

They also reveal some of the dreams he had for himself. London, he wrote, was enthralling. "The places of interest one can visit are endless in number; I shall only be able to visit a small number during my stay; we have been to St. Paul's, Westminster, the Zoological Gardens, Houses of Parliament, up and down the Thames on the swift little steamers of which there are at least a hundred . . ." He took a trip on the new underground railway, the world's first, which had begun service four years previously in 1863; New York wouldn't get such a system until the beginning of the twentieth century. He wasn't as impressed as he had expected to be with the English trains: "I can't say that you ride any smoother than on the Penna. Central RR," he wrote, enclosing a little drawing of the rail bed. "The large brewery of Perkins & Barclay did not quite come up to my expectation. I thought I would see horses as large as elephants, whereas they are no larger than our own mill horses." Barclay, Perkins, as it was more usually known, was one of the great sights of London in the nineteenth century, on the bank of the Thames at Southwark; Washington may have confessed himself disappointed, but the place was, just then, producing nearly a half million barrels of beer a year.

London's bridges were carefully noted. "The numerous bridges across the Thames are exceedingly interesting; two of them are in process of construction at present," he wrote to his father. "From my room I can overlook the construction of the Blackfriars Bridge . . . of which I have sent you a description. Above the Charing Cross Bridge they are sinking the cylinders for another roadway bridge over the Thames . . ." The foundation stone for the new Blackfriars road bridge, an arch bridge designed by Joseph Cubitt, had been laid in July 1865; it would be finished in 1869, the year construction on the East River Bridge began. A few days later Washington made his way to the bridges at Lambeth and Chelsea; the latter, designed by Thomas Page, he found "very handsome, especially the end view; sideways not so much because there is not the slightest camber to the floor." Both of these bridges have long since been demolished and replaced.

They sailed to Paris at the end of the month and spent a week there. The chief reason for their visit was the Universal Exposition, a world's fair on an unprecedented scale that had opened on the Champ de Mars in April that year. The official catalog claims there were forty-two thousand exhibitors (later sources put the figure even higher) on a site that exceeded one hundred twenty acres in size. Everything the world had to offer was on display; there were ten categories of presentation, from works of art to books to mathematical instruments and musical instruments; from "the apparatus and processes used in the common arts" such as mining, forestry, chemistry, and tanning, not to mention engineering, public works, and architecture; food was on show, and techniques of preserving it; there were livestock and farm buildings for all to see—including "Bulls, Buffaloes, &c.; useful insects."

Also in Paris was Abram S. Hewitt—the Roeblings' competitor in Trenton who would come to play an enormous role in the building of the East

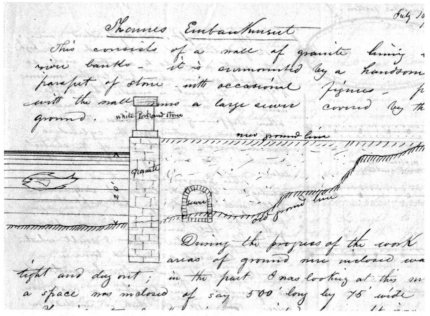

*Washington's drawing of the Thames Embankment at Blackfriars, complete with fish, in a letter to his father, July 18, 1867 (detail)*

River Bridge. Hewitt was there in his capacity as a United States commissioner to the fair; his published report is scathing regarding American representation at the Exposition. It was of "so meagre a character that foreigners, judging from the lessons of the Exposition, would have come to the inevitable conclusion that the iron and steel industry of the United States is not entitled to the rank which it undoubtedly occupies in the metallic production of the world . . ." But Hewitt noted that the Exposition marked a critical transition in the world's metal manufacture—a shift that would be of no small import in the building of the bridge over the East River. "By common consent it seems to be agreed that the most striking feature of the industry of the present day is the marked advance in the manufacture of steel and its progressive substitution for iron, in all cases where strength must be combined with lightness. Notice has already been taken of the enormous masses of steel in the Exposition, but it was only by observing the infinite variety of forms and purposes to which it was applied, that the intelligent observer was compelled to admit the transition which is taking place from the age of iron to the age of steel."

Washington spent three days at the Exposition, which was "ample to see it all and get tired of it; to me it seems like a great advertising show, as everything there is for sale." He was, furthermore, pretty keen to leave Paris altogether for other reasons: "We should have stayed longer than one week, were it not for the frightful expense attending it; my money would not hold out two months in Paris; moreover one lady costs as much as two men in the way of traveling abroad."

In this letter, too, is a careful little drawing of a caisson, the chambered foundation deep underground that was beginning to transform the building of large bridges. He met with a young English engineer called James Dredge—whose father was also in the business—who "gave me to day the bare idea of the manner in which Mr, Ordish laid down his piers for the bridge he is building over the Moldau." Rowland Mason Ordish was an eminent engineer, with work on the awesome roof of London's St. Pancras Station to his credit; he was now constructing a pedestrian bridge in Prague that used this innovative construction method. Washington's illustration shows the river, four stone piers or columns, with the caisson and its rim—the cutting shoe—beneath. He described a strongly built platform with three cylinders cut through it. "The platform is at first lowered by block & tackle . . . the brick and stone work being at once built on it as it sinks. The cylinders are worked by pneumatic apparatus forcing in air; through the two smaller cylinders the dirt is hoisted & through the other the water is pumped, men descend, etc . . . The pier is built at once as the foundation goes down; the greater the depth the greater the weight to force it down . . ." And there, quite succinctly, is an essential description of what would come to work, on a very much larger scale, under the river in Brooklyn and New York. The matter-of-factness of the description here is characteristic. He is in no doubt of what could be achieved. A little later he would head to Runcorn, on the River Mersey, twenty miles from Liverpool, where William Baker's "immense" bridge—still in use to this day—was nearly completed. "The foundations were very interesting," he wrote to his father, including, again, detailed drawings of that bridge's caissons.

He made "flying visits" to Thomas Telford's Menai Suspension Bridge, connecting the island of Angelsey to Wales, and to Isambard Kingdom Brunel's famous Clifton Bridge, in Bristol—both enduring icons of engineering, and greatly admired in their own day too. But Washington was never

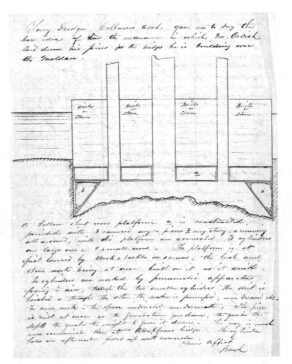

*Washington's explanatory image of a caisson, drawn in a letter to his
father while traveling in Europe, July 29, 1867*

one to be swayed by others' opinions. "As far as picturesque and romantic
setting go, Telford's bridge certainly has the advantage of the two," he remarked. Of the Clifton he said that "the towers are very ugly in design and
still more in execution, they being built of small uneven cellar stone scarcely
fit for the meanest cellar. My sketch I find looks better than the reality."

After London he headed north—where he had been eagerly awaited.
While still in London he had had a visit from John Thewlis Johnson, "the
nephew of 'Johnson & Nephew' "—who had supplied the wire for Niagara
and Cincinnati. Johnson had heard of Roebling's visit to Europe, "and came
on purpose to see me. He was terribly afraid that I would not go to Manchester and made a special point that I should stay with them 3 days. The poor
man didn't know that that was all I came to England for." Fourteen closely
written pages describe everything he saw at the Johnson and Nephew plants,
which were at the forefront of their industry. Just a few years previously, the

company's technical manager, George Bedson, had patented the world's first successful continuous rod rolling mill, which sped up the process of wire-making considerably; Abram Hewitt, for one, had been deeply impressed by Bedson's innovation when he saw it on display in Paris: "Bedson's machine has . . . the double merit of producing a better article, at a lower cost, than has hitherto been obtained; and it is a matter of regret to those who have become familiar with its novelty and its merits, that it received only the recognition of a silver medal, when it so justly deserved the highest prize."

Washington and Emily went back to continental Europe—and to his father's homeland. By early October he would write to John Roebling of all the places he had been: Gotha, Frankfurt, Homberg, Darmstadt, Würzburg, Nuremberg, Przibram (now Příbram), Prague, Dresden, Freiberg, Leipzig, Erfurt, and home again. "It presents quite a formidable list of names, but is only a small journey, after all, as most of the places are accessible by railroad." He listed numerically the objects of interest, including transmission of power by a wire rope at a cotton mill in Oberursel near Homburg; he sketched the transmission ropes down the long edge of the paper. At Prague he visited "the new suspension bridge nearly finished on Ordish's system"—the caisson system that had so caught his eye in Paris. Now he was able to see it for himself.

Emily—now heavily pregnant—had settled in Mühlhausen, taking rooms just opposite his aunt's house at "the English Hof" before Washington set out on his travels. Nonetheless, Washington's letters from John Roebling's place of birth remain focused on the job at hand, mainly concerned with observation of technical and engineering developments in Europe. Personal matters are left largely aside. "Most of the middle aged men, who were boys of 10 or 12 at your departure, remember nothing concerning you except a wonderful pair of yellow breeches you wore at the time; they want to know whether you still have them," he wrote to his father. He would send another fourteen-page letter from his visit to the Krupp factory in Essen—in Paris Krupp had taken the grand prize for a breech-loading cannon weighing some 110,000 pounds—the factory boasted a hammer that alone weighed forty-eight tons. He traveled at night into Bohemia to see the silver mines there, where the ore was brought up from underground using wire rope; he would visit the factory, too, that made the ropes. But for all he had not spoken English himself until he was quite grown, his boyhood tongue was suddenly

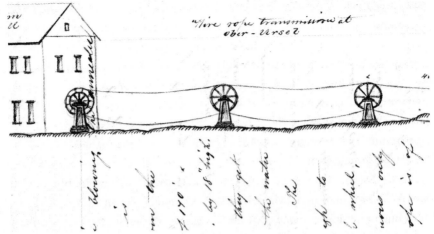

*Washington's drawing of power transmission by wire rope at Oberursel,*
*Germany, in a letter to his father, October 2, 1867 (detail)*

of little use. "I found myself a stranger in a strange land, my German was not understood by the people who speak nothing but the native Bohemian or Czechish; the only Germans there are the Austrian officials at the mines and they seem to be cordially hated."

The dual monarchy of Austria-Hungary had only been established earlier that year, in what came to be called the Compromise of 1867. Washington found the country he traveled through very poor, and the people bigoted; there was universal dislike of Austria. "Business is very much depressed and suicides frequent," he told his father. "As the train left Hořovice for Prague a well-dressed stranger laid down in front of the locomotive, which cut his head off instantly; the head rolled down the bank and the body remained lying between the rails; he was not recognized." In this same letter, down the side of the next page, is a detailed drawing of the Franz Joseph Bridge in Prague—Ordish's bridge. He wrote that he had sent his father a copy of one Mr. Loeffler's book on the foundation of the new pier of the Pregel Brücke at Königsberg; "the account is very interesting and is accompanied by excellent drawings giving all the details of the machinery for compressing the air in the cylinders, and for dredging out the dirt besides. It is the only definite work on the subject I have been able to get hold of so far, notwithstanding my enquiries by letter to all the scientific booksellers . . ." Clearly the

Roeblings were operating—or planning to operate—at the leading edge of technology. As the autumn progressed into winter, he got samples of steel billets and rods at his father's request. "Bessemer steel is by far the cheapest and they are constantly improving the quality of it," he reported.

Washington and Emily's only child, a boy, was born in Mühlhausen on November 21, 1867. Emily—left with Washington's relatives while he made his way through the Austro-Hungarian empire at his father's bidding—had apparently had a fall in the last weeks of her pregnancy that affected her badly. Washington told the story not to his father, but to Charles Swan. "My dear Mr. Swan," begins the letter he wrote just before Christmas that year. Emily, he said, had had "a bad time" on account of the fall; "she has been bleeding more or less all the time for the last 4 weeks, which made her excessively weak; it has stopped however and she sat up for the first time yesterday, and is getting strength again quite rapidly. I must say however that I dread the sea voyage back on her account; I have a notion at present of leaving the first week in February, either from Bremen or Hamburg—To go over to England is too bothersome." In the end, they did indeed go back to England, for all the bother that entailed. At the very bottom of the letter, he adds a postscript: "The little boy is growing fast and promises to be a bright little chap; he looks like me all but with a pointed chin."

The baby was baptized John A. Roebling II in the big church on the Untermarkt in Mühlhausen—its claim to fame that Bach had once played the organ there—in the first days of the new year. After the baptism, "the usual coffee party took place, without which such ceremonies are not complete here in Mühlhausen," Washington told his father. Emily wrote a note to her father-in-law. "The name of John A. Roebling must ever be identified with you and your works but with a mother's pride and fond hopes for her first-born, I trust my boy may not prove unworthy of the name though I cannot hope that he will ever make it famous as you have done." She thanked him, too, for a note he had written to the family in Mühlhausen to express his gratitude for the care they had taken of her; her German, she said, wasn't up to the task, and "you know Washy is rather indifferent in matters of courtesy."

WASHINGTON'S EUROPEAN TRIP was hard work—and hard proof of his dedication to his father's business, and the task John Roebling had set him.

*Emily with baby John, 1868*

As David McCullough has written, "the value of such a son was not lost on the father"—but that value had its price, and Washington was well aware of it. While on his grand tour, he thought he saw a way out from the life his father had planned for him, and he determined to make a break from John A. Roebling's Sons.

For it was during his travels through Germany that Washington encountered the invention of Eugen Langen, a German engineer four years older than himself—"a smart, ingenious, shrewd and able man," he called Langen in a letter to his father. Langen had been working on a small engine, powered by gas rather than coal; Washington believed there might be a future in Langen's work. After their encounter in Frankfurt he wrote: "This little engine supplies a want, which has long been felt in Germany, and to some extent in America also, that of a cheap & economical motor for small mechanics who cannot afford to have a steam engine or whose work is too intermittent to require the steady use of one." A close description of nearly six pages followed. A steam engine, with its external boiler, is a cumbersome machine; coal is an inefficient fuel, but liquid fuel—kerosene—had been developed in the 1850s. A few days later, however, he went farther in describing to his father what he had seen: "The motive power of the machine depends upon the principle that 95% of air mixed with 5% of common coal gas (by volume) produces an explosive mixture of considerable power . . . His idea is that a company should be formed to buy his patent and then manufacture the machines under the patent in the United States . . . Langan [sic] has put the refusal of the thing at my disposal for a certain length of time. It is an entirely new thing and I have not the slightest doubt will take well in the U.S."

The letter to his father was cautious, especially as regarded his own involvement in this new technology. And yet he was thinking of what awaited him back in the States. At thirty, he believed—married, more independent after his trip—that he might have a choice about what he would do with his life. He'd written to his brother Ferdinand about an idea for a business using scrap metal from the wire mill in Trenton: "I will have to go to work at something as soon as I get back and may as well do that." But when he had announced his engagement to Emily to his father, John Roebling had made his feelings clear regarding the future of the family business. No "strangers" for John Roebling: Washington was "expected to step in & help forwarding the interests of the family as well as yours individually."

Washington saw the gas engine—small, flexible, portable—as an even better bet than scrap metal. In mid-September he wrote to Charles Swan about the engine. "Suppose you mention that Gas Machine to Charley Carr and see what he thinks of it . . . If I were only home now I think I could make the thing go and make something of it." And to Ferdinand back in Trenton he proposed a real business plan—one that had nothing to do with building bridges or manufacturing wire. This was a bid for a whole new life. "I want to make you a proposition in regard to the Gas Machine. You say they want $100,000 for the patent, which with gold at 40% perm & Exchange 10% would bring it up to $150,000. The thing would then be to form a company to buy the patent for $200,000 which would leave $25,000 for each of us without spending anything. The Co. could then either dispose of the rights to make money out of it or by making a small assessment of the stockholders start the manufacturing of the engines. There could be no better time than now to start such a thing. There have been so many boiler explosions all over the country for the past year (averaging about one a day with more or less loss of life) that the news papers are taking up the matter very generally," he wrote. This was the moment to be seized, and Washington knew it. But money—a lot of money—would be required for the initial investment, and the one rich man they knew didn't think much of the whole idea. "Father says he won't have anything to do with it, if he had I should want to be out.

He does not seem to care much how we get along and the sooner we can be independent of all others the better it will be for us . . . I wish you would write to me as soon as you get this letter and let me know what you think."

This letter was written on November, 12, 1867. It crossed with a letter from his father, written from Trenton two days previously. John had been speaking to his in-laws about Washington's ideas. John's older sister, Frederike Amalia—of whom he was extremely fond—had married a prosperous Mühlhausen merchant called Meissner; the family had gradually all relocated to the United States, following the Roeblings. John Roebling wrote in very plain language to his son: "The Meissners enquired what I was doing about Langens Machine. I told them I had nothing whatever to do with it, and at the same time informed them that Langen did not want to have anything to do with their firm. They feel sore about [it] and I wish you have never concerned yourself about it. There is no money in it, I am satisfied."

There would be no more discussion of engines. If John Roebling thought there was no money to be made in bringing Langen's newfangled invention to the United States, then there was no arguing the case. He reminded his son of the bad luck the Swedish engineer John Ericsson—who had designed the *Monitor* for Lincoln during the war, and also developed the screw propeller—had had with his "caloric engine," which used hot air, rather than high-pressure steam, to move pistons. The innovation had never caught on, and the ship that bore his name, the *Ericsson*, had sunk in a storm off New York in 1854.

John Roebling was mistaken regarding the future of the gas engine. By the time Washington encountered Langen, the latter had already joined forces with Nikolaus August Otto, whose four-stroke engine would make the modern automobile possible; their company, N. A. Otto and Cie, in Cologne, was the world's first engine factory. Together they had won a gold medal at the Paris Exposition; in later years the company—which still exists today as Deutz AG—would employ Gottlieb Daimler and Wilhelm Maybach. There would be money in gas engines, for sure; but it would not be Washington Roebling who made it. We don't know what he would have said to Emily when his father's letter arrived; or indeed, whether he would have said anything at all. Perhaps his resignation was wholly private. In the Union Army he had followed orders. Now he was his father's lieutenant, and that was where his work, and the path of his life, lay. His father's wealth was not available to him, he knew that much. When he had written to Swan of the investment the engine business would require, his letter had ended on a plaintive note. "The trouble with me is, that I have no money of my own to do anything on the matter myself—" Yours affectionately, he'd signed off to Swan.

In later years, when the automobile came into wide use—Henry Ford's Model T appeared in 1908—it was always said that Washington Roebling disliked them. He refused ever to ride in one, preferring the Trenton trolley cars, even when he was a very old man.

WASHINGTON—AND EMILY, and now with baby John in tow—returned to the United States early in 1868. Winter weather wouldn't allow an earlier return. "Everything is frozen up around here," he wrote in a letter home before the turn of the year. "I wish you all a merry Christmas and regret that I am not with you—It will be 6 weeks before I can think of returning; it is not safe for so young a child during the stormy winter months."

Back in New York, not everyone shared John Roebling's unwavering confidence regarding a bridge over the East River. A letter in the *Journal of Commerce* was headed "Bridge to Brooklyn." "Messrs. Editors—That a bridge, such as proposed by Mr. Roebling, can be built, there is but little doubt; but will it be traveled? It would be a great work, and there would be a fine prospect: but I doubt its utility. By the calculation made by Mr. R., even if ships would strike their upper masts, passengers would have to ascend 180 feet. But to avoid these topgallant masts, they must ascend 200 feet from our streets. Brooklyn Heights, at Columbia and Willow Streets, is but 66 feet above the level of the water; how is this elevation of 200 feet to be overcome twice or more times each day, by a resident of Brooklyn? I am under the impression that, unless the boats were abolished by law, very few inhabitants of Brooklyn would pass the Bridge more than a few times each year. Signed yours &c 'Heights.' "

In January 1869 the bill before the Albany legislature to approve the bridge was mentioned almost in passing by the *Brooklyn Daily Eagle*, sandwiched between news of a committee to report on the postwar reconstruction of Georgia and a pension bill. In the same issue ran an article headed "The Bridge Scientifically Discussed," reporting on a meeting held at the Cooper

Institute in lower Manhattan. Mr. Wright, an English engineer, proposed a combination arch and suspension bridge: "It would be constructed of iron ribs, and rest like a girder, almost level, against two massive abutments on each side of the river." Mr. Dow suggested constructing huge piers out from each side of the river, "narrowing the channel very materially; at the same time he would build a causeway from the Battery to Governor's Island." Dr. Rich "thought all plans for a bridge were but mere chimerical experiments, and that the desideratum to be attained would be better accomplished by the erection of a causeway, as the level of a bridge of sufficient height to permit ships to pass under it would be at least three-quarters of a mile from the river front on each side, and the distance would be too great for such a busy people as we, impatient of delay . . ." Mr. S. F. Shelbourne said that the whole discussion seemed to be "unprofitable."

And yet the work moved forward. The next month, John Roebling met with the "consultants" who would approve his plan: they included Horatio Allen, who many years before had driven the first locomotive in the United States—the Stourbridge Lion, purchased from Britain's great railway engineer, George Stephenson. There was William Jarvis McAlpine, president of the Society of Civil Engineers, and James Pugh Kirkwood who, along with Colonel Julius Adams, was responsible for the beautiful Starrucca Viaduct in Pennsylvania, which remains a monument to their achievement.

Adams and John Roebling had history: a dozen years earlier, they had clashed over the construction of a railroad bridge over the Kentucky River. Washington's account says the pair had come to blows. By the time this team of "consultants" was gathered, the influential Brooklyn contractor William C. Kingsley had made some effort to patch things up between the two men; the effort worked well enough that Washington formed a friendship with Adams after his father's death. So there were, apparently, no hard feelings; though Adams had made his own proposal for an East River crossing, north of where the Brooklyn Bridge now stands, and not long before the opening of the bridge he wrote a long letter to the *Engineering News-Record* when they reported that "the originator of the scheme was J. A. Roebling." "In 1866–67," Adams wrote, "I was brought in contact with Mr. Wm. C. Kingsley, then a successful contractor of large experience, and accustomed to meet and overcome difficulties in the way of his enterprises; and while I cannot say that it was through his suggestion, but it certainly was through the encouragement received from him, that I prepared plans in detail for a suspension bridge between the two cities of New York and Brooklyn, and which were exhibited during the whole time of the fair of the American Institute in New York in 1866." Adams did not wish his place in history to be lost.

And so Adams was among a group of nearly two dozen influential men who accompanied the elder Roebling on a tour, in the spring of 1869, to see the work that had made his name: a "Bridge Party," largely organized by his son, to Pittsburgh, Cincinnati, and Niagara, where his great works stood. In this way men of influence, men of technical expertise, could see what John Roebling was capable of. One of the men along for the ride was Thomas Kinsella, the *Eagle*'s editor; the light-hearted tone of much of what appeared in the paper speaks of his confidence in the whole endeavor. A brief story on April 17, 1869, from a mysterious correspondent calling himself "O'Blique" reported on the doings of the party. "Roebling who is going to build our bridge, built some bridges out West, and at Niagara. A party of military officers, editors and other scientific men have gone to see if the Bridges are still there, and if they can get one that will fit the East River to buy a ready made bridge and bring it in with them."

But as he looked at the Allegheny River Bridge in Pittsburgh, the traveling correspondent addressed the chief question that arose in many minds. "Well, it may be asked what assurance have we in the existence of a suspension bridge whose greatest span is 350 feet that one of 1,600 will stand? The principle is the same. The span of the Niagara Suspension Bridge is over 800 feet; that of the Ohio Bridge, at Cincinnati, is over 1,000; the East River Bridge will be 1,600 and in the opinion of Mr. Roebling a span of 2,000 feet presents no difficulties that cannot be overcome." A few days later the party left the "great sooty blotch" of Pittsburgh behind and headed for "Porkopolis." There, the vision of what the Roeblings had built "broke upon us all at once—the stateliest and most splendid evidence of genius, enterprise and skill, it has ever been my lot to see," as Kinsella wrote. "There is no appearance of fragility or danger about it. We walked to and fro upon it for an hour, and it seemed as solid and stable beneath our feet as the earth on either side of the river. Yet as we stood under it, and saw it swing in the air so high, so graceful, and yet so secure, it was indeed a work to excite amazement and wonder."

Last stop on the tour was John Roebling's great feat at Niagara, a delicate bridge that supported the weight of a thundering steam train rushing over the great chasm below. In these turbulent times it still seemed as if anything was possible: a bridge seemed an expression of how the new world and the old world might meet, and indeed, how a wounded country might be reunited.

ON MAY 19, 1869, the *Eagle* announced the project "finally sanctioned by the Government Commissioners." In June, Brigadier General and Chief of Engineers A. A. Humphrey conveyed the secretary of war's approval of the plan. John Roebling's estimate for the final cost of the bridge had been precisely $6,675,257; in the end, the final cost of the structure would be more than twice that sum. Senator Henry Murphy was the president of the New York Bridge Company, but the driving force behind the project had been contractor William Kingsley. "Mr. Kingsley may be regarded as the real projector of the East River Bridge," the *Eagle* would eventually write. Just what that projection involved would be called into question when the Tweed Ring was exposed years later and "Boss" Tweed stood trial on corruption charges that have hardly been equaled since.

Decades later, as he composed his memoir, Washington would vent his spleen regarding William Kingsley and his machinations; he was, Washington wrote, "a plausible political contractor of scotch ancestry . . . tall lank and lean, with red curly hair . . . He was a man towards whom I felt an intolerable aversion, like I feel towards all men who are not free and above board." Washington describes a striking antipathy between Henry Cruse Murphy and Kingsley—despite how closely the two worked together to promote the bridge. Murphy, Washington wrote, "hated Kingsley and had more than one life and death struggle with him—On one occasion Kingsley in a drunken fit had old man Murphy down in a corner pounding him with a club, when his yells fortunately attracted some people on the side walk who rushed up and saved his life." A startling story, one that never made it into the *Brooklyn Daily Eagle*. As Washington would reflect all those years later, "It was not for

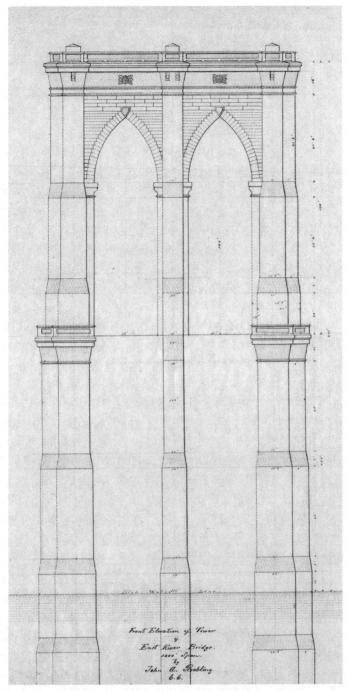

*Full elevation of a tower of the East River Bridge,*
*drawn by John A. Roebling*

a considerable time afterwards that I learned all the facts connected with the promotion of the enterprise."

THE BRIDGE TOUR behind him, in the spring and early summer of 1869, Washington began to build his new life in Brooklyn. He and Emily were now settled in a house on Hicks Street in Brooklyn Heights. "Our house is in a state of chaos yet," he wrote to Ferdinand back at the mill in Trenton, "just like the East River Bridge." His correspondence shows that at every stage there were obstacles to be overcome, yet his tone remained sanguine—even cheerful. "We have had very exciting discussions with consulting engineers— some are stubborn as mules," he wrote to Ferdy. "The papers are full about the Bridge all the time." Washington set about looking for the men who would work alongside him. "The trouble now is getting good assistants." Hundreds of men would build the bridge; but six men worked most closely with him: Colonel William Paine, with whom he'd served in the war; C. C. Martin, an RPI man who had worked at the Brooklyn Navy Yard and as chief engineer of Prospect Park; Francis Collingwood, who had also been a student in Troy. Sam Probasco had been born in New York but traveled to California when still in his teens; a year in the mining camps of the great gold rush had been enough for him, and it is said he walked back home to the East Coast, where he took up engineering. George McNulty, age twenty in 1869, had been at the University of Virginia; and Wilhelm Hildenbrand, not long arrived from Germany, was a consummate draughtsman who had designed the roof of the Grand Central Depot. The little notebooks Washington kept in his pocket show a record of the likely candidates he met, mixed in with lists of suppliers, divers, bids, and prices:

"C. P. Ladd, 31 Woodhull Street, Brooklyn
Offers as an assistant, or foreman—has worked at Victoria Bridge etc. also at some Susp. Bridges—he is familiar with iron work. May make a foreman.
Mr. McNamee in Col. Adams' office, formerly Laws' 1st Asst. on water works is recommended by Mr. Kirkwood as a tolerably good man.
A. M. Howell, Timber & lumber inspector, No. 2 South 4th St.
A. Williamsburgh—Wants to get a job on the caisson in regard to lumber"

Astonishingly, the bridge over the East River was not the only bridge he had to concern himself with. Ferdinand was keeping him completely up to speed with the many other structures John A. Roebling's Sons were undertaking at the time; certainly they were very much smaller projects, but all the same they had to be well made and built to last. "I have told every man that I will build them right off at a fixed price, and I am wondering many will bite," Ferdy wrote to his elder brother. There was a plan for a bridge at Hancock in Delaware County, New York, and for one at Bushkill in Pike County, both to be built for $21,000. "At Pond Eddy, above Port Jervis, the chance is very certain . . . The Bridge for that man at Union Broome Co will cost about $50,000, at least it is worth that to build it; that is probably more money than is in the county."

And then in late July John Roebling had his toes crushed in the pilings of a Brooklyn ferry slip; a little over a week later, he was dead. At thirty-two, his son Washington was the man who understood what his father envisioned; had traveled across Europe to familiarize himself with what was, then, the cutting edge of technology; had made his father's bridge over the Ohio a reality. The minutes of a meeting of the directors of the Bridge Company dated August 3, 1869, formalized his position. "On motion of Mr. Jenks, it was resolved that Col. Washington A. Roebling be appointed Chief Engineer; that the Executive Committee have power to fix his compensation, and that he have power to employ such assistance as he may deem proper, subject to the approval of the Executive Committee."

This moment of responsibility would be the great turning point in his life. "He would be the one with the final say," David McCullough has written. "However much staff help he got, every important decision would be his in the end, and at this stage there was seldom any sure way to know which decisions, of all that had to be made, might turn out to be the important ones in the long run . . . Roebling would be the one to answer the sort of criticism from other overnight experts that the newspapers and a few of the professional journals liked to give space to . . . As the work progressed, he would have to account for each and every step along the way . . . If things went wrong, if materials proved shoddy, if equipment broke down, if the work fell behind schedule, if some part of the structure itself failed, if accounts were juggled or costs got out of hand, if there were mistakes in judgment by any of his subordinates, if there were accidents, he would be the one held accountable."

It is easy enough to imagine him, on a warm, late summer evening, or perhaps an early morning, walking by the sweep of the river on the Brooklyn
shore, this man of medium height, his light topcoat slung over his arm, a
cigar in his pocket. His level gray gaze watching the ferries leave white wakes
behind them, ghostly paths on the water. He stops, and then goes onward.

# "Good enough to found upon"

THE TOWERS AND CABLES of a suspension bridge, its stiffened roadway, are its visible wonders, strength and flexibility hung over water, hung in the air. But all this rests on a hidden foundation, buried deep underground. You can't see what underlies the structure. But it's there—and everything else depends on what has been sunk in the darkness and depth.

This is true of everything we build. Every action and every decision have their roots in a foundation—usually, as far as individual human beings are concerned, in the foundation of character. The building of the Brooklyn Bridge tested Washington Roebling as he had never been tested before, even in battle; certainly, it would alter his life far more than the war ever had. Those tests are the hinges of a life: moments when the choice to carry on, or how to act in a given situation, throw one's personality into sharp relief. These moments don't offer proof of any kind of heroism. They are about endurance and persistence. For Washington Roebling, the first of these great tests is marked in the sinking of the caissons beneath the towers.

At the time of his death, John Roebling was still drawing up his plans for the bridge that would, one day, cross the East River; according to his son, John never expected to be the engineer who would see it through. He was, after all, not a young man. "He at no time expected to see the great work upon which he was engaged finished, but did desire to live long enough to see it fairly launched upon its way towards completion." It was his son who had worked, day after day, on the cables in Cincinnati; and it was his son who had traveled across Europe examining the cutting edge of engineering technology. When it came to sinking the foundations of the towers under the East River, it was Washington who was, by far, best equipped for the work to come.

"The caisson of the Brooklyn foundation is a large inverted vessel or pan resting bottom upward, with strong sides. Into this air is forced, under a sufficient pressure to drive out the water. Entrance is had to the large working chamber thus formed underneath, through suitable shafts and air locks." Washington wrote this simple description in a lucid, compelling account published in 1873 of the building of the foundations of the bridge. Hamilton Schuyler, the Roeblings' first biographer, would expand the description a little. "A caisson is practically a box of timber open at the bottom, which rests on the river bed," he wrote. "The water is kept out by pumping compressed air into the caisson. While the men working in the compressed air dig out the earth underneath, others in the open air on top of the caisson lay the masonry of the towers. As the towers rise, their weight, together with the removal of earth from underneath, causes the caissons to sink gradually to their final resting place at the bottom of the river. Every foot that the caissons sink necessitates an increase in the air pressure in which the men are working, and there is always danger, unless the utmost precautions are taken, of what is known as the dreaded 'caisson disease,' or 'bends,' developing among those employed in working in the high-pressure atmosphere."

This was technology in its infancy. John Roebling's papers show that solving the crucial problem of how to support the great towers of his bridge was one that had tested his powers of reckoning. An undated drawing for the bridge shows a caisson and tower; there are notes below that read: "Principle of Timber Platform/Piling or no Piling?/ If no Piling, how deep?/How to sink it? . . ." James Buchanan Eads, a self-educated engineer who had made his fortune with a salvage company on the Mississippi, was using caissons on the steel truss bridge he was building in St. Louis. He had made use of compressed air in diving bells for salvage work; Washington followed news of his progress carefully. While in Europe he had read and seen all he could about this new development. He copied out plans for building a bridge over the Rhine proposed in 1850; he also made sure to bring home a carefully copied essay on the foundations of the railroad over the Jumna at Allahabad, East India, from the *Civil Engineer* of December 1863; and he copied out and translated an article on a "Lattice Bridge on Cast Iron Cylinders over the Seine at Argenteuil," which had appeared in Opperman's *Nouvelles annals des constructions* in January 1864, as well as a contemporary treatise on the construction of a caisson in Brest.

Both of the caissons for the Brooklyn Bridge were built by Webb & Bell, shipbuilders at Greenpoint—for the structures would be built and launched like ships, albeit ships that were upside down. Each was nearly 170 feet wide and more than one hundred feet long, a massive timber foundation, with a roof fifteen feet thick, to support a tower weighing nearly seventy thousand tons. Men working in the caisson's chamber—which had a headroom of just over nine feet—dug down into the river's bed, so that the caisson would sink; meanwhile stone was piled on top. The compressed air pumped into the chamber kept the river out. Shafts through the chamber's roof allowed material to be put in and taken out, using a clamshell dredge—"the only known instrument which possesses the precise action of the human hand in picking up things." Each shaft was equipped with a locking door at top and bottom, so that the pressure inside could be maintained. Men entered and left through an air-lock chamber and climbed down an iron ladder into the caisson itself, which was divided into six rooms by dividing frames. The main question was: how deep would the caissons have to be sunk? A trial boring had been made in the river in 1867; gneiss rock, the city's bedrock, was struck at ninety-six feet below high water, but at fifty to sixty feet the material encountered was extremely compact; the borehole was sustained without tubing for weeks. But there was little consistency in what lay below the surface. "It is well known that the drift formation of Long Island presents a great variety of strata in comparatively short distances." In New York City's Municipal Archives, Washington's drawing of the results of that test boring can be found: a striking image, layers of earth and rock down into the river.

The Brooklyn caisson was launched from the shipyard at Greenpoint on March 19, 1870. It weighed three thousand tons and was made from 111,000 cubic feet of timber and 250 tons of iron. The launching, Washington wrote, "was a success in every respect; as soon as the last block was split the caisson began to move . . . neither the battering rams provided to start her, nor the checks to hold her back, were needed . . . The whole of the launching arrangements, as well as the responsibility of the entire launch, rested with the builders, Messrs. Webb & Bell, who deserve the greatest praise for the successful manner in which they carried out so novel a work. They accomplished the result by simple common sense arrangements; no money was wasted upon elaborate precautions or fancied contingencies."

Such success was by no means assured. In 1857 the great Brunel had attempted to launch his tremendous steamship, the *Great Eastern* (then still called *Leviathan*) from her cradle at Millwall in London; she got stuck in the ways, however, on more than one attempt and wouldn't make it into the water until January 1858. *Mechanics' Magazine* called the failed launch "the greatest and most costly example of professional folly that was ever seen." The smooth launch of the Brooklyn caisson was due, declared the *Eagle*, to the skill of the engineers. "Where Brunel was repeatedly baffled, Roebling succeeded promptly and instantaneously." If the Chief Engineer had his own doubts—if he thought of Brunel and his vast ship as the caisson waited in the slipway—he did not speak of them. There was no time for that.

Work down in the caisson was like nothing the men—who were paid two dollars a day for an eight-hour shift; as the caisson went deeper, this was raised to $2.25—had ever seen. Master mechanic Frank Farrington described the atmosphere in a lecture he gave, years later: "The temperature in the caisson was about 80°, and the workmen, with half-naked bodies, seen in dim, uncertain light, brought vividly to mind Dante's 'Inferno.'" In the summer of 1870 the *Eagle* sent one of its reporters down into the chamber. Walking into the workmen's yard above the entrance to the caisson, he wrote, "our attention was soon diverted by the appearance of a gentleman with a very red face slowly emerging from the shaft, and quietly but forcibly asserting that he didn't want to try that again." The intrepid journalist went down all the same. Below, he wrote, "small calcium lights throw streaming luminous jets into the corners where the workmen are busy, and sperm candles stuck in sockets fixed on iron rods serve as torches for visitors treading the planks laid down for paths in the sub-marine cavern . . ." He described "the mud and silt underfoot, the dark wooden walls at the sides, the chambers dim with their scattered lights, and the laborers moving like good-natured gnomes in the shadows." A remarkable drawing by Washington shows the men working in the caisson, the stone being piled on top as they dig below. There are five active vibrant figures down in the chambers, the cutting "shoe" digging down in the river. The *Eagle* reporter noted that "most of the laborers say that 'no inconvenience is experienced from the density of the atmosphere.'"

Nothing was simple. In the Brooklyn caisson, huge boulders impeded the caisson's descent; they were "so solid that a steel pick had no effect," Washington wrote. Once again, his keen interest in and knowledge of

John A. Roebling, circa 1865.
*Special Collections and University Archives,*
*Rutgers University Libraries*

Johanna Herting Roebling, circa 1860.
*Image provided courtesy of the Roebling*
*Museum's Ferdinand W. Roebling III Archives,*
*Roebling, New Jersey*

Washington as a student at Rensselaer.
*Roebling Collection, Institute Archives and Special
Collections, Rensselaer Polytechnic Institute, Troy, NY*

381—Blondin's Rope Ascension over Niagara River.

Stereopticon card of Blondin crossing Niagara, with John Roebling's bridge behind;
note Washington's annotation on the right. *Roebling Collection, Institute Archives and Special
Collections, Rensselaer Polytechnic Institute, Troy, NY*

Washington around the time of his enlistment in 1861. *Roebling Collection, Institute Archives and Special Collections, Rensselaer Polytechnic Institute, Troy, NY*

Puppet drawn by Washington in the back of a notebook he kept around the time of his marriage, 1865. *Roebling Collection, Institute Archives and Special Collections, Rensselaer Polytechnic Institute, Troy, NY*

Emily and Washington at the time of their marriage. *Special Collections and University Archives, Rutgers University Libraries*

Stereopticon card of the board of experts taken to see John Roebling's bridges in 1869; taken by Charles Bierstadt at Falls View Suspension Bridge, the narrow pedestrian and carriage bridge completed in 1869 that gave a view of the falls and John Roebling's Niagara Bridge. John Roebling at right front; Washington is apparently on the left, in front. *Roebling Collection, Institute Archives and Special Collections, Rensselaer Polytechnic Institute, Troy, NY*

Brooklyn caisson before
launching, 1870. *Roebling
Collection, Institute
Archives and Special
Collections, Rensselaer
Polytechnic Institute,
Troy, NY*

New York tower under
construction, September
1872. *Roebling Collection,
Institute Archives and
Special Collections,
Rensselaer Polytechnic
Institute, Troy, NY*

Given from
to Bed Rock
under N. Y. Town
East river Bridge
& 1872
79' feet below H.W.

COUNTERCLOCKWISE FROM TOP LEFT: Bedrock from below the New York tower of the Brooklyn Bridge, labeled by Washington Roebling. *Erica Wagner*

Wire rope for the Brooklyn Bridge. *Roebling Collection, Institute Archives and Special Collections, Rensselaer Polytechnic Institute, Troy, NY*

Men walking on the temporary footbridge used during the construction of the Brooklyn Bridge. *Photographer unknown/ Museum of the City of New York, X2010.11.8463*

The Brooklyn Bridge under construction. Photograph by J. A. LeRoy. *J. A. LeRoy/ Museum of the City of New York, X2010.11.8439*

*The East River Bridge*

*will be opened to the public*

*Thursday, May twenty fourth, at 2 o'clock.*

*Col. & Mrs. Washington A. Roebling,*

*request the honor of your company*

*after the opening ceremony until*

*seven o'clock.*

*110 Columbia Heights,*

*Brooklyn.*

Invitation to the Roeblings' reception
after the official opening of the
Brooklyn Bridge. *Roebling Collection,
Institute Archives and Special Collections,
Rensselaer Polytechnic Institute, Troy, NY*

Bird's-Eye View of the Great New York
and Brooklyn Bridge and Grand Display
of Fireworks on Opening Night, May 24,
1883. *A. Major/Museum of the City of New
York, 29.100.1752*

BIRD'S-EYE VIEW OF THE GREAT NEW YORK AND BROOKLYN BRIDGE,
AND GRAND DISPLAY OF FIRE WORKS ON OPENING NIGHT.

The Brooklyn Bridge in 1925. Photograph by Irving Underhill. Washington wrote "keep this for my album" on the back. *Roebling Collection, Institute Archives and Special Collections, Rensselaer Polytechnic Institute, Troy, NY*

Emily Roebling in later years.
*Roebling Collection, Institute Archives and Special Collections, Rensselaer Polytechnic Institute, Troy, NY*

Colonel Washington A. Roebling.
*Roebling Collection, Institute Archives and Special Collections, Rensselaer Polytechnic Institute, Troy, NY*

John A. Roebling II, while at Rensselaer, 1888. *Roebling Collection, Institute Archives and Special Collections, Rensselaer Polytechnic Institute, Troy, NY*

FACING PAGE: Washington's mineral collection at his home in Trenton. *Roebling Collection, Institute Archives and Special Collections, Rensselaer Polytechnic Institute, Troy, NY*

BOTTOM: Emily Roebling on the staircase of the Roeblings' Trenton home, showing the Tiffany window of the Brooklyn Bridge above the staircase. *Roebling Collection, Institute Archives and Special Collections, Rensselaer Polytechnic Institute, Troy, NY*

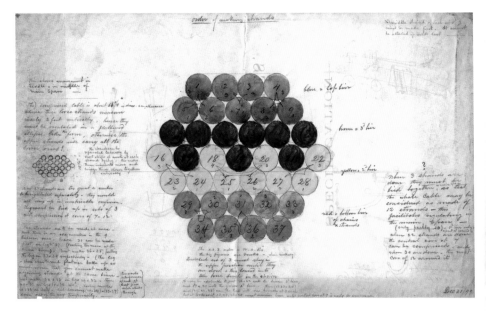

Washington's drawing of a cross-section of a cable for the Williamsburg Bridge, dated December 31, 1899, illustrating the arrangement of cable strands. *Roebling Collection, Institute Archives and Special Collections, Rensselaer Polytechnic Institute, Troy, NY*

Charles Swan, the business associate of John Roebling who showed great warmth to Washington as a young man. *General Research Division, New York Public Library, Astor, Lenox and Tilden Foundations*

Charles Roebling, circa 1908.
*Special Collections and
University Archives,
Rutgers University Libraries*

Ferdinand Roebling, circa 1913.
*Special Collections and
University Archives, Rutgers
University Libraries*

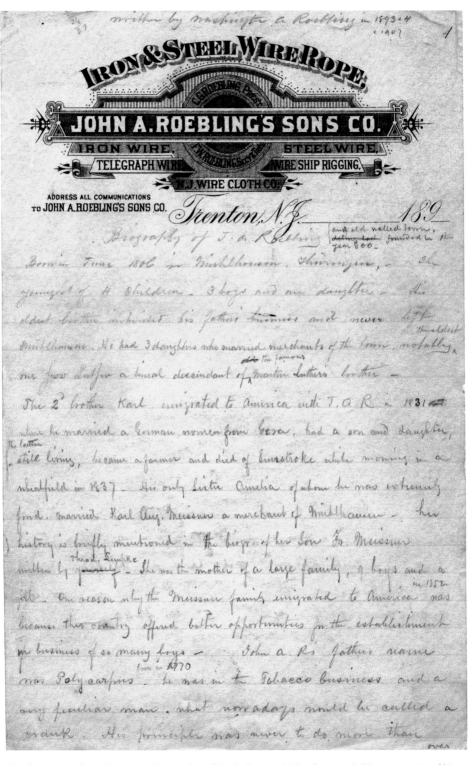

written by Washington A Roebling in 1893 & ~1907

## IRON & STEEL WIRE ROPE,

### JOHN A. ROEBLING'S SONS CO.

C. G. ROEBLING, PRES'T

IRON WIRE     STEEL WIRE

J.A. ROEBLINGS SONS CO

TELEGRAPH WIRE     WIRE SHIP RIGGING,

N.J. WIRE CLOTH CO.

ADDRESS ALL COMMUNICATIONS
TO JOHN A. ROEBLING'S SONS CO.

Trenton, N.J. _____ 189_

Biography of J. A. Roebling and old walled town, founded in the year 800 —

Born in June 1806 in Mühlhausen, Thüringen, the youngest of 4 children. 3 boys and one daughter. His oldest brother inherited his father's business and never left Mühlhausen. He had 3 daughters who married merchants of the town, notably the eldest one, Frau Luther a lineal descendant of the famous Martin Luther's brother —

The 2d brother Karl emigrated to America with J. A. R. in 1831 where he married a German woman from Saxe, had a son and daughter, the latter still living, became a farmer and died of sunstroke while mowing in a wheatfield in 1837 — His only Sister Amelia of whom he was extremely fond, married Karl Aug. Meissner a merchant of Mühlhausen — her history is briefly mentioned in the biog. of her Son Fr. Meissner written by Theod. Luyke — She was the mother of a large family, 9 boys and a girl — One reason why the Meissner family emigrated to America in 1852 was because this country offered better opportunities for the establishment in business of so many boys — John A. R's father's name was Polycarpus — born in 1770 — he was in the Tobacco business and a very peculiar man. what nowadays would be called a crank. His principle was never to do more than

over

The first page of Washington's biography of his father—which often reads like a memoir of his own life. *Special Collections and University Archives, Rutgers University Libraries*

THIS PAGE: Emily Warren
Roebling at the
coronation of Czar
Nicholas in 1896.
Carolus-Duran
(Charles-Auguste-
Émile Durant) (French,
1837–1917). *Portrait
of Emily Warren
Roebling, 1896.* Oil
on canvas, 89 x 47
1/2 in. (226.1 x 120.7
cm.). *Brooklyn Museum,
Gift of Paul Roebling,
1994.69.1*

FOLLOWING PAGE:
Washington Augustus
Roebling by Théobald
Chartran. Théobald
Chartran (French,
1849–1907). *Portrait
of Washington A.
Roebling, 1899.* Oil
on canvas, unglazed,
framed weight is 166
lbs.: 79 1/8 x 53 1/8 in.,
166 lb. (201 x 134.9 cm.,
75.3 kg.). *Brooklyn
Museum, Gift of Paul Roe-
bling, 1994.69.2*

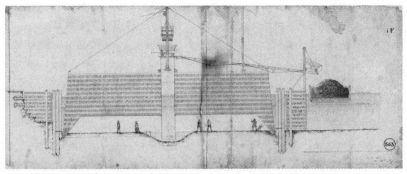

*Washington's drawing, dated September 28, 1869, of men working in the caisson, the stone being piled on top as they dig below. There are five active, vibrant figures down in the chambers, the cutting "shoe" digging down in the riverbed*

geology is evident in his record of the descent. "Nine tenths of the boulders were trap, transported hither during the drift period from the Palisades of the Hudson. Owing to their hardness they had resisted the wear of time the longest. They occurred in all sizes, from one cubic foot up to two hundred and fifty. Boulders of quartz and gneiss rock occurred more rarely. Two large boulders of red sandstone were also found. The softer varieties of rocks had all been worn out to pebbles. A collection made of the various specimens encountered during the descent of the caisson presents a complete series of the rocks found for a hundred miles to the north and northeast of Brooklyn." Just months before, he had gone up to Maine to see for himself the quality of the stone from the Fox Island granite quarries; he was impressed. "The general quality of the stone is a coarse crystalline compact granite—The Feldspar forms nearly 50%—quartz 30% and hornblende 20%—The Feldspar predominates and imparts the general color to the stone. The general appearance of the islands is a low rounded outline, no bald projecting points. Glacial action everywhere visible—The rock in its place is more or less split up into sheets nearly horizontal or following the carved outline of the hill. Deep natural fissures cleave it in other directions," he wrote, with a drawing, once again, alongside.

Danger was never far away. One September Sunday, the water in one of the shafts got too low—below the rim of the shaft; this water level effectively sealed off the caisson from the unpressurized atmosphere above. "The result was that all the air in the Caisson rushed out instantly," Washington recorded in his private notes of the event. "No one was inside. For the previous 3 weeks no water came in from the river under the shore and all the necessary water was supplied artificially though pipes, from the City water . . .

"It took about 2 minutes for all the air to escape. The water, stone & mud were sent up with terrific force to a height of 500 ft, accompanied by a dense cloud of fog owing to the expansion. The whole neighborhood was covered with mud. The watchman was standing on the center of the caisson. He found it impossible to escape, owing to the rush of air towards the watershaft from without, which threw him down. While lying there he was hit on the back by a stone. The Dredge, which weighs 5 tons, was lifted like a feather and broken to pieces in its fall—Two steel wire ropes which held it, prevented it from being thrown into the river. The noise resembled the explosion of a large powder magazine, prolonged for 2 minutes. The whole neighborhood for several squares around took to flight, running in every direction in the wildest confusion, as if an earthquake had happened."

It took a half hour before the air locks could be closed and water poured back down the shafts of the caisson, which remained perfectly sealed from the river—no river water at all leaked into the chamber. When the caisson was repressurized, Washington went back in "with considerable misgiving," he wrote in his later report. Without the pressure of the compressed air, the whole weight of the masonry had weighed on the caisson's roof, walls, and frames. The chamber had settled by ten inches; all the timber courses had been compressed by two inches, squashed by the great burden put upon them. Washington calculated that the structure might have borne as much as eighty tons of weight per square foot in the moment of impact; when it settled, it bore a weight of twenty-three tons per square foot. When the bridge was finally finished, it would need to bear only five tons per square foot; his private notes at the time show a quick run of calculations. The caisson's extraordinary strength had been proved.

There had been no one in the caisson that Sunday morning, of course, Sunday being the day of rest; but Washington was down in the chamber on the day the workmen accidentally dropped first one load of bricks, and then another right on top of the first, down a supply shaft. The bricks were used to build piers inside the chamber—there would be seventy-two of them—"their ultimate capacity was just sufficient to support the whole weight above in case the air should blow out." The proper procedure was to seal the door at the bottom, drop down the bricks, and seal the door at the top; only then would the bottom door be opened, to prevent any compressed air from escaping. "To say that this occurrence was an accident would certainly be wrong, because not one accident in a hundred deserves the name," he had written about the water shaft blowout. Carelessness—"overconfidence in supposing that matters would take care of themselves"—was once again the cause. The double weight of bricks caused the door in the chamber to thud open. The result was dramatic.

"As soon as this happened the air commenced to rush out of the caisson with a great noise, carrying up stone and gravel with it. The men above ran away, leaving those below to their fate. Anyone with the least presence of mind could have closed the upper door by simply pulling at the rope.

"I happened to be in the caisson at the time. The noise was so deafening that no voice could be heard. The setting free of water vapor from the rarifying air, [produced] a dark, impenetrable cloud of mist, and extinguished the lights. No man knew where he was going, all ran against pillars or posts, or fell over each other in the darkness. The water rose to our knees, and we supposed, of course, that the river had broken in. It was afterward ascertained that this was due to the sudden discharge of the columns of water contained in the water shafts. I was in a remote part of the caisson at the time; half a minute elapsed before I realized what was occurring, and had groped my way to the supply shaft where the air was blowing out. Here I joined several firemen in scraping away the heaps of gravel and large stones lying under the shaft, which prevented the lower door from being closed. The size of the heap proved the fact of the double charge. From two to three minutes elapsed before we succeeded in closing the lower doors." The brick pillars and other supports were examined to see whether the weight of the stone suddenly piled on top of them—thirty thousand tons—had done any damage: there was no sign of yielding.

"The question naturally arises, what would have been the result if water had entered the caisson as rapidly as the air escaped," Washington wrote. "The experience had showed that the confusion, the darkness, and other obstacles were sufficient to prevent the majority of the men from making their escape by the airlocks, no matter how ample the facilities. If the water entered as rapidly as the air escaped there would then be the same pressure of air during the whole time of escape. Now it so happens that the supply shafts project two feet below the roof into the air-chamber: as soon, therefore, as the water reaches the bottom of the shaft it will instantly rise in it, forming a column of balance and checking the further escape of air. The remaining two feet would form a breathing space sufficient for the men to live, and even if the rush of water were to reduce this space to one foot, there would be enough left to save all hands who retained sufficient presence of mind." In his notes Washington remarked that "the only way to avoid it for the future is to make some part of the door shutting machinery the means of turning on the compressed air"; and to his brother Ferdinand, back in Trenton, he was brisk in dismissing the blowout as a matter for concern. In a note tucked into the corner of a letter on New York Bridge Company stationery that is mostly about accounts, he said simply: "Had a little accident this morning, all the air blew out of the caisson while I was in it. She took 6 minutes to blow out. No harm done beyond a scare."

MEANWHILE, UP IN the daylight, huge blocks of stone, weighing as much as eight tons, were hauled up to the top of the tower on derricks supported by iron guy ropes; one day in October one of the guys snapped under the weight of its load, sending stone and derrick crashing down.* "The accident we had yesterday was very dreadful," Washington wrote to Ferdinand. "Three men are dead, and the rest are so scared that you cannot get them to work again. Two derricks are completely destroyed, and other damage done which $20,000 will not cover. Stone setting will be stopped for the season." But such things, as Washington knew, only happened if they had a right to happen: "It was all the fault of the Trenton blacksmith. The weld in the handle of the socket was merely a cold shut hammered over on the outside to hide it. No part of the iron was welded, and the great wonder is that it did

---

*The masonry of the Brooklyn Bridge is composed of granite and limestone. The towers are mostly limestone below the water, and granite above. The anchorages are mostly limestone— with 650 cubic yards of granite blocks placed on top of the anchor plates.

not break long before. The strain on it was only 1/5th of what the iron should bear if it had been sound."

And in the archives at Rensselaer is a remarkable document, a note written first in pencil, crossed out, and rewritten in ink beneath in a hand that is still almost completely legible, no matter in how much haste this aide-mémoire was composted. "Accidents," is the heading, and it begins with one word: "Fire." Throughout his life, Washington Roebling would write on any available scrap of paper, on the back of old stationery, on small slips found in the backs of drawers. Here is one, evidence both of attention and exhaustion, of the need to keep every detail in his mind. It reads, now, almost like a kind of urgent poetry. "Several small Fires. Leaks in seams—a caulking of oakum catches easily—some easily put out. One over shoe . . . slightly serious—5' wide and 6" burnt away—had to flood caisson—water entered easily as pressure was let off—only 3 courses masonry on at the time—water pumped out—4 hours by air.

"Increased watchfulness henceforth. Water supply from outside getting scarcer, finally all cut off—Necessity of flooding caisson from outside if fire got started. Danger of doing it. Increased caution—water pipes, hose— steam hose from outside. Fire on night of Dec 1 . . . Candle, pointing with cement. Bad place, would not be seen. Burnt appearance—living coals . . . no smoke—amount of water applied . . . Above frame no chance to stop up leaks. Steam no good—apparently out. Digging out—boring commenced— No fire at first, smell outside—danger of letting out air, settling of caisson— over night. Risks. Ultimate decision. Boats, 38 streams, 5 hours time—contents = 1,350,000 gallons—Fire not out until roof reached.

"Flooded how long—water pumped out in 6 hours—forced out over water shafts . . . smell, suspicion of fire—reasons against it—How long it lasted—heat in wood—no air bubbles."

On the timber roof of the caisson the whole structure rested: and then, that timber caught fire. Everything could have been lost thanks to a moment's carelessness. "The immediate cause of the fire must be owing to a candle held in the right hand of the man who had his coat or dinner in a candlebox which was nailed up over the door close to the roof. He could only reach the box by stepping up on a frame brace, when he would hold a candle with his right hand and reach into the box with his left. He must have held the candle there at least a minute," Washington surmised. The man, the *Brooklyn Daily Eagle* reported, was called McDonald; once he had seen the hole burned through the wood, he filled it with plaster "to conceal his blunder"; he soon after disappeared, "and has not been seen since." But in the oxygen-rich atmosphere of the caisson the wood behind the patch job kept burning, "living coals" as Washington described them. Buckets of water, carbon dioxide from fire extinguishers had no effect; a "desperate remedy" had to be tried. There was nothing for it but to flood the caisson from above; but such a plan was more than just risky. "If the air should all be out before the water had reached the roof, the result would be a sudden drop of the caisson, and the destruction of all supports by the weight of twenty eight thousand tons, besides running the risk of causing the caisson to leak so badly as to render its reinflation impossible . . ."

Now the hard work, and more crucially the weight of responsibility, began to take its toll on Washington. "All these considerations had to be carefully weighed, and the risks looked at from both sides, before giving the order to flood it. Outside of Farrington, who had gone to bed, there was no intelligent mind to consult with, as all of my assistants make it a point to live 3 miles away from this work so as not to be on hand in case of an emergency," he wrote with not a little bitterness. "In the mean time I had been down in the Caisson for 7 hours and began to experience that peculiar numb feeling in the small of the back & lower limbs which precedes paralysis . . ." Fireboats were called; 1,350,000 gallons of water were poured down through the caisson's shafts, and the caisson remained flooded for two and a half days; it settled by two inches. When the water was eventually pumped out, the damage was painstakingly repaired: at first cement was pumped into the blackened caverns left by the blaze, but this turned out to be a bad mistake when it was discovered that the wood had burned into a layer of soft, friable charcoal, which would have to be cleared out and replaced with new wood. The work of digging out the cement took eighteen carpenters, working day and night, two months. "The work was extremely disagreeable and unhealthy," Washington wrote. "Men had to lie for hours in confined spots, without room to turn, and breathing a foul mixture of hot candle-smoke and cement-dust combined with powdered charcoal, and under pressure at that, the temperature being eighty degrees." Eleven courses of timber had been damaged; more pine was forced into the breaches, and more cement, once the charcoal had been cleared away; iron straps were bolted to the chamber's roof.

After those seven hours down in the caisson, Washington had had to be taken home and "rubbed for an hour on the spine with salt & whiskey." He had tried to rest, but at any moment expected "to hear the door bell ring with a message that the Caisson was burning yet"; and indeed that message had come. "It seems there was the usual delay in executing my orders (all ordinary human beings, including myself"—those two words inserted above with a carat—"delight in delays and procrastination)."

He recovered enough, clearly, to write up his notes; whether it was the salt and whiskey that did the trick, or simply being away from the caisson, we don't know. We don't know if Emily tended to him, or how much he would have seen his three-year-old son, or his youngest brother, Edmund, who was now living in the Hicks Street house. But here again is this quality of practical rigor, "the rigor of the open world where uncertainty is profound." The construction of the caisson, the blowouts, the fire—the most careful plans could be made, and yet every day brought new challenges and new uncertainties. Washington Roebling might call all this simply doing his job—but considering the strength of mind and feeling required is what draws us into that Hicks Street room, still scented with smoke and whiskey, to find Washington back at his desk. There were still no end of solutions to be found.


AT THE END of December the masonry reached the level of ordinary high water; in early March 1871 the Brooklyn caisson was filled in. Twenty thousand cubic yards of material, by Washington's account, had been dredged up in buckets; work that should have taken a month took five times as long, thanks to the boulders beneath the river's sleek surface, "five months of incessant toil and worry, everlasting breaking down and repairing, and constant study to improve if possible." Sometimes the stones below the shaft had to be dislodged by hand: "When the lungs are filled with compressed air a person can remain under water from three to four minutes with ease." So spoke the man who had measured the length for a bridge by swimming across the swift-flowing waters of the Shenandoah with the measuring tape in his mouth. Before long, he would begin to pay the price for his unceasing attention to the work.

In May, the New York caisson was launched; it was roughly the same as the one across the river, though there were seven more courses of timber in the roof, and it was lined with a light iron skin to keep it airtight—and protect it from the risk of fire. The air locks were much bigger, and they were built into the roof of the caisson so that the men didn't have to climb in the compressed air. "The idea of placing the air lock at the bottom of the air-shaft . . . in masonry caissons, is not new, having been proposed in England as long ago as 1831 by Lord Cochran, and again by Wm. Bush in 1841, and still later in 1850 by G. Pfannmuller, of Mayence," Washington wrote in his report on the construction of these remarkable structures. "It, nevertheless, remained for Captain Eads, in his St. Louis caissons, to make the first practical application of the same on a really large scale in this country." Washington's notebooks make it clear that he had gone to St. Louis to see the work that Eads was doing on—and under—the Mississippi; just how much he owed to Eads would shortly become a terrible bone of contention between the two men.

There were, Washington wrote, "vexatious delays" in obtaining possession of the ferry slips on the New York side; but the material they would encounter once the digging began was wholly different in nature from what had been found in Brooklyn. Past a layer of disgusting dock mud, the decaying remains of anything that had ever been thrown into the river ("the character of the work at this particular time was more disagreeable than at any subsequent period"), the caisson descended quickly through yard upon yard of sand, which could easily be thrown up to the surface by a system of pipes that used the available force of the compressed air—the same force that had occasioned the great blowout in Brooklyn. "At a depth of sixty feet, sand was discharged through a three and one-half inch pipe continuously for half an hour at the rate of one yard in two minutes. This represents the labor of fourteen men standing in a circle around the pipe and shoveling as fast as their strength would permit." The material, Washington wrote, was sucked out of the chamber with "tremendous velocity, stones and gravel being often projected at least four hundred feet high . . . In order to deflect the sand at the top of the pipe at right angles, both wrought and cast iron elbows were used at first. The sand blast would generally cut through these in an hour or two, sometimes in a few minutes, the thickness of iron being one and a half inches." Finally, heavy granite blocks were placed over the mouth of the pipes. "Several minor casualties occurred from the discharge of stones, such as a boatman having his finger shot off and a laborer being shot through the arm by a large fragment."

Down and down the caisson went. Clever Paine rigged up "a mechanical telegraph" to facilitate communication between the world below and the world above. But at sixty-eight feet, the material under the workmen's feet changed to coarse gravel and stones, which often blocked the sand pipes; up until this point the caisson had descended by up to a foot a day, but now two feet a week was considered good progress, and often it was much less than that. And more than the material of the riverbed was problematic. In Brooklyn, the caisson had come to rest forty-five feet below the river. Atmospheric pressure is the force exerted by the entire body of air above the specified area; at sea level, "one atmosphere" exerts 14.7 pounds of pressure per square inch; forty-five feet below sea level the pressure in the caisson was twenty-one pounds per square inch. The *Eagle* reporter who had ventured down into the caisson in 1870 had reported that the men working below suffered "no inconvenience": but this was no longer the case. By January 1872, the New York caisson had reached a depth of fifty-one feet, and the pressure was at twenty-four pounds per square inch; by this point, Washington wrote in his report on the building of the caissons, "scarcely any man escaped without being somewhat effected by intense pain in his limbs or bones or by a temporary paralysis of arms and legs."

And so early in that year Washington hired a former army surgeon, Andrew H. Smith, to attend to the men working down below, just as Eads, in Mississippi, had brought his personal physician, Dr. Alphonse Jaminet, to look after his workers. Neither man, however, would finally grasp the cause of what is now called decompression sickness, a disease of the modern age, brought about entirely by human interaction with developing technology.

As early as the seventeenth century scientists and natural philosophers began to examine the nature of air as a medium through which we move, as much as swimmers do in water. Evangelista Torricelli, a seventeenth-century Italian scientist who first developed the barometer, remarked in a letter that "we live submerged at the bottom of an ocean of air." In 1648 the French scientist and philosopher Blaise Pascal demonstrated the effect of altitude on a column of mercury: the height of the column was three inches lower at the top of the Puy-de-Dôme mountain in the Auvergne than it was at sea level in Paris. Before the century was out Edmund Halley had begun his experiments in the other direction, creating one of the first diving bells; in 1691 Halley and a few companions spent an hour and a half sixty feet below the surface of the sea off the south coast of England.

The first real caisson was constructed in a coal mine in the Loire Valley by Charles-Jean Triger; and here, in 1840, the first medical observation of "the bends"—so called because of the twisting agony it caused in its victims— were made. Two physicians, B. Pol and T. J. J. Watelle, noted the "severe pains in the arms and knees" suffered by some of the men; they would also observe that *"on ne paie qu'en sortent,"* "one only pays on leaving." It wasn't being down in the caisson that caused the problem: it was coming back up into normal atmosphere. They were also the first to suggest that a return to the compressed air would effect a cure. Decades later, in Brooklyn and New York, Washington too would note that this "heroic mode" of returning straight down into the pressure would alleviate the "intense pain" the condition caused. "The fact remains, however, that the effects of compressed air on the human system constitute the principle difficulty attending deep pneumatic foundations," he wrote in the annual report for 1872.

Smith would publish the results of his attendance at the New York caisson in 1878. It was the prize essay of the Alumni Association of the College of Physicians and Surgeons of New York; in its opening pages, it traces the early use of caissons by men like Triger, and the effects on the workers who were sent down into them. It is carefully observant of the strange effects of the disease: pallor of the skin, caused, he surmised, by decreased circulation to the surface of the body; for the same reason the fingertips would appear shrunken, as if the hands had been soaked in water. The men suffered pain as if they had been "struck by a bullet"; "as if the flesh were being torn from the bones." Washington had written in his notes for the caissons that everything must be built "strong enough for a pressure resulting from a depth of 100', to which it may be necessary to go." But Washington Roebling and Andrew Smith were both aware of how much worse the problem had been in St. Louis, where the caisson for Eads's bridge was sunk to ninety-three feet below the river; fifteen men died in Eads's caissons, and Washington had no wish to replicate this kind of casualty figure. In late April a German called John Myers died; and then a fifty-year-old Irishman, Patrick McKay, after he was found "sitting with his back against the wall of the lock, quite insensible." Washington would note that while the deaths were regrettable, "the number has been much less in proportion to the number of men employed than upon any similar work."

At the end of his 1878 report Smith would describe something like a modern "hospital lock," a chamber constructed on the surface to enable the men to return to a pressurized atmosphere without having to go back down into the caisson; it was never built for the men working on the Brooklyn Bridge. It's not hard to understand why that "heroic mode" was so infrequently applied: relief from pressure, rather than the pressure itself, is the cause of the problem, which seems counter-intuitive. Anyone would have wanted to flee from the strange atmosphere far below the river. Neither Jaminet nor Smith would understand the true cause of the caisson sickness afflicting the workmen. At high pressure, nitrogen in the atmosphere saturates the blood and fatty tissue of the body; if decompression is too swift, these stores of nitrogen are quickly released, forming bubbles in the blood that coalesce in the venous bloodstream, blocking the flow of blood and causing terrible pain or worse, depending where the blockages occur. French physiologist Paul Bert had begun to understand this at just the time Jaminet and Smith were considering the problem, and he was the first to describe the real cause of decompression sickness—but Smith disagreed with Bert explicitly. Smith believed, as many at the time did, that the problem was caused by too much time spent inside the caisson, and that longer should be spent "locking out"—he was right about that, but for the wrong reason. He advised five minutes per atmosphere, or for every 14.7 pounds of pressure added; even that was hard to get the men to adhere to, and it bears no relation to the times advised for decompression in modern diving tables. Jaminet, in Mississippi, decompressed in three and a half minutes when the pressure there was at forty-five pounds per square inch, the very deepest the caisson would reach; a modern diver would take two and a half *hours* to decompress from such a depth.

That spring the city's papers kept the public informed about the cost of the work to men in the caisson. Eugene Sullivan, of 23 Roosevelt Street, and

Louis Gifford, of 25 Baxter Street, were taken "suddenly ill" on April 30, 1872, according to the *New York Herald*; Michael McGuire and Dominick Montrass were both "seized with sudden illness" according to the *New York Times* on May 5. In the same week, the men went out on strike after higher pay for what had become dangerous work; on May 10 the Bridge Company agreed to pay $2.75 for a four-hour shift.

But on April 26 the *Herald* had run a short article authored by Francis Collingwood, Washington's able lieutenant. "The pleasant task devolves to me to announce to you that we have this afternoon uncovered the bedrock under the New York caisson," he wrote. At seventy feet Washington had ordered that soundings for bedrock begin—and at seventy-five feet "sharp, thin ridges" of Manhattan's gneiss had been discovered. The bedrock wasn't level—but Washington's knowledge not only of engineering but of geology, too, meant that it didn't need to be; there was no cause to keep digging, and risk the lives of more and more men, in order to lay the final foundation for the New York tower. "No part of its surface shows the rounding action of water or ice," he wrote of these spurs of bedrock in his official report, showing they had never been affected by erosion or movement. They were covered with "a very compact material, so hard that it was next to impossible to drive in an iron rod without battering it to pieces." And so the caisson could come to rest. "On such a bottom no sliding can ever take place, no matter what the average slope might be," he wrote. This surface "was good enough to found upon, or at any rate nearly as good as any concrete that could be put in place of it . . . It was determined to rest the caisson on this material at a depth of seventy eight feet."

In the archives at RPI there is a small collection of Roebling "artifacts," as they're called in the catalog. They're noted at the end of the listing, after all the series and subseries that make up the university's holdings, from John Roebling's "cholera expenses" from 1854, to Washington's copy of G. P. Quackenbos' "An English Grammar," published in 1862. There are not many objects: the doorknobs and door knockers from Washington and Emily's house on Columbia Heights, now long demolished; the marble bust of John Roebling that presides, somewhat eerily, over the library's Fixman Room.

One of these artifacts is a small stone, not quite two inches long, perhaps an inch wide. It doesn't look like anything special. It is labeled, with a piece of tissue carefully cut to fit its shape. On the tissue Washington has written in his careful hand: "Gneiss from the Bed Rock under N. Y. Tower East River Bridge 1872 79 feet below H. W."*

It is very unlikely that he collected this stone with his own hands. For by now Washington himself was badly affected by his constant trips from Brooklyn to the construction site, by the pressure, by the strain of the work. As he would write coolly in the report on the construction delivered in June 1872, "considerable risk and some degree of uncertainty was necessarily involved" in the construction of the towers' foundations. "At the commencement of this enterprise, they constituted the principal engineering problem to be overcome, and not until they were accomplished facts could it be said that the project of the bridge was placed upon a firm, immovable footing. The subsequent building of the Towers and of the superstructure is all work which has been done before on a smaller scale. But upon the Tower Foundations rests the stability of the entire work. The whole enterprise depended on their success."

Perhaps it was Francis Collingwood—three years older than Washington, who had started his career in the family jewelry business in Elmira, New York, before being drawn back into the work he'd trained for at RPI—who brought Washington this simple treasure. Also at the library at Troy there is a long document in Emily Roebling's sloping, confident hand, a short biography, of Washington. "The period of time at the end of the sinking of the New York caisson was one of intense anxiety for Col. Roebling," she wrote; this unremarkable stone offers a physical marker of relief. Washington Roebling was a thoughtful man, and a sensitive man; but he was not a sentimental man. He was not a man for keepsakes, although he would, in later life, carefully preserve the public record of his and his family's lives. This stone, however, he kept.

The digging stopped on May 18, 1872. It was a cloudy day, according to the *New York Times*, which also printed on that day an exchange of letters between President Ulysses S. Grant and former Confederate general Robert E. Lee. Grant assured Lee that "this application of Gen. R. E. Lee for amnesty and pardon may be granted to him"; a few days later Grant would sign the Amnesty Act, lifting voting and office-holding restrictions on most former

---

*The difference of a foot, between the report and the label, may allow for a few inches either way.

members of the Confederacy. Only now was the terrible war coming to its close, those who'd fought in it might think; rebuilding a country meant building bridges where before there had been none. And here in New York and Brooklyn there would, at last, be a bridge that would be, in its own way, a symbol of that reconstruction.

# "I have been quite sick for some days"

ANDREW SMITH'S 1878 ESSAY on "the Caisson Disease" is perspicacious, though not perfect in its understanding of the new syndrome it describes. But the copy held at RPI is annotated: down along the sides of certain paragraphs there are fine pencil lines of emphasis; and then, every so often, words written in the margin. It is clearly Washington's hand. He highlighted certain aspects of the condition's character: the way in which the increased pressure of air in the caisson changes a bass voice "to a shrill treble"; the way in which the heart rate was often increased; the way in which sweat would not evaporate down below; the way the appetite was greater after exposure to the compressed air.

There is no way of knowing when Washington made these annotations. One suspects that it was much later in his life, when he was composing his memoir of his father's and his own life, and when he was combing through his papers in the knowledge that one day others would comb through them, too. And until the East River Bridge was finished, the work—and his health— would allow little time for reflection. But even at the height of his anxiety over the sinking of the great caissons, Washington did not neglect his work on behalf of the family firm, just as, when he was at work in Cincinnati, he had mailed out company circulars all over the country. Now, he sent off plans for a bridge of three spans of 435 feet each to John W. Glenn, the mayor of Austin, Texas, judging that the construction would cost $85,875; he recommended one Charles McDonald to actually build the thing, as he himself was somewhat preoccupied: "He is a reliable man in every respect . . . My own individual time is too entirely taken up with the construction of the East River Suspension Bridge." There are estimates for a bridge out in Ohio; and on May 15—just three days before the New York caisson settled on its final resting place on the East River's bed—he corresponded with Mr. Wilkins, from Wilkins and Davison engineers in Pittsburgh, who was wishing to make alterations to the Allegheny Bridge in Pittsburgh so that locomotives might cross. Washington's reply is a considered analysis of the problem arising from the concentration of the load on the bridge.

His home life, his personal life, is elusive. His son John was a toddler now; his youngest brother, Edmund—whose life became a mysterious tragedy, childhood damage never repaired—was in his charge. Edmund had been sent to boarding school, as Washington had been as a boy, in 1867, where "he was subjected to evil influences of so galling and insidious a nature that he ran away—was caught, brought back, and nearly beaten to death by a brutal father, and sent back." He escaped again, Washington wrote in a private memo written in the 1920s, vanished, and was found in a Philadelphia jail-house. "He had himself entered as a common vagrant to get away from his father, and was enjoying life for the first time." Washington had hired a tutor for his youngest brother for a while; this was apparently not a success. There are records of Washington's household expenses from this time: "Emily's household account, $4185.42" is listed; this is followed a few lines down by "Pittance to Wash" of $897.57; a little later on in the accounts he made reference to "Wash's dole." He was always worried about money; his salary couldn't keep pace with his expenses, and his brothers Ferdinand and Charles, spared the danger he faced daily, were always better off than he was.

He worked as hard as anyone he employed, and he paid the price. In the spring of 1872 he suffered an attack of the bends worse than any he had experienced; he was, apparently "brought up out of the New York caisson nearly insensible, and all night his death was hourly expected." Five days after he decided to stop the digging in the New York caisson, he wrote to his brother Ferdinand: "I have been quite sick for some days, owing to imprudence in remaining too long in the caisson on Saturday last. Saturday night the pains were so intense that I thought I would not get over it. By the plentiful use of morphine however I was finally stupefied sufficiently not to feel them for a time." And yet, his letter ends: "Everything is going well here otherwise."

In Smith's paper there is another fine pencil line drawn beside the physician's observation that while caisson disease may well cause paralysis, such paralysis does not relieve pain: "The patient does not, however, obtain relief from his suffering, since the pain in the limbs will continue after pain from other sources is no longer felt. Thus a leg, for example, may be entirely insensible to pricking or pinching, while at the same time it is a seat of extreme suffering." Attacks like this can last for three or four hours to six or eight days, Smith noted; Washington, even in his coolly composed official report, made a note of his own "severe personal experience" of excruciating pain, which was "only relieved by the complete stupefaction afforded by morphine." And while Smith dismissed Bert's new understanding of how nitrogen could affect the bloodstream, he had his own theories, treating the problem with ergot (an alkaloid derived from a fungus that grows on wheat and rye), which constricts the blood vessels; it is still used to treat migraine. Ergot, Smith wrote, would "correct the want of tone [of the capillaries] which I considered to lie at the foundation of the difficulty": apparently he had seen severe pain "completely relieved" in this way. By this passage Washington has simply written "doubtful" in the margin.

In the depths of winter, when the ferries were pushing through the ice in the river, Washington felt he could no longer go down to the bridge. Afraid that he too might not live to see the work completed, hour after hour he wrote and he drew, leaving precise directions for every aspect of construction. He was too sick to leave his room; but he could not abandon his—and his father's—work. His eyesight began to fail, and he could not even carry on long conversations with his assistants, whose help, whose belief in the wisdom of his directing hand, were even more vital now. He thought the seclusion must have done him some good; in early February he let his brother know that he might be able to go up to the mill. "Everything is quiet here, frozen up or snowed under—I am improving very slowly, and have an idea that I may be able to get to Trenton in the course of a fortnight."

Eventually, though, he feared complete collapse. He requested a leave of absence from his work at the bridge. His skilled and devoted assistant engineers knew their tasks, thanks to his instructions; he would head to Europe—to the spa baths at Wiesbaden—for his health. It seems an extraordinary choice, given the scorn he would eventually express for John Roebling's douches and spritzes and sitzes when it came to giving an account of his father's life. But no other treatment was on offer.

He left for Europe in May 1873. "I find that I am a little too early for the German baths and shall therefore take a preliminary douse at the Maison de Santé of Dr. Beni-Barde at Auteuil," he told Ferdinand. At the time, Auteuil, then a suburb of Paris, was a renowned center for hydrotherapy and the treatment of nervous disorders. Alfred Beni-Barde's clinic had a wide roster of famous clients: Du Maupassant, Dumas, and Manet were his patients at one time or another. But when he and Emily and Edmund—still under their charge, for the time being, though he would shortly skip off on his own to Vienna—landed in England, rest and recuperation were surprisingly far from his mind. The first four pages of the letter remarking on the proposed trip to Auteuil are a characteristically detailed account of a visit to the Hooper's Telegraph Works in London's Millwall Docks, where submarine telegraph cable was manufactured. He was especially interested in the method of splicing cable used, and was evidently thinking hard about how the method could be adapted for the thousands of miles of wire that would shortly be crossing the East River. "The splices are rather new—in splicing the No. 13 Bessemer, they file it off like a bridge splice and wrap it around with pure tin wire," he wrote, and drew a neat, delicate image. The tin was then brushed with acid and soldered, and then the acid was neutralized. "It makes a nice splice . . . it goes very rapidly, and I think can be used by us to advantage," he wrote to his brother. The contract for the wire for the East River Bridge had not yet been awarded, but there seemed little doubt, at this stage, that it would go to John A. Roebling's Sons.

He and Emily returned to Paris, and viewed the paintings in the annual Salon, admiring the figurative works on show. "After seeing them you come to the conclusion that most of the pictures sold in New York are trash," he wrote to his brother; "the French however are poor hands at a landscape." (Perhaps it's just as well they arrived a year too early to take in the Impressionists' first breakaway Salon in 1874.) He was tempted to purchase a painting but was once again fearful his money would run out: "Emily is spending more than I expected." She was clearly glad to be back in the City of Light.

He enjoyed his time at Auteuil. "I think the douche has done some good," he wrote hopefully. "They put you in a pen and squirt cold water on you with a steam fire engine and a 1" nozzle. It is a capital thing when you are well." When the weather warmed they finally moved on to Germany and

Wiesbaden; its waters are mentioned by Pliny, and ever since, Europeans had traveled to bathe in its dozens of thermal springs. Later in the summer he received news of the birth of Ferdinand's first son; he must have hoped his own letter of congratulation would be received before the boy was named: "The selection of a name for a child is one of the difficult choices one has to make in life—Don't give him such a long handle as I have to my name. You waste a power of ink during a lifetime writing it." Ferdinand and his wife, Margaret, named their son Karl Gustavus.

Washington and Emily sailed home from Southampton on October 13, 1873; nineteen-year-old Edmund had gone his own way early on, and would sail home from Bremen. But Washington's early optimism had been dashed: the trip had not done any good. "I shall leave Europe no better than I came— If anything worse," he wrote to Ferdy. Yet he had been visiting the Millwall Docks, and the Paris Salon. It is as hard to fathom the true nature of his condition now as it was for his physicians in his own day. Retrospective diagnosis is impossible. But the violence of Washington's childhood, the brutality of his education, the terrors of the long war, and the hard work and uncertainty that had followed nearly every day since—all these factors make it hard to avoid the conclusion that there was some psychological element to what was afflicting him. Recall the little joined puppet he had drawn in that notebook eight years before, its hanging limbs and pained face. A great burden was still on his shoulders, his father's shadow still cast over him.

So he sailed back toward the East River, and the tasks that awaited him there.

ILLNESS WOULD HAUNT him for the rest of his life. He would be an "invalid," a term prevalent in the nineteenth century, covering a myriad of conditions, and allowing for some of the mystery of what exactly might have been wrong with him. But he had grown up in a household where great—and peculiar—emphasis was placed on health and sickness, and the balance between the two. For John Roebling, good health had a moral aspect: he claimed to have escaped a cholera epidemic during the building of the bridge at Niagara by simply resolving not to get it. Fifteen years before Washington had ever been down in the pressurized air a sequence of abdominal attacks had nearly floored him; that he kept such close track of them, and of his self-administered treatments, says something about the origins of his concerns about his health.

While in Auteuil he had written to Ferdy back in Trenton, who had sent him a clipping from a technical journal. "I received some days since your letter with enclosure from RR Gazette containing Capt. Eads' article. It made me mad for several days. I have never read an article that contained so much venomous envy and the same quantity of lies in the same space. He is manufacturing public opinion in his favor because engineers decline to pay him royalties until our suit is decided."

James Eads—who was also using caissons to build his great bridge across the Mississippi—had decided to sue Washington Roebling for an infringement of the patent of his caissons. The argument—which would not be settled until 1876—became decidedly personal. The dispute between the two men would escalate into a lawsuit, and the two great engineers would continue to trade insults. Coincidentally, the same year this argument began, a board of Army engineers began to investigate whether Eads's bridge was an impediment to navigation on the Mississippi; one of the members of the board was none other than General G. K. Warren, Washington's brother-in-law. In the end, nothing came of the Army engineers' report; and the lawsuit would be settled out of court, Washington handing Eads five thousand dollars to be rid of the matter. Washington maintained that the design of the caisson and the pattern of the air lock were technical innovations that could not be subject to a patent; different patents might be granted in different countries based on separate developments in those countries. But Washington didn't help his cause by answering explicitly Eads's personal attack. His own work, he was certain, was superior. "Where would you go to find an easier material to sink [a caisson] through [a river] than at St. Louis, or a more difficult one than in the East River?"

Clearly Washington was strongly influenced by what he had seen in St. Louis. In 1870, at the end of a description of the launching of the Brooklyn caisson, Washington wrote plainly that "since my visit to St. Louis it has become evident to me that I made a mistake in not providing more supply shafts—the two that are in are perhaps ample to supply the material as fast as it is wanted; but the great drawback will be the distance it has to be hauled below to reach the remote parts of the Caisson." And while Eads's main suit was about the placement of the air locks—at the bottom of the caisson rather than at the top—he took issue with the placement of the supply shafts, too. Washington's observation of what he saw in St. Louis would seem to put Eads in the right. Eads contended that his patent number 123,685, "Improvement in constructing subaqueous foundations"—regarding water supply pipes, sand pipes, and air pipes—meant that the system used in the New York caisson was an infringement.

But Washington was nothing if not a man who did his homework. The date of that patent was February 13, 1872, and Washington noted: "The Brooklyn caisson was commenced in October 1869 and launched March 19, 1870; the NY caisson was commenced in Nov. 1870 and launched May 1871. All the pipes complained of were built into their caissons previous to the launch." He denied, in any case, that the placing of a pipe was a patentable idea.

The conflict with Eads would haunt him, nonetheless. Ironically, perhaps, Eads's own work would also be affected by his failing health; but the public perception of Eads was different than that of Roebling. Eads was a self-made man, something of a heroic figure. By contrast, the dispute played into one perception of Washington Roebling that he himself was all too well aware of: that the younger Roebling had had everything handed to him on a plate. Not only was he just following the plans his illustrious father had left, but the inference was made that he would even stoop to stealing another man's work. It was more than Washington could bear. "The whole Engineering world in the U.S. sided with Cpt. Eads, most unjustly so . . . These feelings were actuated entirely by jealousy—the Brooklyn Bridge was by far the largest and most prominent Engineering work of its day—Throughout its construction the attitude of the Engineering fraternity towards me was one of hostility, envy and hatred."

*Envy*—used both in the letter he wrote to Ferdinand at the time and then, much latter in his life, when he was setting down his memoirs—is a powerful word, *hatred* even stronger. His language reveals the personal and moral cost of this work to Washington Roebling. The mid-nineteenth century was a time of such rapid technological change that innovations often overlapped—the line between invention and adaptation very rarely absolutely clear. What was clear was Washington's desire to protect his good name.

When Washington returned from Europe in late 1873 he would no longer live in Brooklyn. He and Emily bought a new house there, on Columbia

Heights, but he repaired almost immediately to Trenton. The towers and great anchorages were finished without his ever being near them; the cable-making machinery was made ready, too. His assiduous correspondence with the loyal men who worked under him—Paine, Collingwood, Farrington, and the rest—enabled him, always, to ensure that his plans were enacted to the letter. Washington Roebling could not see the work. But he knew what was happening all the same. His assistants served his faithfully—and what's more, they *knew* him, as he knew them.

Just before Washington and Emily had left for Europe in 1873, the Brooklyn anchorage had been started; the massive structure, which would keep the bridge's great cables fixed to the earth, was completed just over two years later. In 1875 the Brooklyn tower was completed—as the great stone numerals atop it proclaim—and its New York brother was finished just over a year later, and the New York anchorage was finished by then, too, in July 1876.

But when he had left for Europe there had been good reasons for him to wish to put distance between himself and the bridge—reasons that had nothing at all to do with his health.

# "Now is the time to build the Bridge"

By 1870 the reign of William Magear Tweed—"Boss Tweed"—was at its
height. Tweed's name has become a byword for political corruption. His ex-
ploitation of private and public finance as New York secured its position as
one of the world's industrial, financial, and cultural capitals in the years after the Civil War remains unrivaled. Born in 1823 on Cherry Street—now
the site of the Manhattan approach to the Brooklyn Bridge—his real ascent
began with his service as a volunteer fireman with one of the many rival companies that served the city in those days. Engine Company Six, its ensign a
roaring tiger, was a finishing school in local politics for Tweed. The one-time
bookkeeper and brush-maker learned how to command men and gain their
trust and their loyalty, and he began to learn, too, how to manipulate systems
of local government. Charismatic and charming, his huge girth a seeming
demonstration of the space he felt he was entitled to occupy in the world, he
was elected to Congress in 1852 at the age of twenty-nine; by 1858 he was a
member of the general committee of Tammany Hall—the Democratic political machine that dominated New York City politics well into the twentieth century. Tweed had what a contemporary described as "an abounding
vitality, free and easy manners, plenty of humor, though of a coarse kind,
and a jovial swaggering way which won popularity for him among the lower
and rougher sort of people."

He picked up the nickname "Boss" once he had become "grand sachem,"
or leader, of Tammany Hall in 1863. The title of grand sachem, like the
name of the society itself, had been adapted from Tamanend, a chief of the
Lenni-Lenape tribe, when "the Society of Saint Tammany" had been founded
in 1788. Tamanend, legend had it, greeted William Penn when he landed in

the New World in 1682. Writing about the cult of Tamanend in Philadelphia, John Adams told his wife, Abigail, "The people have sainted him and keep his day." But Tweed was no saint; as he rose up the city's political ladder he created an extraordinary machine whose true purpose was to line his own pockets and those of his cronies.

BY THE TIME the bridge was under way, Tweed's "Board of Audit"—consisting of New York's Mayor A. Oakey Hall, Comptroller Richard Connolly, and Tweed himself—collected all outstanding bills payable to the county Board of Supervisors. This looked, on the surface, like good government reform: but the bills ran into tens of millions of dollars, and Tweed and his gang added a 50 percent markup, divided among Tweed, Connolly, Hall, and Tammany strategist Peter Sweeny, plus the clerks who knew about the payment padding: this was the great Tweed Ring. Especially choice cuts, of course, could be taken from inflated bills for public works projects, including the construction of the grand new courthouse on Chambers Street, and the construction of Central Park. Debts owed by New York—both the city and New York County—rose from $36 million in 1869 to $73 million in 1871.

The grandiose wedding of Tweed's daughter Mary Amelia in the summer of 1871 marked the beginning of Tweed's downfall. The *New York Times*'s report of the occasion is a model of the reporter's craft, the careful placement of adjectives and details of description more damning than any editorializing. From the wedding guests who waited in Trinity Chapel in "speechless astonishment" for the ceremony to begin, to the note of Tweed's "palatial mansion" on Forty-Third Street and Fifth Avenue, to the practically biblical list of wedding presents including forty sets of sterling silver and forty pieces of jewelry. Fifteen of those were diamond sets; and "a single one of the latter is known to have cost $45,000." In another five weeks the *Times* followed up with evidence of the ring's corruption, and Tweed's fate was sealed; a warrant was issued for his arrest on October 26.

In his glory days, Tweed would have wanted a piece of the gigantic undertaking growing up between New York and Brooklyn. Since the time of John Roebling's funeral in 1869 his name had been associated with the project— he had been to the first meeting of the new executive committee of the Bridge Company. When he eventually stood trial he would claim that Henry

Murphy had handed over at least fifty-five million dollars to the ring, a little something so that the New York Board of Aldermen could be brought on board; Murphy strongly denied the story. Tweed would swear under oath that William Kingsley had been on his payroll, and indeed at the time it had been plainly impossible to do business in the city without doing business with Tweed.

Washington was an engineer, not a moneyman, and as such he had managed to keep his distance from the whole sordid business, and would continue, at some personal cost, to do so as the work progressed. Years later he intimated that his father, certainly, had understood all too well the lay of the land as far as financing public works in New York City was concerned. Washington exchanged letters with the sculptor William Couper in 1907, when Couper was modeling the statue of John Roebling that still stands in Trenton's Cadwalader Park; discussion stretched beyond the matter of statuary. "You may not be aware that this bridge was started by the infamous 'Boss Tweed Ring' for the sole purpose of using it as a means to rob the cities. When this fact began to dawn on my father's mind he made up his mind to get out. This ring was overthrown in 1871." Washington, especially as he got older, clearly liked to say things for effect. But the statement sums up much of his dismay at the financial shenanigans surrounding the financing of the bridge.

John Roebling's original estimate for the cost of construction had been just shy of $6.5 million; costs had risen far above that, and following the downfall of Tweed, serious questions were being asked as to whether the sort of extortionate pricing for goods and materials that the ring had facilitated might not be affecting work at the bridge, too. Tweed's place on the executive committee of the New York Bridge Company had been filled by Abram Hewitt, who immediately asked Roebling to let the committee know whether prices paid for stone, lumber, and other materials for the work had been "just." Washington could not have been clearer in his reply.

"It has been alleged that supplies have been furnished by members of the Company, at prices prejudicial to the interests of the Bridge. In all such cases I know that the supplies have been furnished after a reasonable competition, and at rates lower than those of any other bidder.

"I can further say that every dollar's worth purchased for the Bridge has been expended in a legitimate manner, and for the proper purpose for which it was designed, and nothing whatever has, to my knowledge, been diverted into any outside channel. I am in daily attendance at the Bridge, give it my whole time and constant superintendence, and am therefore in a position to give an honest judgment on this question." But there was no doubt that costs had risen: Washington estimated in the summer of 1872 that the final sum would be $9.5 million.

Tweed was not the only one to fall into difficulty in those years: 1873 was the beginning of what became known as "the Long Depression," five years of financial slough that followed the prosperity of the Civil War and the burst bubble of the Gilded Age—a term coined by Mark Twain and his coauthor, Charles Dudley Warner, in their novel of the same name, which was pre-sciently published that year. "Railroad mania"—over-investment by both banks and individuals in tens of thousands of miles of new railroad, many of those miles proving far less profitable than their investors hoped they would be—affected the steel and iron industry badly. Great banks failed—even Jay Cooke and Company, the country's premier financial institution, which had largely financed the Union Army during the war. Not long after Washington and Emily returned from Europe, he wrote gloomily to Ferdinand that "work on the bridge has progressed very slowly during my absence, neither are the prospects for the future very bright at present. City bonds are almost unsale-able, even when offered below par."

And costs were still increasing—while Washington's remuneration was not. Money worries continued to haunt him, but he can hardly have failed to know the way the political game was played, in those days, both in New York and Brooklyn. He wanted no part of it.

Even though work on the structure slowed, the salaries of the assistant engineers still had to be paid, as well as rent and fuel and taxes and all the other expenses of construction. By 1875 he knew that the bridge was going to cost at least $13 million, and he wrote a forthright plea to Henry Murphy. "Such increase of cost will always attend an irregular supply of funds . . . Now is the time to build the Bridge. At no period within fourteen years have the prices of labor and material been as low as at present. A rise of 10 percent in these items during the year is within the experience of all, and is but little thought of; but a rise of 10 percent means over a million in the cost of the Bridge. To build now is to save money!"

All the while, terrible and mysterious pains continued to afflict him. He came to depend, above all, on Emily. In these next years, as the towers and anchorages were completed, as wire began to be worked back and forth across the river, as the roadway would eventually be laid, their marriage evolved into an extraordinary practical partnership that swiftly became the subject of speculation, rumor, and eventually myth and legend.

IN MAY 1912, some nine years after Emily died, Washington received a letter at his home in Trenton from Mr. M. D. Maclean of the *New York Times*. It seems to have been sent to the offices of John A. Roebling's Sons, and from there made its way into Washington's hand. "Gentlemen: We write to you thinking that you may be able to answer a question someone asks us. After the death of Mr. John A. Roebling, did his wife take an active part in directing the work of completing the Brooklyn Bridge? We know, of course, that Mr. Washington Roebling was in charge of the work."

Washington made a careful and thorough answer. "I have your letter of May 14th asking whether 'after the death of Mr. John A. Roebling, did his wife take an active part in the directing the work of completing the Brooklyn Bridge,'" he wrote. "In this I would reply that Mr. John A. Roebling died in 1869 before actual work had been done on the bridge—His widow (his 2d wife) did no work whatsoever on the Brooklyn Bridge. As a matter of fact I never saw her again after the estate was wound up—

"A few years after the work was actually commenced I was stricken with a severe illness, and my own wife the late Emily Warren Roebling who died in 1903 became of the greatest assistance to me in the conduct of the routine matters of the work—An eye trouble (long ago healed) prevented me from doing much reading or writing, so that her services as amanuensis became invaluable to me, and this led to other duties in the way of interviewing people, avoiding personal friction by her tact, smoothing over difficulties which were naturally inherent in a work somewhat political in its conduct and which was done by days [sic] work and not by contract thereby throwing vast amounts of additional work on the engineering staff—The responsible body in charge of the work consisted of a board of Trustees, ten from each city whose persons usually changed with each change in politics, each new man being of course saturated with suspicion of the old management—Here again her remarkable talent as peacemaker came into play, and her thorough knowledge of the work and the plans carried conviction to the heart of each new member—Being assisted in these ways for fourteen long years with the various phases of the work she earned a well-deserved recognition as well as my everlasting gratitude."

He signed off: "W. A. R.

Engr. Brooklyn Bridge

Hale & Hearty at the age of 75"

Looking back at Washington and Emily's exchanges during their courtship and in the early years of their marriage—through Washington's surviving letters—it is clear that her vision of what it meant to be a wife was one of equal partnership. But she could not have known what that partnership would entail, or just what would be required of her as her husband was gripped by the maladies that would plague him all his life.

In this period, Emily and Washington were bound together more closely than most married couples ever are. Reports persist that in the mid-1870s he was paralyzed, blind, deaf, and mute; but while he was badly affected, his mind was as sharp as ever, and he was never not in control of the work. His engineers followed his plans closely and reported the details of the work directly to him. At the end of December 1875, Francis Collingwood wrote his account of the finishing of the Brooklyn tower, carefully addressing it to "Col. W. A. Roebling, Chief Engineer of the New York and Brooklyn Bridge." The great saddle plates, over which the main cables of the bridge would ride, had been hoisted into position at the end of the previous year. They weighed thirteen and a half tons apiece; four were mounted atop each tower. "The finishing off of the masonry was a work of considerable difficulty and danger," Collingwood wrote to Washington, "as many of the roof stones weigh from 9 to 10 tons each (the key stones weighed 11 tons each)." There was danger, too, from the hoisting rope used to lift the stones: the rope was made from steel wire, one and a half inches in diameter; vibration along its great length, as the motor powering the hoist pulled upward, was difficult to control. But all the same, nothing went awry: "It is a great satisfaction to be able to state that during the building of all that portion of the work which is above the roadway (by far the most difficult part) there has been no accident to life or limb . . . The result has only been secured by incessant watchfulness on the part of the engineer and foremen." A total of 497 yards of masonry had been set; the falsework inside the gothic arches of the tower had been removed.

Emily would have read this report aloud to Washington. After the middle of December 1875, the notebooks containing the drafts of letters to his assistants, to Henry Murphy, to anyone, are all in Emily's slanting hand. The words flow smoothly, his swift dictation proving he was certainly neither deaf nor mute.

Just before Thanksgiving he had received word from Henry Murphy that funds were running out; Murphy wanted to suspend the work altogether, and suggested halving the pay of the assistant engineers. On December 6, Washington had replied with his characteristic calm. "I do not myself think it a fair arrangement to put any of these gentlemen on half pay—They are not like day laborers that can be picked up at any time—they have to devote a great deal of thought to this work when they are away from it as well as when actually present at it." He had in fact written twice to Murphy that day: the second letter concerned the awarding of the contract for the iron links that would attach the cables to the plates at the bottom of the anchorage. Each of these anchor plates weighed forty-six thousand pounds, twenty-three tons; these enormous plates, buried beneath masses of stone in New York and Brooklyn, would hold the cables in tension and support the weight of the bridge. There were two bidders, the Keystone Bridge Company and the Passaic Rolling Mill; the latter was the underbidder—but they had never made anything like this before. "I do not think the East River Bridge is the proper kind of a structure to be experimented on. Lastly I desire to say that if the rule is to be invariably adhered to of giving all contracts for supplies to the lowest bidders irrespective of all other considerations I hereby absolve myself from all responsibility connected with the successful carrying on of the work," Washington wrote. He noted that he had no connection to the Keystone company "financially, politically, socially or in any other way." That he had to make this assurance is not only evidence of the political climate in the two cities at the time, but also a foreshadowing of trouble that was to come—trouble that Washington, despite his incapacity, seems to have already been anticipating.

Now it was Emily who would facilitate his close attention. These would have been long days, pencils sharpened and resharpened, coal burning low in the grate. Their son John—educated first at the Collegiate School and later at Brooklyn Boys' Preparatory School—had just turned eight, but whether the sight or the touch of his little son was any comfort to either Washington or Emily, we don't know. What was private was private. Or at least, almost. For there is a final paragraph in the letter to Murphy—in Emily's handwriting.

"My health has become of late of so precarious a nature, that I find myself less and less able to do any work of any kind. I am therefore reluctantly compelled to offer my resignation as Eng. of the East River Bridge. The hopes that time and rest would effect a change have been in vain, rest being simply impossible." There is no signature. "Less and less" replaces "incapable," which has been crossed out. It seems this paragraph was never sent to Murphy; one or both of them had second thoughts.

Some believe that, during the years to come, Emily became her husband's engineering equal—and there is little doubt that sitting with him day after day, going over plans and taking down letters, she became extremely well acquainted with the work. Just as the bridge was finished, in 1883, a piece appeared in the *New York Times* headlined HOW THE WIFE OF THE BROOKLYN BRIDGE ENGINEER HAS ASSISTED HER HUSBAND, with a dateline of Trenton, New Jersey.

"While so much has been written about the Brooklyn bridge and those who have had a share either in planning or building it, there remains one whose services have not been publicly acknowledged . . . 'Since her husband's unfortunate illness Mrs. Roebling has filled his position as chief of the engineering staff,' says a gentleman of this city well acquainted with the family. 'As soon as Mr. Roebling was stricken with that peculiar fever which has since prostrated him, Mrs. Roebling applied herself to the study of engineering, and she succeeded so well that in a short time she was able to assume the duties of chief engineer. Such an achievement is something remarkable,'" the story ran. The gentleman remained anonymous. But the article went on to note that when bids for steel and ironwork requiring wholly new shapes to be made—those linking anchor bars—the bidders found themselves meeting not with Washington, but with Emily. "Their surprise was great when Mrs. Roebling sat down with them, and by her knowledge of engineering helped them out with their patterns and cleared away difficulties that had for weeks been puzzling their brains."

The story caught on all over the country and was reprinted in the *Atlanta Constitution*, the *Decatur Herald* in Illinois, the *Times-Picayune* in New Orleans, to name a few. At a time when it was unusual for a woman to speak in public, it was unheard of for a woman to assume the duties of a chief engineer. A few days later, the *Boston Weekly Globe* went even farther, noting how, at the grand opening ceremony for the bridge, Emily Warren Roebling had been accorded "the rather empty honor" of driving the first team of horses over the bridge. "It only illustrates how the work of a woman becomes a lost force in the world," ran the *Globe*'s piece. "She is frequently the power which makes possible the achievements of brother, husband, son. But her work flows into his, her life is absorbed in his, and so, while she furnished much of the motive power, he gets all the credit."

But the honor of driving her carriage across the bridge was not an empty one. She was invaluable to her husband: whether it is fair to call her an engineer is another question. Her empathy and courage and wit were surely just as valuable to him as her understanding of the work. As his good right hand Emily also combed through the papers of New York and Brooklyn for any news of the work, and preserved two detailed scrapbooks of clippings. Articles from the *Brooklyn Daily Eagle*, the *New-York Tribune*, and the *New York World* were clipped and glued by her to make a striking record of the bridge's construction and the often wildly varying opinions held by the public, and by newspaper editors, on that construction. There is no way of knowing which articles were chosen by Emily, and which by Washington—except, perhaps, for a letter, from W. H. Francis of the Edge Moor Iron Company, dated October 28, 1879, regarding a contract for steel for the superstructure of the bridge. It is simply addressed "Dear Madam," having been sent to Mrs. W. A. Roebling, Columbia Heights, Brooklyn.

In 1875, in New York City, the *New-York Tribune* built itself a grand edifice on Nassau Street and Printing House Square, its finial a majestic 260 feet high—just sixteen feet shy of the final height of the bridge's towers—which soon enough would cease to soar over everything else in lower Manhattan. Another two years would pass before Washington and Emily returned to Brooklyn, to the house they had bought on Columbia Heights overlooking the work, the river, and Manhattan across the water. "The Colonel has been absent from Brooklyn for nearly two years," wrote the *Eagle*—in another clipping carefully saved by Emily in her book. "His health remains about the same, and although he gives the closest attention to the details of the Bridge construction, he is still unable to give the work his personal supervision."

In CINCINNATI, JUST after the war, Washington had personally supervised the laying of the great cables over the Ohio entirely, climbing up and down the towers day after day, the sweat soaking through the shirt on his back. Now, despite being confined to his room, he worked just as hard, just as attentively, as he had a decade before. He was pleased with the work, and confident that the greatest obstacles had already been conquered—and there is something of his father's swagger in his description of the work so far. "Both towers and anchorages have been completed as far as may be during this time," he wrote in his report of 1876—the draft copy of which is another document in Emily's hand. "In all suspension bridges the masonry usually forms about one half of the total work to be done. We may therefore congratulate ourselves that we have this much behind us.

"The work henceforward will be of an entirely different character. Our towers and anchorages as they stand challenge comparison with any masonry in the world, as regards their solidity, the material of which they are built and the careful manner in which the material was laid . . . The machinery for delivering the stones on top of the towers was so adequate that it cost no more per yard to lay the top courses, than the bottom ones. This implies something beyond what at first might well appear; since the summit of the New York Tower is 345 feet above the foundation." There had been some delays in the construction of the towers: but these, he said, "are not to be wondered at, when it is remembered that the stones were cut, over six hundred miles away, and gathered from at least twenty different quarries, with boisterous seas intervening between them and the place of delivery." Thousands of loads of stone had been delivered: only one had been lost.

While he could still write, he drafted for his engineers a forty-four-page treatise on the manufacture of the cables of the New York and Brooklyn Bridge. This had become its legal name, for it was no longer a private enterprise but a public work, "to be constructed by the two cities for the accommodation, convenience and safe travel of the inhabitants" of the two cities it would soon permanently link. The New York Bridge Company had been dissolved in May 1875, and now the cost of the building going forward would be borne not by stockholders but by New York and Brooklyn.

Every possible contingency of manufacture was covered in this closely written document. Material—"cast steel wire Hardened & tempered & galvanized"—the strength of that material, how it would be attached to the anchor chains. This work was not so different from what had been done over the Ohio, but it was on a far greater scale. Each of the four great cables would be made from nineteen "strands" of wire; the wire within the strands was all laid parallel, and then each strand—composed of 332 parallel steel wires, making 6,308 wires with a total ultimate strength of 10,730 tons—was bound together with a clamp. "Spinning" the cables, the task was called, which implies the strands were twisted around each other, as the smaller, diagonal stays are. But this isn't the case for the main cables. "All four cables will be made at once," Washington wrote. "By so doing you reduce the number of bosses, overseers, watchmen, engineers, splicers, cradlemen etc. In fair weather the strands can perhaps be made quick enough to employ one special gang to let the strands off & regulate them, which takes 7 or 8 days here. After the tedious preliminaries of getting started it is just as easy to make 4 cables as 2 at a time, especially as all machinery is double—Finally 2 years of time are saved. Although cable making will take 2 years in any event." The cable making in Cincinnati had taken nine months; he had extrapolated this figure from that.

The work in the caisson had been a process of discovery—to some extent, a leap in the dark. But of the wirework he was, despite his physical frailty, confident and sure in the briskness and clarity of the descriptions he set out. "The wires are carried over the river from anchorage to anchorage by means of an endless wire rope called the working rope. It is made of steel, 3/4" diam., No. 17—At the Br. anchorage the rope passes around a driving wheel, say 15 in diam. and 2 leading sheaves for getting width . . . At the N. Y. anchorage the rope passes around 2 sheaves, 4' diam. arranged in a transverse modern frame setting on horizontal slides and held by block & fall which keeps the rope tight & hauls up the constant stretch. The rope must not be hauled so tight as to jump out of the sheaves." After the first twelve of the nineteen strands were completed and regulated, they were compressed to form the central core of each cable, around which the remaining strands were placed; then the whole would be wrapped with wire.

In the autumn of 1876 he would have occasion to write to Henry Murphy that "one mind can in a few hours think out enough work, to keep a thousand men employed for years." "The great weight and size of all the materials handled required apparatus of the strongest kind combined with constant exercise of care and judgment," he assured the board that same year, making plain that it was his care and judgment that were indispensable. The two factors concerning the strength of the anchorages are "granite and gravity," he wrote: "The first a material whose very existence is a defiance to the 'gnawing tooth of time.' The second the only immovable law in nature. Hence when I place a certain amount of dead weight in the shape of granite, on the anchor plates, I know it will remain there beyond all contingencies."

On August 14, 1876, the first wire rope was taken across the river from Brooklyn to New York and raised from tower to tower. In the morning, at slack tide, a scow had been towed across the river, paying out a rope that had been passed over the Brooklyn tower; all the work was supervised by C. C. Martin, George McNulty, and master mechanic Frank Farrington. The distance was 1,599 feet 6 inches, according to the *Brooklyn Daily Eagle*; as the scow let out the line, the rope sank to the bottom of the channel, 65 feet below the surface at its deepest point. When the scow reached the farther shore, a hoisting engine pulled the rope over the New York tower. "But the main body of the cable still lay at the bottom of the East River," ran the *Eagle*'s report. "The next business was to hoist it into position." But for that a full stretch of clear water was needed, and as the engineers and the gathering crowds watched, it seemed this might never come. "Swift steamers from Harlem dashed to and fro. Lazy lighters and heavily laden schooners passed back and forth. Two barges and an excursion steamer moving out into the stream from Jewell's Dock finally got under way, bound for Oriental Grove, on the Sound, laden with a picnic party."

Finally, at eleven thirty-five in the morning, with no boats or ships or schooners between the towers, a cannon was fired, the lashing holding the rope to the Brooklyn dock was cut, and a hoisting engine was set in motion, drawing the rope, at first invisibly, up out of the water. Wilhelm Hildenbrand published an account of the day the following year, in *Van Nostrand's Science Series*. According to Hildenbrand it took four minutes for the rope to come out of the water; and in six minutes it hung freely clear over all the river's traffic.

The rope, Hildenbrand wrote, formed "the first connecting link between the cities of New York and Brooklyn, destined never more to be broken." He described in understated style the mood of the spectators that bright Monday morning. "Simple as this operation was, it created considerable excitement and interest among the population, probably caused by a feeling of historical importance for the day, which practically should unite two cities. Thousands of spectators lined the shores, who greeted with loud cheers the appearance of this little rope, as it rapidly ascended high up in the air. It seemed that all doubts, hitherto entertained as to the erection and subsequent safety of the bridge, had vanished from that moment. The East River Bridge was considered an established fact."

There would be even bigger crowds in ten days' time, when Farrington, clad in a linen suit and wearing a felt hat, became the first person to cross the river between the towers—on the " traveler," or endless wire rope loop (much like a clothesline), which would, in the years it took to make the cables, carry the thousands of miles of steel wire necessary back and forth between the anchorages on either shore—there are over 14,000 miles of wire in the finished cables. He sat in a boatswain's chair—just a plank attached to a couple of ropes—and made the twenty-two-minute journey accompanied by "wild shouts" in appreciation of "the boldness of the voyager," who "kissed his hand to the populace in response to the cheers," waving his hat as he went, and all in all was "very cool and collected in his perilous position."

As the work continued, Farrington was at pains to assure Washington Roebling that everything was under control. "I have anticipated much trouble in instructing inexperienced men in the details of cable making, as we have only one man who has ever seen anything of the kind; but the intelligence and skill displayed by the workmen in these preliminary operations, had me to think that, I may have overestimated the difficulties to be overcome," Farrington wrote. The Chief Engineer had, however, anticipated such difficulties. Not only was the work novel to all of them, but in a city of immigrant labor many different languages would have been spoken on site; and literacy, certainly English literacy, could not have been taken for granted. And so a model of all the elements necessary to spin the cables—the machinery, the working ropes, the temporary footbridge—was built and kept in the engineering office. The *Brooklyn Union*, in revealing the existence of the model, had credited it to Farrington—which Farrington, in a printed reply, declined to take. "The idea of erecting the model originated with Col. W. A. Roebling, our engineer-in-chief, who gave instructions by which it was made. Owing to the complicated nature of the work incident to the situation, and the precautions necessary to be used to prevent damage from passing vessels to the work in progress, such a model was indispensable, though the plan for making the cables varies but little from that successfully followed on the Covington and Cincinnati Bridge. Colonel Roebling has, however, during his long illness, conceived a design whereby a great amount of labor and material required by the old plan may be dispensed with. Probably the amount saved will not be less than $25,000, besides the advantage of causing but little if any obstruction to navigation, a most important point to all concerned."

Unsurprisingly, Emily preserved this letter in her scrapbook, as she preserved another cutting from the *New York Sun*, which, while remarking on the Colonel's long illness—he "has not seen the structure in two years"—noted carefully that "every step that is taken on the bridge is controlled by the chief engineer, Mr. Roebling." Washington's assistants were devoted to both his service and his reputation. In his report on the cable making, Farrington assured Roebling he hadn't been after any glory when he swung out over the river in August. "It is a matter of history that the river was first crossed on one of the travelers, Aug. 25," he wrote, "and I wish to state here, that the transient notoriety attending the affair somewhat annoyed me. I had a natural desire to cross first, but my principal object was to show my workmen who were to follow under more dangerous conditions, my own entire confidence in the strength and security of the works." He had perfect confidence in the Chief Engineer. "I have carried out your instructions to the letter . . . and from my perfect familiarity with your plans, and my own experience, I shall expect the cables of this bridge to equal, if they do not excel, the best that ever were made."

But there was a long way to go before Farrington's expectations were fulfilled. Everything depended on the wire that would actually be laid to form the cables; bids for the contract would only be entered at the end of the year. "As may be supposed the question of the main cable-wire has formed the subject of much speculation and thought; especially in preparing the specifications for the same. Tests have been going on at intervals for the past six years and constantly during the last two years," Washington wrote in his

1876 report. "The principle which served as a guide in these determinations was to obtain a certain amount of strength for the least amount of money."

STEEL IS STRONGER than any other material in tension. The cables of the Brooklyn Bridge were the first made of this new material, but from this point forward steel cables would stretch between ever more distant shores. The unprecedented span of this bridge made the use of steel a necessity rather than a choice: to use iron, Washington wrote, "would necessitate a cable of such weight and size that it would become unmanageable, and involve the greatest difficulties in making it. I, at least, would not be willing to undertake it."

By the middle of 1877, the cable spinning over the East River was well established. Washington and Emily were now installed in their house on Columbia Heights; if Washington could not go down to the construction site himself, he could observe the work closely through a telescope installed in a bay window. C. C. Martin, Washington's assistant engineer, had assured his chief at the end of the previous year that it had not been hard to find good men to undertake this work: in the work on the cables there had been "a spirit of rivalry among our best men as to who should be assigned the most dangerous and difficult tasks."

A temporary footbridge, made of light wooden planks, had been strung across the span, rising from shore to tower, over the river, and down to the ground again. A sign was posted by the wooden steps that led up to the footbridge: "Safe for only 25 men at one time. Do not walk close together. Nor run, jump or trot. Break step!" It was signed off *W. A. Roebling, Eng'r in Chief.* Early in the year the cities' newspapers had sent adventurous young reporters to make a crossing on the footbridge—a terrifying experience by all accounts. Emily's clippings record that the *Eagle*'s man was frozen with fear on the ascent: "Peter Cooper's glue"* kept him fixed to the spot; he only got moving when the master mechanic remarked offhandedly that his own

---

*Peter Cooper was a famous industrialist and glue manufacturer; his daughter Amelia was married to Abram Hewitt, with whom he worked closely. The partnership survives in the institution of the Cooper Hewitt, the Smithsonian Design Museum in New York.

daughter had made the crossing that very morning. C. C. Martin's girls made the trip, too: "One of them stepped out briskly, leaving the rest of the party behind," ran the *Eagle*'s story, headlined BEAUTY ON THE BRIDGE.

Some still doubted the wisdom of the whole project; one of the first cuttings Emily saved concerns a man called Abraham B. Miller, who in May 1876 attempted to stop the bridge entirely by bringing a lawsuit against the cities, "giving notice of a motion for injunction restraining them from progressing with the completion of the East River Bridge. The complainant sets forth that the projected bridge will be a nuisance, and is being built without lawful authority, and that it will be injurious to the business of the complainant, he being the owner of warehouses facing the East River where vessels discharge their cargoes." The case would drag on until 1880, when it was finally dismissed—a comparative frivolity which, in 1877, was hardly a distraction from Washington's real, and very alarming, concern.

On December 3, 1877, he dictated a letter to his trusted lieutenant, Colonel Paine, regarding samples of the bridge wire that had been sent to him. A wire had snapped ten days earlier. "I have examined the piece of wire which broke on Thanksgiving day and you no doubt have done so also," he wrote to Paine. "It is as brittle as glass. That this particular piece happened to break was mere accident. The first question which arises is how much of this same brittle wire has been going into the cable without our knowledge and secondly what steps must be taken to prevent its recurrence. Is it down to a wrong system of inspection? Is it down to a laxity in inspection or what is the reason in your opinion?"

The same day he wrote to Paine he dictated another letter, to Henry Murphy. "I send you herewith some pieces of the wire which we had to cut out of the cable Thanksgiving day . . . It is brittle as glass, and the most dangerous material that could be employed. Its character was revealed accidentally. How much of this poor wire has been going into the cables I do not know. Can I be held responsible for this? It is scarcely right that the Engineer should have to be acting as a detective. I see but one way of preventing such wire being run out and that is to double the number of inspectors at the contractor's works. They had had too much work put on them all along. It is impossible that any two men can test so much wire carefully."

The wire being spun across the East River was not made by John A. Roebling's Sons. When it was announced that the bidding for the wire contract would be opened, in the autumn of 1876, most people taking an interest assumed that the contract would go straight to Trenton. It was a Roebling bridge; surely it had been designed with Roebling wire in mind. But, after the fall of Tweed, that was just the kind of impression that the trustees, and the cities of New York and Brooklyn, wished to avoid. Further roiling the waters, back in September, when the trustees had met to discuss the awarding of the wire contract—an enormously significant step in the construction— Abram Hewitt had made a surprising announcement regarding the wire. The bridge's towers and anchorages were complete; there was no reason that construction should not be swiftly completed. He had, reported the *New York Times* and other papers, "examined the specifications for the great cables and found them satisfactory." He thought it was unwise to allow the Roeblings, one of whom was engineer, to receive any further contracts for wire or other material. The *Times*'s report was straightforward.

Mr. Hewitt offered the following resolution, which was unanimously adopted:

> Resolved: That bids from any company in which any trustee, officer, or engineer at the bridge has an interest, will not be received or considered, nor will the successful bidder be allowed to sublet any part of his contract to any such persons or company.

The meeting was adjourned.

What the paper did not report was the statement made by one of the trustees, John Riley, who was concerned at the rumor he had heard that Roebling himself was far too ill to concern himself with bridge building; that his wife was apparently in charge. Surely the time was right to replace Roebling? Thus, in the contest over the contracts for the wire began a real attack, in private and public, against Washington Roebling's leadership and decision making, against his ability to see the job through to the end.

Washington was outraged that his integrity, and that of his brothers, should be publicly questioned. But he knew the way of the world now. Writing his memoir, years later, he said simply that "the safest plan is to take it for granted that every human being is a thief who tries to get the best of you." In 1876, when Hewitt's resolution appeared in the paper, Washington determined to wash his hands of the whole affair: he would resign from the bridge. He had suffered too much, he wrote to Henry Murphy, and he had neglected his own business affairs in the years he had given to the bridge. But most of all, "I am obliged to resent the gratuitous insult offered to me by the Vice-President of the Board of Trustees . . . a man whose designs upon the cable wire and the work of the superstructure are only too transparent and whose nominal connection with the Board of Management has had from the first no object but his own personal advantage."

Washington Roebling knew that Hewitt's lofty disinterest was not what it seemed; and when Henry Murphy declined to accept his resignation, Washington replied that the board would do well to keep its eye on Hewitt. "As you seem to be deeply impressed with Mr. Hewitt's action in declining to become a competitor for this wire, I desire to say that his magnanimity is all a show, as the firm of Cooper and Hewitt have no facilities whatever for making the steel wire, and if you receive a bid from a Mr. Haigh of South Brooklyn, it will be well for you to investigate a little."

The correspondence reveals Washington's keen eye, his ability to pay attention to every detail of the work no matter how much he suffered—and his almost maddening integrity. For shortly after he wrote to Murphy, ads were placed in the papers announcing that the bidding for the wire was open. The *Brooklyn Daily Eagle* carried the notice on its front page on September 30:

Proposals will be received up to December 1, 1876, by the trustees of the New York and Brooklyn Bridge, at their office, 21 Water street, Brooklyn, for the manufacture and delivery of 3,400 net tons of No. 8 galvanized steel cable wire, to be used in the construction of the main cables of the East River Suspension Bridge. Printed Specifications containing the full information will be furnished upon application to this office. W. A. Roebling, Chief Engineer.

Washington knew that Haigh was indebted to Hewitt, who held the mortgage on his Brooklyn wireworks. Abram Hewitt's business sat alongside John A. Roebling's Sons in Trenton, and Washington and his brothers made a point of knowing what their rivals were up to. Roebling and Hewitt were "what might be termed commercial rivals . . . This act of Mr. Hewitt's I look upon as a contemptible trick perpetrated solely with a view to gratifying his private revenge."

He sold all his shares in his own company. Three hundred shares—he was out of the business. Now that John A. Roebling's Sons had nothing to do with him, the Chief Engineer of the New York and Brooklyn Bridge, his brothers were free to bid, if they so wished. He would call Hewitt's bluff. And yet he was happy to acknowledge that the interests of the engineer and the interests of the business were not necessarily the same—despite what the papers and the gossips of the two cities said. "The great question is, is it expedient to [bid]?" he wrote in an undated letter to his brother Ferdinand. "The bulk of bidders will put in only one bid and not say a word about the quality of steel they propose to use . . . What I want—and you know this as well as I do—is . . . a uniform steel that will not crack . . . The danger of making the cables out of poor steel is so great even with the best of intentions on the part of the makers . . . only the most rigid inspection will serve."

In the end, Haigh won the bid for the cable wire; notes in pencil on the stationery of the Trustees of the New York and Brooklyn Bridge show the results of testing wire from different firms. J. Lloyd Haigh "presented several samples of wire apparently cast from steel of three different stocks, tensile strength exceeding the requirements. The elongation very good, the elastic limit up to the mark . . . this wire is very straight, galvanizing smooth." John A. Roebling's Sons presented "two rings of wire said to be cast steel and two rings of Bessemer stock. The shortest of the cast steel wire indicated a high tensile strength while the long tests fell considerably below the requirements. The elongation was deficient in the long tests . . . the bending tests unsatisfactory, galvanizing indifferent." It would seem reasonable that the contract should be awarded to Haigh; but Washington and his engineers knew that supplying wire to be tested for a bid was a different matter than supplying wire on a consistent basis, day after day, week after week—indeed, Washington had said plainly to Murphy that tests were never a real indicator of a supplier's long-term ability to reach a certain standard. "I know that nothing can be done perfectly at the first trial; I also know that each day brings its little quota of experiences, which with honest intentions, will lead to perfection after a while." He knew the contract would go to the lowest bidder; it was his "duty to guard these specifications with tests and restrictions, which,

if faithfully followed out, would fully warrant the Trustees in giving the contract to the lowest bidder."

So Washington—via his doughty lieutenant, Colonel Paine—determined
to investigate what was going on. He insisted that every ring of wire Haigh
supplied be tested, instead of every tenth ring; he had suspected that Haigh
would try to bribe the inspectors examining the wire, which indeed he had
tried to do, albeit unsuccessfully. Instead Haigh had concocted an elaborate
system by which wire that had been rejected by the inspectors was put into
storage in his Brooklyn warehouse—but then slipped back onto wagons for
transport to the bridge. It took all of Paine's cunning to discover the trick;
what Washington described, in a letter to Henry Murphy, sounds like an
escapade from their days during the war. "A watch was set . . . and the trick
discovered," his letter described. "The wagonload of [good] wire as it left
the inspector's room, with his certificate, in place of being driven off to the
bridge, was driven to another building, where it was rapidly unloaded and
replaced with a load of rejected wire, which then went to the bridge with the
same certificate of inspection."

Washington's letter to Murphy was written in the summer of 1878; but
the board of trustees took no action. Haigh was called before the board and
asked to justify himself; he denied his intention had ever been to deceive. At
the same time, as Washington wrote Murphy, the cables of the bridge were in
no danger, despite the bad wire that had gone into them—221 tons of it, by
his reckoning. This was no small matter, he wrote to Murphy, for "if a weak
wire is spliced to a strong one and stretched from tower to tower, the strength
of the weak wire is the measure of the whole length." (The splicing system
alone—the means by which individual coils of wire were made into one continuous wire, with every join as strong as the wire itself, had taken two "harrowing" years to develop, Washington had told Henry Murphy.) But he had
made allowance for the cables to be six times as strong as they would ever
need to be, knowing that so many variables had to be taken into account; he
reminded Murphy that he had written, in his report of January 7, 1877, that
this margin already allowed for "any possible imperfection in the manufacture of the cables." Now that Haigh's subterfuge had been detected, he felt
there should be no cause for concern: it was possible to add an additional
fourteen and a quarter tons of wire to each cable without having to enlarge
the cable bands or alter the wrapping machines. J. Lloyd Haigh of Brooklyn

would continue to supply wire for the bridge, "on such conditions and terms as he deems proper under the circumstances," the trustees noted after reading the Chief Engineer's assurances. Washington made it clear that no harm would befall the bridge so long as his directions were followed. "There is still left a margin of safety of at least five times, which I consider perfectly safe, provided nothing further takes place."

Haigh agreed to supply the additional wire needed at his own expense. None of this—the engineers' investigations, Haigh's reparations—was revealed to the world. Perhaps there was enough in the papers that year to distract readers from the business of steel wire; Tweed had died in April—the *New York Times* reported "a peaceful death-bed scene" in the Ludlow Street Jail after contracting pneumonia; he died "expressing the belief that the Guardian Angels would watch over him." In the autumn the Edison Electric Light Company was incorporated; the Bell Telephone Company, too. Business in the city was booming again, and on October 5, 1878, there was real news from the river. THE LAST WIRE OF THE CABLES OVER THE RIVER THIS AFTERNOON, ran the story in the *Eagle*.

"By Monday at the very furthest, the last wire of the last strand of the last cable will have been stretched," the paper's correspondent wrote. "The automatical wheel, which since August of 1877, moving across the river, high in air, apparently of its own volition, has been so familiar to passengers on the ferry boats, will, after Monday next, be missed. Its mission has been accomplished. It bore the wires across from Brooklyn to New York, and in thirteen months has crossed the river nearly 23,000 times. Nearly 3,400 tons of wire have been used in this stage of the work . . . The end, then, is near at hand. It was early in the spring of 1870 that the caisson was launched. As the season for work of 1878 closes down, we find the engineers ready to enter upon the last stage. All the great engineering problems have been solved, all the difficulties have been surmounted, one prediction after another of antagonizing and quibbling engineers have been proved to be false. For so extensive a work, employing so many hands, with duties so hazardous and perilous to be performed, there has been comparatively little loss of life. It is to be hoped that in the work yet to be done the same good fortune may attend the enterprise." In the matter of the manufacture of the wire itself, the paper noted that "it was a matter for congratulation . . . that the fortunate bidder should be a Brooklyn gentleman, Mr. J. Lloyd Haigh, whose works are located in this

city, on Red Hook Point, and give employment to Brooklyn workmen . . .
Each coil was tested by Colonel Payne [sic], of the Bridge Engineers, and his
assistants, so that the quality of every wire that went into the cables was
known and determined beforehand."

The *Eagle* went on to let its readers know that "the next stage in the work
will be gathering the mass of strands together in compact cables, and wrapping them from end to end, with galvanized iron wire . . . this work will
begin this week." The contract for that wire, as it happens, was awarded to
John A. Roebling's Sons. In December 1879 the *New York Times* carried a
report that—despite the upward turn in the economy—one Brooklyn business had failed: that of J. Lloyd Haigh. EMBARRASSED BY HIS WIRE CONTRACT
WITH THE BROOKLYN BRIDGE TRUSTEES, ran the headline on December 30,
1879. "Much surprise was caused in the iron and steel trade yesterday by the
announcement of the failure of J. Lloyd Haigh, manufacturer of wire, at
No. 81 John Street . . . The main cause of his failure is ascribed to the bridge
contract, as unadjusted matters with the parties who furnished him the steel
caused a pressure on his regular business." According to the *New York
Times*, the trouble was caused by the steel supplied by Anderson and Company, of Pittsburgh; their steel "did not fully meet the requirements of the
contract," and Haigh was burdened with its cost. Mr. Haigh was "expected
to extricate his affairs if allowed time." It was not to be; in Emily's scrapbook
a clipping records the Brooklyn wire merchant's imprisonment and hard labor at Sing Sing, New York State's notorious jail. Bankruptcy had led him to
forgery; forgery had led him to the loss of his liberty—and while in prison,
he had tried to bribe his guards.

Emily Roebling did not save—perhaps she did not see—the notice in the
paper that marked the death of Haigh's wife, Eliza. "Mrs. Haigh was a most
remarkable woman, and her devotion to her husband through the long period of his disgrace, trial and imprisonment never wavered or faltered."
Mrs. Haigh was thirty-six years old; she would have been thirty-seven had
she lived to her next birthday, the report was careful to note, and left behind
seven children, the eldest "a young miss of 10."

# 13

## "Trust me"

IT SHOULD HAVE BEEN smooth sailing from there on in. For some of the papers the completion of the work was a forgone conclusion: "It may almost be said that the beginning of the end of the East River Bridge has been reached," one paper wrote, as the cable work was finished. In the middle of 1878, when Abraham Miller was still arguing his case to have the bridge taken down, Washington had summarily dismissed the extraordinary statement made by a former judge, E. M. Sherman, that a study of bridge structures had shown that no wire bridge would stand six years unless supported by guys or stays below the roadway. The *New York Sun* described the Chief Engineer's blunt reply to this ridiculous assertion. "He still suffered from caisson fever, and as he lay on a cot in a front room in his home on Columbia Heights, Brooklyn, he said [of this charge] '. . . To an engineer this is absurd on its face, the precedent having shown to the contrary, and such an opinion shows the legal gentleman referred to to be profoundly ignorant of the rudiments of engineering. The bridge crossing the Ohio River at Cincinnati is similarly constructed to the Brooklyn bridge. Already it has stood thirteen years without showing vulnerability . . . There is no reason why the Brooklyn bridge should not remain intact for centuries.' "

Yet there had been problems. Just a few days after Washington had assured the public that the bridge would last forever, the first accident involving the cables shocked the city. Around noon on Friday, June 14, "people along the line of the bridge upon the New York side were startled by a loud report, closely followed by a furious rushing sound, as if a tornado had struck the city," the *Eagle* reported. "It was a strand of the No. 4 cable of the Bridge which had been released in some manner . . . from the New York anchorage." The strand's great weight, suddenly released, had sent it hissing through the air "with frightful rapidity"; when it struck the river it made a splash fifty feet high, and nearly struck one of the ferries traversing the river: had the flailing cable come down on the boat, "a longer list of dead and wounded would have been printed in the *Eagle* tonight." As it was, two workmen were killed: Thomas Blake, a laborer, was killed outright when lifted by the cable and flung down on the stones of the anchorage; Henry Supple, a foreman of the cable gang who had been invaluable to the work, was also lifted and tossed, though his fall was partially broken by the other strands. But, said the paper, "his forehead is broken in and his brains protruding"; he was not expected to live more than a few hours.

Rumors began to fly about that the quality of the steel was to blame—though the engineers' report found that it was not. For the strand had not snapped—rather, the fall-rope, or hoist, which was being used to lower the strand into its final position, had broken. The fall-rope had been damaged by the edge of the pulley on which it ran; when that rope had broken, the strand it was supporting had been released with catastrophic results. The coroner's jury recorded a verdict of accidental death in the case of Blake and Supple. But the great cost of the bridge, the length of time it was taking to build, the invisibility of the Chief Engineer—and the whispered presence of his wife on the scene in his stead—did the work no favors in the public eye, and in the eye of some of the influential men of Brooklyn and New York. Money was still very tight, and men were laid off from the work, "because of the failure of funds caused by the refusal of the New York Controller to pay over any more money." John Kelly, Tweed's replacement as "boss" of Tammany Hall, "says he has been assured by some persons that the Bridge will not be of great use when finished, and on the other hand, he thinks New York never should have paid anything toward the work for it will benefit Brooklyn exclusively." So reported the *Brooklyn Daily Eagle*, whose editorial begged to differ: "No man in his senses can possibly dream of abandoning such a work as this at this stage of its progress, and every day's unnecessary delay in the completion of a work in which one or ten million dollars is invested, must necessarily add largely to the cost of the enterprise."

Washington knew that if the bridge itself was not under threat—it was nonsense to argue that with the towers and cables complete, the structure should not be finished—his position certainly was. Throughout 1877 and 1878, rumors continued to circulate in the papers that the Chief Engineer was insensible, incapable. But the worst battles were to come. Newspaper gossip was one thing: a move against him from the bridge's trustees, quite another.

Things began to get stiff in the middle of 1879. The truss work for the bridge floor was soon to be built; originally it was to be have been made from iron, but following a report from Paine, Washington decided that it would be better to use steel, making the New York and Brooklyn Bridge an all-steel construction. However, at a meeting of the Trustees in early May of that year, this decision was called into question: one of the trustees, General Henry W. Slocum, said that he had heard on good authority that Paine had been taking bribes from steel manufacturers. He also questioned the involvement of John A. Roebling's Sons in the enterprise altogether. "I want to say right here that I think it is indelicate that the brothers of the Chief Engineer should be engaged in furnishing us materials." He went on to bluntly accuse the company of shady practices: testing steel for use in the bridge, apparently condemning it as unfit to be used—and yet not returning it to the manufacturer.

Washington and his brothers, especially Ferdinand, were outraged. This was the same General Slocum who he had encountered before the great battle of Chancellorsville; according to Washington's account of that battle, Slocum had called him a god-damned liar—and threatened to have him shot. Ferdinand was called down from Trenton to appear before a special investigative committee; all the charges were finally revealed to be utterly false. But Washington's anger was vivid. "I hope I have heard for the last time your oft-repeated remark that you think it indelicate in me that I should allow my brothers to do any work for the bridge while I am the Chief Engineer," he wrote to Slocum. "Did it ever occur to you that my brothers act independently without consulting me and I have no control over them even if I wished to prevent them bidding on any contract for the bridge? Or did you ever consider that the John A. Roebling's Sons Company hold the first rank in this country as manufacturers of wire rope—and the word 'fraud' has never been coupled with their name save in your board?. . . The course of a true gentleman would have been to come to me first with a lie that had been whispered behind my back and at least heard what I had to say, whether you believed it or not."

But then there had been moves against him long before. Lloyd Aspinwall, who had been made a director of the Bridge Company with Hewitt, in 1872, when Tweed's Tammany men were turfed out, had called for a consulting engineer as far back as 1876, concerned as he was for "the precarious condition of Colonel Roebling's health." In those days, Washington had been willing to explain his reasons against such an appointment in a temperate manner. "I consider the motion of Mr. Aspinwall quite natural," he had written to Murphy then. "Neither he or some of the New York Trustees know me save by reputation or have even seen me. It is known that I have been sick for the two years these gentlemen have been members of the Board and even for some time previous. It is therefore quite excusable that he should feel a little alarmed both as to the future management of the work and the troubles that might possibly arise in case I die before the Bridge is complete." He went on to argue, however, that Aspinwall and the others "must agree, that up until this time my illness has not in any way interfered with my successfully directing to the smallest details, the prosecution of the work." From a "common sense" point of view, too, the appointment of a consulting engineer was "quite opposed to the spirit of American engineering." Consulting engineers "usually have nothing to do with the actual construction of the work." A "quiet talk with my foreman or assistant engineers produced all the necessary mental friction for eliciting new thoughts or modifying old ones" better than the appointment of a consulting engineer.

"Continuing to work has been with me a matter of pride and honor!" he had emphasized to Murphy; and he had made assurance that he would give ample warning if he could not carry on. And in anticipation of a later battle, he set out the problem, as he saw it, of appointing a consulting engineer. If the board were to appoint someone superior to him then clearly there would be no need of him, and he would leave; if the consulting engineer was his equal, "then you would see the remarkable example of a man learning in five minutes what it has taken me seven years of hard study and thought to acquire. My feeble light would be so pale before such transcendent genius I should again feel constrained to step out!" And then, if he were Washington's equal—well, then, there was simply no need for the appointment. And his present assistants have "grown up with the work . . . they expect to remain on the bridge until it is completed, and they are as familiar with the remaining work before us as it is possible for anyone, but myself, to be . . . My counsel in this matter is to let the subject rest, and trust me." For evidence that the Chief Engineer was working as hard as ever at the time, we can take a note to Farrington, dispatched in the autumn of 1876: "P. S. Send me my big Webster's Dictionary, I am studying Danish and Swedish and need it." (The Danes and the Swedes were in the vanguard of steel production in this period, giving the English a run for their money.)

As Washington had noted, "the building of the whole Bridge is a matter of trust." And so the work progressed. Careful calculations had been done to ensure that each of the nineteen strands of wire making up each cable were correctly regulated, or set in precisely the right place—not a simple task, as heat and cold caused the steel to expand and contract. By the beginning of 1880 the cables had been wrapped, the strands bound with galvanized iron wire that spooled out from a drum traveling the length of the cable, the workmen tightening the wire wrapping by means of a spoked wheel, turned and turned again. "These operations, though simple in themselves, acquire a special interest from the circumstance that they are carried on at such a gigantic scale and such an enormous elevation above the river," said *Scientific American.* The papers continued to send reporters to ferret around the work. "The four great cables have been covered over with stone at the anchorage, so that they now project out of a solid mass of masonry, as though carved out," noted the *New York Sun* in June 1879. "A reporter, yesterday afternoon, in peeping down a hole in the anchorage, saw a half nude man with a sponge dripping with white lead, going over the different strands of the cable at the point where they are fastened to the anchor plates." And the hot summer had no effect on the atmosphere in the great caverns of the anchorages. Under those deep arches "it is as cool as in an ice cellar."

In December, Frank Farrington gave a series of public lectures—in places such as the Brooklyn Music Hall—on the construction of the bridge. They were very well attended. "Doubtless you are all interested to know when the bridge will be completed. I have my own idea about the matter, and I can say that if no more idiotic opposition is offered to its progress, and the funds are forthcoming from the two great cities which have assumed control of the work, it will be completed in about 18 months." He described the floor system, made from six longitudinal steel trusses, connected by the floor beams and suspended from the cables by suspender ropes, which were made from twisted galvanized wire, capable of sustaining a far greater load than they would ever be called upon to bear, Farrington assured his audience. "The great weight which will ever be brought to bear on them is 10 tons a piece, yet I have seen them tested with a weight of over 140,000 pounds without giving way. There are no such things as rotten wires in this bridge." The construction of the superstructure had been delayed because producing the steel required had presented such a challenge—the trusses called for larger steel bars than had ever been produced in the United States, and the different behaviors of iron and steel during manufacture were still in the process of being understood. But by the beginning of 1881 *Scientific American* stated confidently, "All these engineering and mechanical difficulties have now been surmounted."

But such confidence was not reflected in the board of trustees—especially among its newer members. Seth Low was the energetic new mayor of Brooklyn; at thirty-two he was the same age Washington Roebling had been when he took over as chief engineer after his father's death. A reformist Republican and son of the wealthy merchant A. A. Low, he was notable for his hostility to patronage. Even before Low attended his first meeting of the trustees at the beginning of 1882, there had been suggestions that one way of reducing the cost of the bridge would be for the Chief Engineer to reduce his salary; and for the assistant engineers to take a reduction, too. The suggestion had apparently come from Kingsley, but it was Murphy who had written to Washington, and so to Henry Murphy Washington made his definitive reply. "I do not know that any of the engineers would consent to remain with you if their pay should be reduced," he wrote in April 1881. The work still needed the attention of his trusted lieutenants; and in any case, "you would find it very difficult to fill their places with other men at any price." As to his own salary—ten thousand dollars per annum—"I should not be willing to take any less pay than I have received since the beginning of the work. For me to consent to a reduction of salary now that the bridge is so nearly complete would look as if the Trustees could get along very well without me now, but

I feeling anxious to stay, make them a bid to keep me by offering to work for anything they see fit to pay me."

It seems as if he knew very well that worse trouble was brewing. In the last months of 1881, the board had questioned his decision to add one thousand tons of steel to the superstructure to enable it to carry Pullman cars—should such plans present themselves. But the board had interpreted the chief engineer's alteration as vacillation, and had called upon him to explain (which he had done, with some patience) his decision. Washington aimed to future-proof the bridge: despite his opposition to the idea of Pullman cars, he had deepened and strengthened the floor beams of the bridge, "fearing that [Pullman cars] might be used in spite of all my opposition." Now it seemed as if the board were considering dispensing with him altogether. "If the Board of Trustees decides by vote that they can no longer afford to pay me my present salary, they need not pay me anything. If I live long enough to direct the important work still to be done, I know it will be finished cheaper & better than it will be if left to some engineer who has not had my experience in constructing suspension bridge floors." However, there is a sentence in the letter that was not included in the final draft; it is crossed out, but perhaps best expresses the truth of what he felt: the dedication to it and the price he had paid and knew he would go on paying. "I have given my whole life to finishing the bridge, and never expect to do any more work when it is successfully completed."

AND NOW ANYONE could see that there would be a bridge over the East River before long. In the spring of 1881 the floor beams were stretching across the water; John Roebling had imagined they would be made of iron, but Washington had decided that steel was a better choice. It had first been used in quantity for railroad, but by the 1870s mills began rolling structural shapes suited for use on a bridge such as this one. In 1879 the first all-steel railroad bridge—a truss bridge—had been erected in Glasgow, Missouri. "The manufacturers have been able to make a steel which meets all the requirements of the work, and the tests have been very severe," Washington wrote to Henry Murphy. "There will be nothing experimental, therefore, now on our part in the use of steel . . . as a steel in every way suited to bridge construction can be produced." The Edge Moor Iron Company of Wilmington, Delaware, would supply nearly six thousand tons of steel for the bridge's floor beams and stiffening trusses. In the Chief Engineer's report to the trustees dated December 1881, Washington laid out in detail "all material of every description that will be required to complete the Bridge and its approaches." Two-hundred twenty-four tons of steel rails for the tracks had been ordered, as well as 15,330 railroad spikes; ten thousand bricks would be delivered to complete the approaches of the Bridge.

But Washington's position was not as secure as the progress of his construction. In 1876 he had tried to resign from his work at the bridge; Murphy had refused his resignation. Now, with the bridge so near completion, resignation was the farthest thing from his mind. It was about to be forced upon him.

He had left Brooklyn for Newport, Rhode Island, on the advice of his doctors; Emily kept a clipping from the local *Knickerbocker*, dated July 13, 1882: "Colonel W. A. Roebling, chief engineer of the Brooklyn Bridge has taken the Meyer Cottage for the season, and proposes to remain there, far from the madding crowd." Emily's brother G. K. was stationed there, too, as district engineer. Emily also kept a cutting from the *Star*, dated around the same time: ROEBLING HOPELESSLY SICK: A NERVOUS DISORDER WHICH BAFFLES ALL SKILL. One of his "intimate friends" apparently told the paper, "At times he loses all control over his mind." There is no record of Emily's thoughts, but there is surely a sense of irony in the preservation of this tall tale. And while he was at Newport, some trouble arose among his lieutenants. It seems that engineer C. C. Martin and Farrington, the master mechanic, had never got on well; now, something flared up and in 1882 Farrington simply walked away from the work. Washington wrote in some irritation to Martin that he was "only surprised that Farrington has staid as long as he has, all . . . of you have been trying to get rid of him for the past ten years."*

Martin had served him well; but, five years older than Washington Roebling, perhaps he felt his time had arrived. He must have known the board

---

*Farrington, evidently, did not hold his grudge eternally; when the Covington and Cincinnati bridge was overhauled under the supervision of Wilhelm Hildenbrand in 1895–99, the master mechanic returned to help with the project. He was fatally injured when he was struck by a fast-moving cable car as he worked on the bridge. *New York Times*, April 16, 1898.

favored his work as "active" engineer. And Washington, even before he had moved to Newport, knew that the board wished him to appear in person before them. "I am not well enough to attend the meetings of the Board. I can neither talk or listen to conversation. I came here by the advice of my physicians who hope that living out of doors, away from the continual noise of a city may lessen the irritation of the nerves of my face and head. I am glad to say I am much better this spring and able to walk out occasionally. The journey down here was a very painful effort, and I have been in my room most of the time since I have been here," he told Henry Murphy. He had informed the board he would not appear before them in a five-word telegram: CANNOT MEET THE TRUSTEES TODAY—which seemed, to say the least, offhand. But Washington was exasperated by what he saw as the trustees' presumption. "I did not state in my telegram that I was too sick to attend your meeting," he wrote to the trustees, "because I am sick and tired of hearing my health discussed in the newspapers though there is not the slightest objection to any trustee knowing my condition. I believe there is not a day that I do not work for the bridge and I think that those Trustees who are familiar with the work will agree that my assistants work with the perfect confidence that they can always refer to me and get any advice or assistance they need . . . I shall be most highly honoured to be present at meetings of the Board as soon as I am well enough to be of any use there."

Darkness crowded around him—and around Emily, too. In early August, as the demands of the board grew increasingly shrill, Emily's beloved brother, Washington's former commanding officer and the man who had brought them together, died at the age of fifty-two. The cause was ostensibly diabetes, complicated by liver failure, but many would say he died of a broken heart, having failed to clear his name over the Five Forks affair; tragically, the military court would find in his favor just a few months later. And on August 22 the trustees appeared at a meeting called by Mayor Low, who said forcefully that the Brooklyn Bridge should not be in the hands of a sick man. He had traveled to Newport to see Roebling; Washington would say in his notes of the meeting that Low's reasons for wishing to remove him were "weak and childish." Indeed, he had been reduced, finally, to insult: "Mr. Roebling, I am going to remove you because it pleases me."

Low called for Roebling to be replaced; once again, the dreaded words "consulting engineer" were used. A resolution was proposed. Roebling was "an invalid"; thus, "this board does hereby appoint Mr. Roebling Consulting Engineer, and Mr. C. C. Martin, the present First Assistant Engineer, to be the Chief Engineer of the New York and Brooklyn Bridge." The board, of course, wished to express its "regret at the necessity of such a change"—but deemed it a necessity all the same. On the day of the meeting a piece in the *New-York Tribune* praised Low's initiative. "The day for sentiment and dawdling has gone by . . . this is the day for sharp decision and vigorous action . . . Let Mr. Roebling be retained on a pension if necessary." More vigor was required, and an engineer who would "bend every energy to pushing forward the work." For the *Brooklyn Daily Eagle*, while acknowledging that "the people of [New York and Brooklyn] are weary of the existing conduct of the work," blame didn't lie with the Chief Engineer. Whatever "mistakes and blunders" there had been, those errors "are not chargeable to the professional men employed upon it." Yet even the *Eagle*'s praise of Roebling had a memorial air. "No generous man would willingly degrade Roebling, on the eve of the completion of an enterprise which, to him and his, will be as the victorious flag of his country to the soldier whose frame has been shattered in its defense."

Some on the board robustly defended the chief engineer. Ludwig Semler, German-born like John A. Roebling, had arrived in the United States in 1851 with five dollars to his name and had risen to be comptroller of Brooklyn. If Roebling had been found at fault for having delayed the work on the bridge, said Semler, he himself would have been "in favor of retiring him . . . but there is not a shadow of a charge upon which to base such action." James Stranahan, who had been with the enterprise from the beginning, stalled for time, hoping the resolution could be deferred; "I cannot help but feel that there is something yet due to the Chief Engineer."

And so the decision was deferred, at which point Emily Roebling took a tactical decision. She wrote to Semler; neither she nor Washington had ever met him, for he was a new man on the board. She had heard of his remarks at the meeting; "as you are a stranger to Mr. Roebling all that you said was doubly appreciated." She invited him to Newport, so that he might see for himself whether Washington was in any way impaired; he set off to Rhode Island on September 5, and was most impressed by the man he met there. "If his intellect has been impaired, I should be a happy man if I had what he lost," he would say. "He spoke to me with clearness, and exhibited a memory which was something astonishing." Washington made it plain to Semler that he would never accept the post of consulting engineer. Roebling had said, "If they want to remove me, let them do it absolutely," Semler reported.

Semler told the reporters who greeted his return from Newport that he was certain none of Roebling's men—such as his first assistant, C. C. Martin— would accept the post of Chief Engineer anyhow. Washington himself was not quite so sure—he had written a warning note to Martin before the trustees' meeting in August. He knew, Washington told Martin, that many on the board wished to be rid of him; but if he were to be dismissed, he would withdraw absolutely. "Whoever takes the place of Chief Engineer must do so with an entire understanding that my interest in the bridge ends the day I am ruled out of the position of Chief Engineer," he wrote to his lieutenant. "I do not see that you can as Chief Engineer render the board any more efficient service than you have as first assistant but it certainly makes no difference to me who is put in my place, when I am through with the Bridge . . . I have worked for the bridge when I could scarcely eat speak or move, and I shall leave the work in the knowledge that I have done my duty in every way as far as my strength would let me."

Washington depended absolutely upon those men who served him; but he would have known, too, that there can be moments in a man's life in which ambition overrules loyalty. His reminder, to Martin, of what his work on the bridge had cost him is a clear rebuke.

Washington's anger now led him to error. He had already expressed himself in no uncertain terms to Henry Murphy after the trustees' meeting in August: "I cannot see why Mayor Low in his resolution took the trouble to offer me the place of consulting engineer as I *positively* refused it in the interview . . . I should like some proofs of the judgment that my absence from the post of active supervision is necessarily in many ways a source of delay. I have made every possible arrangement for many years that a work of such magnitude should not be exposed to any of the vicissitudes of my health . . . The final claim of Mayor Low's resolution"—in which he expressed his regret of the necessity of this change—"is particularly offensive to me . . . You will either allow me to finish the bridge to the best of my ability or dispense altogether with my services in the future." Now, astonishingly, in the wake of Semler's visit to Newport, he spoke to a reporter from the *New York World* who had come nosing around in search of a story—which he duly got. He spoke to Washington, who told him that despite the efforts of the board to remove him, "he proposed to stick." "He suggested that one reason why perfect harmony did not exist in the board was that there were four candidates for Governor in it." By introducing politics, Washington had just shot himself cleanly in the foot.

The piece appeared on Friday, September 8, in the paper owned by Jay Gould—who had nearly ruined the nation's financial system in 1869, and who now had a big stake in the Manhattan Elevated Railway, a rival route of transportation; the article's mischief was to a purpose. Certainly, it caused damage; after it appeared Emily Roebling wrote to William Marshall, one of the original trustees, for she knew he was well disposed to her husband. She had, she said, believed the reporter from the *World* when he told her he would not publish a word he had heard in their house. She was full of regret. "I am sorry as there was no need of Mr. Roebling unintentionally offending those who were hoping to help him . . . There is no doubt of Mr. Roebling's perfect sanity and ability as an engineer, but he certainly is unfit to be on a work where so many political interests are involved for he has no idea of doing anything for the sake of policy." And Washington himself wrote to Ludwig Semler, sounding rather abashed. "Unexpected complications have arisen which may probably defeat your efforts on Monday, to see justice done me as a man and as an engineer. But whatever the result I must ever feel most profoundly grateful to you for the kindly interest you have shown in me."

Monday, September 11, was the day of the vote—the day that would decide Washington's fate. The weather was forecast to be cloudy, with rain; some summer heat still lingered in the air as the trustees gathered in Brooklyn. A detailed account of the meeting, printed the following day, was front-page news for the *Brooklyn Daily Eagle;* its report put readers in the room with the seventeen out of twenty trustees who were gathered there. Only three were unavailable, and, as the paper remarked, only the older members would have "any recollection of seeing so many of their colleagues at one time." Before the vote, the business of bridge-building was discussed. A total of $13,883,168.28 had been expended so far; C. C. Martin assured the assembled men that work on the viaduct, the station, and the

rope-driving machinery for the bridge trains, the boiler house, and the rail-
ings for the bridge approaches was continuing.* But these technical matters
were just a warm-up act for the main show. MAYOR LOW OPENS THE BALL, ran
the headline as the Brooklyn mayor Seth Low set out to finally oust the chief
engineer from his post. The question, to him, was a simple one. "Shall we
have a sick man or a live man—a man who is responsible and with whom we
can come into contact day after day?"

Not for him any assurances made by Roebling, or by anyone else. Henry
Murphy said he had a communication from the Chief Engineer, and it
should be read—the letter Washington had written to him. He struggled
against Low even to get a word in; it's clear the younger man had no wish to
hear from Roebling. But still, he could not avoid listening to that dry, assured
voice. "I am an invalid," Washington allowed, "and certainly it would be
greatly to my advantage to be well, but I should like some facts to sustain his
[Low's] judgment that my absence from the post of active supervision is necessarily a source of delay."

Before the vote was taken, William Marshall was permitted to speak.
Marshall had been born in Belfast, in1813; his parents had emigrated when
he was a boy and they had settled first in Delaware before coming to Brooklyn. He too was a rope man: the cordage business he had helped to found
grew to be the largest in the United States. He was hemp to Roebling's wire,
the old world that came before the new; one might suspect that he would be
no friend to the Roeblings, father and son. But such was not the case. He was
simply astonished, he said, that Low and his cabal—in particular Mayor
William Russell Grace of New York and Comptroller Allan Campbell, those
men with powerful political ambition to whom Washington had referred—
would attempt to drive Roebling out.

"For what?" Marshall asked. "Why? As a bridge builder he has not his
equal on the face of the earth. I defy contradiction! There are two bridges
across Niagara. He built the largest of them and it stands there today—a

---

*Until 1950, the bridge railway transported passengers back and forth, its cars—"models of
beauty and comfort," as Washington wrote—drawn by an endless steel wire cable. The Library
of Congress holds an extraordinary two-and-a-half-minute film of the journey over the bridge
by rail, made by the Edison Manufacturing Co. in 1899.

perfect success. When I say 'he' I mean his father and himself—the father who was sacrificed on this bridge. There are two bridges across the Ohio, one built by Mr. Roebling and one built by a man who is ashamed of his name. The one at Wheeling fell into the river; the other, at Cincinnati, is an honor to the man who built it." Perhaps, for once, obscuring the distinction between father and son served Washington's cause; but Marshall, certainly, knew of whom he truly spoke. "I know this bridge has been kept back time and time again by many, but I never knew that Mr. Roebling has kept it back by one day or one hour," he continued. "For one I would take the arm off my shoulder before I would permit myself to vote against a man standing here without a blemish upon his character and ability. But our friend Mr. Low goes down to him and demands his resignation: by what authority? You gentlemen form yourselves into a delegation of representatives of this board! By what authority? Have you any law for it? If you have I should like to see it. I should like to know by what parliamentary usages you get the power to resolve that three of you shall do the work of twenty members. If I had no other objections to raise to these resolutions these would be enough for me to stand up against them for forty years, if I lived so long. If there has been any fault in the board for the last ten years, for one I am willing to assume the responsibility for it, but I don't want to sneak out and place it on the shoulders of the chief engineer. It would be mean and contemptible for me to do that, and I don't propose to do it."

Silence for a moment or two. The rustle of papers. Sounds from the street below, the shouts of men, the clatter of horses' hooves. Who knows whether the first to speak will influence the others who will follow? When William Marshall died—suddenly, at his dinner table, in 1895—another eminent Brooklyn Democrat, William C. De Witt, would say that "I have seen him, solitary and alone, shake his fist in the faces of fifty ring politicians." He was, said de Witt, "keen, sagacious, energetic, tough as a gnarled oak." Now treasurer Otto Witte—born in Prussia the very same year John Roebling set out for the United States—spoke too. He had, he said, never met Washington Roebling; but he disliked the speculative nature of Low's argument. "We are called upon to act upon inspiration derived from a more or less venturesome imagination," he said. But then Campbell stood with Low, against Roebling—not on the basis of any direct evidence, but based on that troublesome piece in the *World*. "The Controller of Brooklyn, I believe, has been to see the engineer," he said, referring to Semler's trip to Rhode Island, "but I saw a communication from the engineer, or rather the report of an interview with a reporter and he doesn't seem to consider that he himself is at all accountable or blameable in the matter. There is where he does us an injustice. I think his fault is in trying to put upon us trustees his infirmities. I think that is very unjust and very unkind indeed."

When the vote was taken, Low, Barnes, Campbell, Grace, Van Schlack, Clark, and McDonald called for Roebling's dismissal. Standing with the chief engineer were Marshall, Kingsley, Semler, Witte, Stranahan, Swan, Slocum, Agnew, Murphy, Bush—even Slocum, who had once said he would shoot Washington Roebling himself. Ten to seven: the Chief Engineer was retained in his post.

Whatever else was said in the Meyer Cottage when the news arrived in Newport, whatever quiet satisfaction and relief there was between Washington and Emily in that house still hung with black—in mourning for Emily's brother G. K.—Washington's first response was to his staff, and to the work. On September 12, the day the *Brooklyn Daily Eagle* splashed its vivid report of the dramatic meeting, the Chief Engineer sat down to write a letter to C. C. Martin. If Washington had ever been concerned about the ambitions of his assistant, he didn't reveal them.

Dear Martin,

Your congratulations reached me last evening. Many thanks for your good wishes and your efforts on my behalf. It gives me fresh courage to find I have such good devoted friends.

In spite of all the anxiety we all felt as to how the vote would go, I find my mind was after all more occupied with how the bridge stood the fierce storm and gale of yesterday. I shall wait with great anxiety a letter from you telling me of how fierce the storm was in Brooklyn and what were its effects if any on the bridge. Here the storm was a fearful one.

Yours very truly, W. A. Roebling

---

THE OPENING CEREMONIES of the New York and Brooklyn Bridge took place on May 24, 1883, two days before Washington Roebling's forty-sixth birthday. The six months since the dramatic meeting of the trustees had passed without incident. Earlier that spring of 1883 Washington had set out with McNulty to see the Brooklyn terminal for the train, at Sands Street; he hadn't felt the need to climb out of his carriage—and so never set foot on the bridge in all the time of its building. A contract was awarded to the U.S. Illuminating Company to supply seventy arc lamps to light the bridge, at a cost of eighteen thousand dollars: it was the first time electric lights would be used over a river. And in the middle of May a light victoria—a carriage fashionable among ladies—drove over the bridge. "Colonel Roebling desired to know the effect upon the structure of trotting a horse over the roadway," ran the story in the *New-York Tribune*. During the war, when she and her husband had first met, Washington had noted Emily's sense of adventure, her love of riding; he had warned her, in those days, to be careful. She was very safe on the bridge he had built. "When Mrs. Roebling drove over on Monday she was accompanied by Engineer McNulty, who carefully observed the effect. No vibration in the suspenders could be seen. In all other European bridges," the story noted, "the vibration is very perceptible."

Now the bridge would be opened to the world—but not before the president of the United States, Chester A. Arthur, along with Grover Cleveland, governor of New York, would make a ceremonial crossing from New York to Brooklyn; not before the bridge was "presented" to the mayors of New York and Brooklyn, Franklin Edson—who had replaced William Grace—and Seth Low, by William Kingsley, acting president of the bridge trustees. Low had declared May 24 a public holiday in Brooklyn; on the New York side, there was a "strong expression of sentiment" in favor of closing the New York Stock Exchange early. Thousands of people would cross the bridge in the wake of the president, all of them in receipt of engraved invitations sent out by the trustees. On both sides of the river the wharves and streets were thronged with eager crowds; the river itself was packed with ships and boats, and the North Atlantic Squadron of the U.S. Navy had been sent specially as a watery honor guard: the *Vandalia*, the *Kearsage*, the *Yantic*, and the *Tennessee*—the latter the flagship of the fleet and the largest ship in the navy. Newspapers all over the country devoted pages and pages to the event, producing special supplements, commissioning grand illustrations.

Washington and Emily would not make the crossing that day; their fine Brooklyn home was decked out for a special reception, organized by Emily, that evening. Washington had been preoccupied by the plans for the opening, knowing how difficult it was to keep crowds under control and fearing what the effect of the planned fireworks might be. He wished to be assured that everyone would be cleared off the promenade by eight P.M. on May 24, the ends of the bridge secured, and no one allowed beyond approaches. "It will I fear be impossible to clear the bridge and I will not be responsible for the consequences if people are allowed to crowd on it. It would be possible for 100,000 people to fit on the main span of the bridge, crowding every available foot of space, cables and tops of trusses. This would make a load three times greater than the live load calculated for. Could you not have the fireworks on the night of the twenty third before the bridge has been opened to the public—it would be very wrong to put it to such an unnecessary test simply to gratify the whims of [an] official." Two regiments of militia were to follow President Arthur, and Washington didn't much like the sound of that, either. If they were to take part in the ceremony, "it must be with the distinct understanding that they are not to march across the bridge either before, after, or during the ceremonies. On no existing suspension bridge are troops allowed to march to music in cross and it should not be permitted here."

He wouldn't get his way; the fireworks went off that night, after many speeches. In particular, Abram Hewitt paid homage to "John A. Roebling, who conceived the project and formulated the plan of the Bridge; Washington A. Roebling, who, inheriting his father's genius, and more than his father's knowledge and skill, directed this great work from its inception to completion, in the springtime of youth, with friends and fortune at his command, braved death and sacrificed his health to the duties which had devolved upon him, as the inheritor of his father's fame, and the executor of his father's plans." He made special mention of Emily Roebling; he had written to Washington to be certain he could describe correctly the role all those involved in the bridge had played. His description of the part Emily had played is, therefore, clear and accurate. "With this bridge will ever be coupled the thought of one, through the subtle alembic of whose brain and by whose facile fingers, communication was maintained between the directing power of its construction and the obedient agencies of its execution. It is thus an everlasting monument to the self-sacrificing devotion of woman, and of her capacity for that higher education from which she has been too long debarred. The name of Mrs. Emily Warren Roebling will thus be inseparably associated with all that is admirable in human nature, and with all that is wonderful in the constructive world of art."

Hewitt had also written to Washington to ask if he could make a comparison between the Great Pyramid at Giza and the bridge just completed— in regard to the "true cost" of the two structures. Washington was not impressed by the notion. "To build his pyramid Cheops packed some pounds of rice into the stomachs of innumerable Egyptians & Israelites. We today would pack some pounds of coal into steam boilers to do the same thing, and this might be cited as an instance of the superiority of modern civilization over ancient brute force—but when referred to the sun our true standard of reference the comparison is naught—because to produce these few pounds of coal required a thousand times more solar energy than to produce the few pounds of rice. We are simply taking advantage of an accidental circumstance.

"It took Cheops twenty years to build his pyramid but if he had had a lot of Trustees, contractors & newspaper reporters to worry him he might not have finished it by this time. The advantages of modern engineering are in many ways over balanced by the disadvantages of modern civilization." But the intervening 4,500 years had, at least, a fiscal advantage by Washington's reckoning. He figured the costings of the pyramid at forty million dollars, based on present stone prices; it would rise to fifty million dollars if profits for contractors were taken into account. The bridge, then, at just over thirteen million dollars, was a bargain.

He might have wondered whether the builder of the pyramid had been as sorely tried as he had been. "For years I have been obliged to possess my soul with all the patience and philosophy that I could muster," he wrote in closing. "And when I have had to yield to the inevitable I have consoled myself by thinking with Pope, 'Whatever is, is right' etc."* Yet whatever animosity had existed between Washington and Hewitt, the latter recognized, in his speech, that the bridge was much more than a mode of transportation, or a

---

*"All nature is but art," poet Alexander Pope had written in his *Essay on Man*. "And, spite of pride, in erring reason's spite, One truth is clear, Whatever is, is right."

demonstration of the scientific and structural advancement of the nineteenth century. It was, Hewitt said, "a monument to the moral qualities of the human soul."*

Emily had gone down to see the Brooklyn festivities in the early afternoon in a train of twenty-five carriages filled with family and friends. She sent out clear instructions to her circle. "I want the ladies to meet at my house at one o'clock on Thursday and go in a procession down to the bridge—sort of opposition to the Presidential procession on the New York side you know!" The police, ringing the entrances to the bridge as Washington had directed they must, had to work hard to clear the road for them to pass.

After the speeches, Emily had returned home to the grand reception she had planned, to a house festooned with flowers, where a bust of her husband was garlanded with the laurel wreath of victory. He was a quiet, pale presence at the reception that followed, and as soon as the president left he slipped back upstairs. It's quite possible to believe that the stream of visitors to the house—a thousand of them, that day—hardly noticed the absence of the nearly silent, gray-haired man who had brought them all together; who had done, at last, what he had set out to do.

From his window, that night, he would have seen the fireworks that had caused him so much worry. As the *Eagle* reported, "a more brilliant illumination within a restricted area has never been witnessed in the city." The bridge had been cleared, the wooden walkway soaked with water to reduce the risk of fire. The president was safely ensconced at Mayor Low's residence. Fourteen tons of fireworks, manufactured by Detwiller and Street, had been ordered for the occasion. The electric lights on the bridge were switched off; and then, at precisely eight o'clock, the daughter of Mr. Detwiller touched the fuse, and the bridge, the river, and the city were lit as never before with an exploding, deafening rain of silver and gold, of red and blue and green.

---

*After Hewitt spoke, an oration was given by the prominent clergyman Richard Storrs, who was careful to note the contribution of immigrants to the great work now connecting the two cities. "It was not to a native American mind that the scheme of construction carried out in this Bridge is to be ascribed, but to one representing the German people . . . the skill which devised, and much, no doubt, of the labor which wrought them, came from afar."

Serpents of fire, flowers of fire, showers of fire: the display could be seen in Westchester, in the mountains of New Jersey, on the Long Island shore. It went on for an hour, and finished with five hundred rockets shooting straight up from the center of the bridge, exploding seven hundred feet above the river in huge blooms of brilliance. As the smoke hung over the river, as the roar of the rockets abated, all the boats on the river let off their horns and steam whistles, every man and woman on the shore and crowding the buildings by the river shouted and howled and clapped; the deafening noise thundered along wharves, out along the bay, and shook the houses all along the Heights. The explosions would have shaken the Chief Engineer's house, too, of course; like many an old solider, he may well have thought of the last time he had been rattled by the sound of bombs and rockets, by the shrieks of men all around.

And then, at the stroke of midnight, the scent of gunpowder still in the air, the New York and Brooklyn Bridge was opened to the public. This was to be, wrote the *New York Times*, "the people's day." "The select ten thousand who participated in the ceremonious opening of the bridge were as nothing when compared with the vast multitudes that swept across the East River structure yesterday." The very first man to cross was Martin Kees, the *Times* reported, "who achieved immortality at the comparatively low price of one cent." He had to borrow a penny to cross back again, but no matter: "pioneers are notoriously improvident." Victor F. Lutz was the first Brooklyn man to cross: and then "there was the first coupé, the first double carriage, the first baby, the first colored person, the first hearse, the first beggar, the first drunken man, the first bag-piper, the first pair of lovers, the first policeman, and the first dude."

None of this fuss was of any interest to Washington. Emily's scrapbooks stop with the completion of the bridge. She did not keep any of the thousands of pages of newsprint that were filled with news of the opening; though her account of her husband's life was printed in the *Brooklyn Daily Eagle* on the day those first crossings were made. This brief, compelling biography—one of the last documents in the letter book for 1882–83—is not a defense, but it was at pains to make clear what he had been through—and to emphasize that the work he had done was very much more than simply a fulfillment of his father's plans. Something in its tone suggests this was not wholly dictated by Washington to his faithful amanuensis, but rather was composed together in those last weeks before the bridge was opened to the public. The biography begins with an arresting description of the "bed of boiling quicksand" through which the caisson of the New York tower descended, and Washington's "bold step" to call a halt to the digging before bedrock was reached, leaving a layer of sand to distribute the weight of the tower.

"It has often pleased the average penny a liner to remark that there is nothing new in the East River Bridge and that Col. Roebling only copied his father's plans," the account runs. But those plans were "most general in character"; "not a detail" had been considered, and there was "scarcely a feature in the whole work that did not present new and untried problems." A long list follows—including the methods needed to get the materials out of the caissons, lighting the caissons, and filling them by the supply shafts. There was the machinery for raising stones onto the towers—all Washington's design. The anchor plates were much longer than his father had intended; and as for the cables, all of his father's bridges had been made with seven-strand cables, while this bridge required nineteen strands for each cable. "This involved new problems in regulating, which under any circumstances is an extremely difficult task." Each of the great cables required two tiers of anchor chains to connect them to the anchor plates: this too had never been attempted before. The footbridge was a wholly new feature; and a new method of splicing had to be devised for the wires in the cable, so that each join would be as strong as the wire itself. This alone "took two years of experimenting before it was satisfactorily accomplished."

For his part, in his memoir, Washington carefully noted the enormous changes that had had to be made to his father's initial design—often in an exasperated, dismissive fashion. There was "too much side walk and too little roadway," Washington realized, when the foundations "had scarcely been finished"; the bridge had to be widened, which also meant widening the tower spaces at the floor line of the masonry. He drastically reduced the amount of timber called for in the caisson roofs, decrying his father's "partiality" to wooden timber foundations: "The precedent of exalted authority should never be blindly followed!" The design of the cables had to be altered once Haigh's bad wire was discovered. Perhaps most critically, the superstructure had to be effectively redesigned. Even at the time the bridge was proposed, Washington states, the other engineers were troubled by the specifics of John Roebling's design, which called for a truss broken up into "little cut up sections" of thirty feet each. This had troubled him "immensely"—and so "I took the bold step (to the amazement of engineers) of cutting the truss in the middle and riveting it together solidly from there to the Towers."

The portrait of Washington in the *Eagle* put it simply. "All the changes that these improvements involved have made the bridge almost a different structure from the one originally designed." It ends with a description of the chief engineer. "Col. Roebling is about five feet ten inches tall. He is a blond of the German type, with large expressive gray eyes. In the early days of the work a newspaper reporter described him as unpretentious in manner and one might scan him for hours and see no traits except his continuous quiet with a firm individuality and strong self-composure. He is a man of very versatile attainments. He is a good classical scholar, a fine linguist, an excellent musician and as a mineralogist he is nearly as widely known as he is as an engineer."

The Brooklyn Bridge is Washington Roebling's enduring monument, an extraordinary symbol not simply of nineteenth-century ideals of progress, but of one man's tenacity in the face of hardship beyond most people's imagining. Looking back on his life years later, he returned to the time when the Niagara bridge was finished; when, still in his teens, he had been pulled out of school to be his father's aide. He remembered all those years when he had listened to his father's tirades and done his father's bidding. He had been, he thought now, too "bashful in disposition, too much given to reading and lacking in aggressive insolence," in those days. "My father had knocked it all out of me, just what I needed most in life."

Strange to say a man is wrong about himself; but people often wish to be other than they are. No one is exempt from that. Washington's lack of "aggressive insolence" is one of the things that marked him out in an age— perhaps that age never passes—in which that characteristic was very much to the fore in public life. But Washington Roebling never needed it. It was not what defined him. What defined him was his work, and his ability to do the job that lay before him. That would never change.

But now, in the summer of 1883, the post of Chief Engineer could finally be filled by another. In one of the Roebling letter books at RPI there is a note pasted on the inside cover, undated and written in pencil:

Dear Sir

The East River Bridge being completed, and open to travel, the Trustees are no longer in need of my services. I therefore desire to relinquish my connection with the work at the end of this month.

W. A. R.

C. C. Martin took over the post, overseeing the bridge's maintenance until 1902, when, with bitter irony, he was removed from his position by New York's commissioner of bridges, Gustav Lindenthal, who changed his title to consulting engineer and reduced his salary by four thousand dollars. As the *New York Times* reported when Martin died—a year later—"It is said that the action of the Commissioner was very deeply felt by Mr. Martin." Just before his demotion, he had been offered the presidency of Rennselaer Polytechnic Institute, but had turned it down, "saying that his life work was the Brooklyn Bridge."

SHORTLY BEFORE THE opening ceremonies of the Brooklyn Bridge, Washington and Emily Roebling allowed a journalist from the *Brooklyn Union* into their home. Despite Emily's certainty that the great bridge would be the last work her husband would undertake, Washington assured the reporter that "if I get well there is lots of big work in the world to do yet." At first, however, it seemed that all this big work might be done by his son, John. At the age of sixteen, a few months after the bridge had opened, the young man had written a poem to celebrate its construction. "Why the thunder of joyous sounds from shore to shore?/ Two cities united high in the air?" It celebrates his father, he "who lived to see the victory," and some of the imagery, personifying the bridge, is really rather fine. "Over the flags of all nations were suspended/ Steel girders, planks and railroad tracks;/ These she holds with a thousand arms, pressed/ Like a lace garment against her breast." The year after the bridge was completed, Washington and Emily moved away from Brooklyn Heights. They settled in Troy for four years while John, following in the apparently inescapable family tradition, studied engineering at RPI. Their imposing house was on the corner of Washington Park, where the city's old grandeur can still be felt. The park is one of only two privately owned urban ornamental parks in the whole United States.* It is a world away from the battered red brick tenement at 97 Ferry Street where Washington spent his student days; and John's life at RPI was rather different from his father's. Emily and Washington opened their home to John's fellow students—"as well as many prominent in society here," the papers reported—at several fine soirées, with music provided by artists celebrated in their day, such as the composer and performer Dudley Buck and the banjo player Reuben Brooks, whose playing met "with deserved favor." There were refreshments and dancing, aspects of college life John Roebling would surely never have imagined.

John's namesake never really took to engineering, in part because his own health was poor: he had a weak heart, a fact that Emily kept from Washington until just before John was due to graduate. Washington himself acknowledged that his son wasn't really suited to the life of an engineer, but sent him to RPI because old man Roebling's grasp was still firm, even long beyond the grave. Five years after John had graduated, after he had worked briefly at the mill and then left, thanks both to his health and a disinclination for the work, Washington considered the enduring strength of that paternal hold. "[Twenty-seven] years ago I read Carlyle's *Frederick the Great*—last week I read it again with such different feelings," he wrote to John. "Then I had the world before me, now it is mostly behind me— Frederick says that no man who has not plenty of hard knocks in his youth ever amounts to anything—that is in allusion to his father's tyranny, he said it in his old age not in his youth," he notes.

Letters between father and son have a kind of scratchy warmth; their relationship was far better than Washington's own relationship with his tyrannical parent had been. To the very end of his life he would unburden himself to John. Carlyle's account of "Old Fritz" clearly chimed with Washington's

---

*The other is Manhattan's exclusive Gramercy Park.

own experience—and he wanted his son to know it. "Further [Frederick] says 'as a young man I was my father's slave—in middle age, the slave of the state, now in my old age I propose to please myself alone,' and then he found he had no health to enjoy anything. It is a book you must read someday."

WASHINGTON TURNED FIFTY in 1887; the world was rushing past, New York and Brooklyn—and the whole United States—bent on "progress." The Rapid Transit Act of 1875 had authorized handing out transportation franchises to private entrepreneurs: the New York Elevated Railway Company, the Manhattan Railway Company, the Suburban Rapid Transit Company. After 1878 the Third Avenue El ran all the way from South Ferry up to 129th Street, and even ran all night: "owl trains," they were called—already New York was the city that never sleeps. After 1880 the Second Avenue line headed north from Chatham Square, and soon the other boroughs were brought within range. By 1891 trains ran across an iron drawbridge over the Harlem River, up Third Avenue in the Bronx, through Mott Haven, Melrose, and Morrisania to 177th Street. South in Brooklyn, passengers left the bridge trains to climb aboard the Brooklyn Elevated Company trains that had been inaugurated by Mayor Low in May 1885. In November 1885, the *Brooklyn Daily Eagle* reported that the Brooklyn Elevated's station at Fulton Ferry was increasing the number of ferry passengers at the expense of the Bridge; "and yet, in spite of this fact, the bridge last week made its biggest daily average, $2,015, or $14,110 for the week." By 1893, thirty million passengers annually used these new lines, most of them heading from their homes in Brooklyn to their jobs in Manhattan.

Pictures of the era show the elevated lines stretching out to infinity, an indication of rapid growth not only in the city but all over the country. In 1878 there were 2,665 miles of railroad track in the United States; four years later, in 1882—the same year that John D. Rockefeller created his Standard Oil Trust—nearly twelve thousand miles of track had been laid. The Brooklyn Bridge was already an icon of that dynamism.

The Brooklyn Bridge stretches out over the river; it rises up, too, in the air, the first real expression of the dynamism and energy of New York City in the years after the opening of the span. The New York Tribune Building and the Western Union Building in lower Manhattan, both completed in 1875, marked the real beginnings of Manhattan's modern skyline; on that island—crowded ever more each day by the thousands rushing over the new bridge—it was time to go up, and up, and up. "Real estate capitalists suddenly discovered that there was plenty of room in the air, and that by doubling the height of its buildings the same result would be reached as if the island had been stretched to twice its present width," the writers of "Sky Building in New York" announced in 1883. The Standard Oil Trust building at 24–26 Broadway was begun in 1884 and headed up to a full ten stories: it was finished within two years. Joseph Pulitzer, now the publisher of the *New York World*, bought French's Hotel at Park Row and Frankfort Street in 1888; he hired the architect George Post to build him what was then the tallest building in the world. At 309 feet, this expression of "true Americanism," as Pulitzer had it, was the first building to overshadow Trinity Church, at the intersection of Broadway and Wall Street. Around this time the term "skyscraper" entered the city's lexicon. Here were the "monsters of the mere market" that Henry James deplored—but then he had his doubts about the city's "league-long bridges," too.

Washington, however, was on the side of the great new towers. He found himself on the nineteenth floor of the Flatiron Building in 1904, not long after it was completed, and was enamored of the extraordinary sensations to be had in the "sky parlor" there. "You feel like jumping into the North River on one side and the East River on the other," he wrote to John. "The view is very extensive . . . The rooms are very warm and the wind produces no effect!" Not all the glories of New York were to his liking: he observed the last stages of the construction of McKim, Mead, and White's Pennsylvania Station—now widely regarded as a lost architectural masterpiece since its demolition in 1963—and didn't like what he saw. The building, he said, "reminds me of a mausoleum."

And the Roeblings' work had to be adapted to suit new times and new pressures. Washington's health, in the years after the Brooklyn Bridge was finished, improved—enough for him to be able to go up to Niagara and oversee the restoration and strengthening of his father's bridge there. He had kept his son up to speed with the work in the days before John was due to graduate from RPI. "The trusses are all now in position and I have about four hundred feet of the new floor in and that much of the old work cut out. This we do at night closing the bridge against vehicle traffic from eight in the evening till about five in the morning . . . If only I had enough men we could finish up the floor system within ten days," he wrote in 1888—sounding almost like the young man he had been in 1869.

A decade later he addressed the panic that had occurred when, on the evening of Friday, July 29, 1898, there had been what sounded very like an explosion on the Brooklyn Bridge—a great cracking bang that brought traffic completely to a halt as fearful passengers and pedestrians fled the bridge. A dray horse had dropped dead in the oppressive summer head. As Francis Collingwood wrote in the *Railroad Gazette*, the stoppage of traffic resulted in a pileup; the bridge had suddenly to bear three times its ordinary load. The explosive sound was made by the buckling of the truss chords of the superstructure. "This article is not intended as a criticism," Collingwood wrote. "Many of us are wiser now than we were 30 years ago. The bridge was designed to transport safely certain assumed loads, but not any such a load as the one of July 29th."

Washington had his own ideas about those assumed loads—and who was to blame for them. In the final years of the construction of the Brooklyn Bridge, certain alterations he had been called upon to make had never pleased him; now he said so plainly in a reply to Collingwood's piece. The draft of his letter is written on the stationery of the Waldorf Hotel, in Manhattan. "In 1880 some of the Bridge trustees, notably W. C. Kingsley"—that name was pointedly inserted above the line—"insisted that I should adapt the Bridge for the passage of elevated locomotive trains, to take the place ultimately of cable-propelled trains—I yielded with reluctance and against my better judgment." The central trusses of the superstructure had to be raised as a result of these alterations, "which is structurally wrong, leaving the outer truss liable to such minor accidents as the recent buckling. But the ongoing evil on the bridge is, that every year since it has been opened to traffic, there have been numerous additions to the dead load, small in themselves but large in the aggregate." A recalculation of the loads was needed "so as to determine the margin of safety in the main parts of the structure . . . I have no fear of the cables. They still have ample strength and could pull up the anchorages with ease." He would make a similar remark nearly twenty-five years later, when it was reported, to great public alarm, that the cables on the bridge were slipping because of the greatly increased loads—elevated trains, electric trolleys, cars, and trucks—it was now being forced to carry.

The two northernmost cables had been observed to slip slightly toward the center of the bridge in June 1922—each by less than an inch, but calamity was rumored despite the reassurance of those in charge. BROOKLYN BRIDGE MUST BE REBUILT, was the headline in the *Eagle* on July 28 of that year; but when the eighty-five-year-old Colonel Roebling was contacted by the *New York Times*, he "scoffed" at the idea of any such necessity. "There is not the slightest danger," he told the paper's man. "All this excitement is for the purpose of hastening the construction of a new bridge across the river, which is proposed, and which is badly needed. There is absolutely no truth in any report that Brooklyn Bridge is in danger.

"They say the cable slipped—this is just what the designers intended it should do. If that big cable had not slipped, one end of the bridge would have fallen down."

"'Rebuild the bridge?' repeated Colonel Roebling. 'Why rebuild it? It is already carrying six to eight times the amount of traffic of its early days. It is carrying elevated lines and a lot of other things that it did not carry in the beginning. There is no necessity to rebuild. It will last 100 to 200 years. Isn't that long enough?'" To Palmer C. Ricketts, president of RPI and a respected engineer, he explained what he believed was the political aim of the story, laying blame at the door of Grover A. Whalen, commissioner of plants and structures for the city—not an engineer, but a businessman and impresario; he would eventually mastermind the 1939 New York World's Fair. "This coup was planned by Commissioner Whalen, who favors a new bridge at the foot of 23rd Street," Washington wrote to Ricketts, "a location which would not relieve the old bridge of a single passenger. By disparaging the old bridge, he hoped to boost the new one . . . The Bridge is functioning all right. Traffic has not been interrupted one minute. My advice is: Let the Bridge alone."

# "She goes everywhere and sees everything"

THE SPAN OF THE Brooklyn Bridge would not be significantly exceeded for another fifty years—when the George Washington Bridge, designed by the Swiss-born engineer Othmar Ammann, was launched over the Hudson River from northern Manhattan to New Jersey. Opened in 1931, in the midst of the Depression, it took the world record with a 3,500 foot center span and four cables, each with a thirty-six-inch diameter—supplied by John A. Roebling's Sons. Washington had been contacted as early as 1895 about a proposed bridge over the mighty Hudson; it seemed a fantastical proposition at the time. Engineer George S. Morison wrote to solicit Washington's thoughts on great cables that might be built "in the form of small strands to be manufactured before they are shipped, fitted into sockets and adjusted under strain to exact lengths." Washington was polite to Morison but told him plainly that he didn't think much of the idea: "an experience of fifty years has demonstrated the fact that when a wire rope is fastened in a socket it will nearly always break in the socket." To Wilhelm Hildenbrand he said simply that "it is the plan of an idiot."

Washington Roebling was an engineer ahead of his time, perhaps even more than he was aware. Experience had shown that the main cables had to be stiffened against oscillatory motions caused by both moving loads and wind. This could be done with either inclined stays or a stiffened floor: Washington used both in combination—but the actual way in which this method worked wouldn't truly be understood until well into the twentieth century. He would never build another bridge after Brooklyn, and a good many of his letters indicate that he is on the brink of death, or nearly; and yet the remaining years of his long life were extraordinarily active and engaged.

His wife, for one, was clearly maddened by her husband's fears for his health. From her surviving letters—most of them to John—it's clear that her patience was often tested to the limit. She wrote to John on her wedding anniversary, January 18, 1896. She and her husband had come to New York, to the Waldorf, to celebrate—though her mood was not entirely celebratory. "Your father has been married 31 years today. I twice that time." Over and again, Emily noted that she was not always entirely convinced by her husband's complaints. Returning from a brief visit away in July 1898, she found Washington in a terrible state. "I found the Colonel had had another severe attack of imaginary indigestion which is a very bad disease to have— particularly in warm weather! He was in bed with the ever faithful Dr. Clark uncomplainingly in attendance." Things improved, apparently, the following morning, as she reported with irony. "Colonel is up and dressed this morning, has eaten about twice as much breakfast as I [have] and is cheerfully planning what he is going to do tomorrow though he had quite made up his mind that it was not worth his while to think of living through last night." Dr. Clark was often called in. "Your father has taken one of his cantankerous spells again, and dies hourly but still manages to eat and sleep like other people. To save Dr. Clark from going wild, I have sent for Dr. Weir to tell us there is nothing the matter." And yet she was at pains to reassure their only child that she still saw the good in him—at least, sometimes: "If you take him as he is, there are enough good qualities in him to make a grand noble man, with many of nature's finest traits," she wrote—"and enough talents of doubtful value left over to make two or three ordinary bad men."

Much of what ailed him would remain mysterious. Only in the mid-twentieth century were the long-term costs of working in compressed air discovered. Washington's time in the caissons might have predisposed him to permanent injury by dysbaric osteonecrosis, when the bones become calcified, thanks to the slowing of blood flow to delicate bone tissue during decompression. This causes permanent damage to the bone, and persistent pain; the condition remains "the single most injurious, potentially permanent, and hazardous aspect of compressed air work."

In the summer of 1889 John married Margaret Shippen McIlvaine— known as Reta—the daughter of Edward Shippen McIlvaine and Anne Belleville Hunt. Their first son, Siegfried, was born in 1890; he got a little brother, Paul, in 1893, and a third son, Donald, was born in 1908. "There is no fear of the race dying out, at least my part of the family," Washington wrote to John just after Siegfried was born, and then warned once again against the family tradition. "You are doubtless cudgeling your brain for a name—don't call him after me or yourself—it makes so much confusion—I have not heard your mother's opinion of the happy event as yet—but she will be delighted," he wrote. Washington always remained close to his daughter-in-law.

All the while Washington kept keenly abreast of the work at the Roebling mill—for although, as its vice president, he drew no salary, in the years to come he would occupy himself deeply with what his father had begun. "At the mill we have never been so busy and in every department—The rush has all been in the spring and we were not prepared. It takes an awful lot of capital to swing so much copper—twice as much as last year—and everyone is behind in payments. That is a way many people have of raising capital namely not paying their debts."

That was never the Roebling way. John A. Roebling's Sons would expand greatly in the coming years, but never by borrowing money to do so. Washington's younger brother Ferdinand was the secretary-treasurer of the company; Charles—whose industry "was indefatigable," according to his oldest brother—ran all the manufacturing and construction operations. Washington was a man to see opportunity when others did not; the year his first grandson was born was also the year of another great panic, caused by speculation in railroads; in some parts of the country unemployment was pushed as high as 20 percent. Demand at the wire mill slackened steeply: "All kinds of juggling were restarted to keep the place a-going," Washington noted; but a depressed economy was also the time "when everything is cheap and therefore the time for laying the foundation for future expansion."

But he had more on his mind that just business: it was around this time that he thought of setting down his thoughts about his father's life. That winter he had got hold of his father's travel journal for the first time: his judgment of its merits is astute. "Written in Sept 1831 it presents an admirable resume of the aspirations of a young colonist who has canvassed pro and con the merits of almost every county in the United States as a place of settlement," he wrote to John. "This masterly production of emigration literature was written by a young man of 25! A year younger than you are and is worthy of being the product of a mature mind of 57 (my age.)" But why the choice of Saxonburg, which "then was a primeval forest where wild pigeons would not even alight"? Fear of "chills & fever," of course. John Roebling had decided that the atmosphere of western Pennsylvania was particularly healthful; it didn't matter to him that the soil was heavy clay and difficult to farm. That clay, however, hid a greater treasure: "Had your grandfather only known what was under his land at that time—The great Golden eagle well that spouted 3,000 barrels a day!" He told the story to Reta, his daughter-in-law, too: "In 1859 when I was in Pittsburg with my father he was offered an interest in the original Rockefeller syndicate for drilling oil wells, refining oil and buying oil land—At the same time some friends who own Oil shale deposits asked him to go in with them—He reasoned that the oil wells would all give out in a short time—that refining was too costly—So he put all his money in shale and lost every cent he put in, but at the same time escaped being a second Rockefeller. Thank God, what would I have done with a billion of dollars! Truth is stranger than fiction."

Rockefeller didn't, in fact, get into the oil business until the 1860s—but one could argue that in this respect John Roebling was ahead of his time. The shale deposits of western Pennsylvania have only been accessed in the twenty-first century. However, whether vast wealth from oil would have stopped Washington from grousing about money is doubtful. "Please be economical," he wrote to his son. "My income is shrinking . . . It takes a tremendous pile just to keep this big house agoing (it ought never to have been built!)—We are certainly no happier in it—" He would later estimate that the upkeep of the house, without living expenses or servants, ran to about eleven thousand dollars a year. Emily was devoted to the house and its upkeep: "More servants have arrived . . . Your mother, for all the world, reminds me of a Prussian drill sergeant, going around with his ramrod keeping order."

Washington and Emily were living in Trenton, now, in a twenty-seven-room mansion that Hamilton Schuyler called the "finest house in Trenton," with good reason. The glass-plate photographs Emily Roebling had commissioned when 191 West State Street was built show a place of striking opulence, with tiger skin and polar bear rugs on the floors, a stunning Tiffany window depicting the Brooklyn Bridge over the entrance hall's grand staircase, a conservatory for Washington's orchid collection, a special "museum" for his mineral collection. From the back of the house there were views of Trenton's Stacy Park, the Delaware and Raritan Canal, and the Delaware River. It was Emily who oversaw the whole design; her carriage can be seen in one of the exterior views of the house, and in the image of the entrance hall she stands on the stairs in proprietorial manner, light shining through the impressive window with its vitreous bridge above her. She entertained often in the house—no doubt even in its bowling alley, for bowling was a fashionable, and slightly racy, sport for women in those days. (It's hard to imagine Washington lacing up his bowling shoes.) Eventually the Roeblings would virtually colonize West State Street: Ferdinand bought number 222 for his large family, as well as buying a row of townhouses on the street as an investment; Charles would end up at number 333.

Washington liked to stay at home; Emily liked to travel. He accompanied her to the World's Columbian Exposition in Chicago in the autumn of 1893 and was admiring of its design—the 630-acre fairgrounds had been laid out by Frederick Law Olmsted. "The vista is so magnificent that the heart threatens to burst its confines with exultant joy . . . to Mr. Olmsted belongs the greatest honor." The exhibits, however, he loathed: "exponents of silly competitive vanity," he wrote to his son; perhaps he caught a glimpse of the "Mammoth Cheese," brought down from Canada, weighing in at eleven tons. But the trip didn't suit him. His digestion was terrible, "due to the vile food which would kill a shark." And the journey home would only make things worse. "Tomorrow evening I am to be dragged back to Trenton straight through without stopping—You can't believe how I dread the journey—my reserve strength is all gone . . . and yet people say how well you look, I feel like killing them."

No wonder Emily wanted to live her own life. And that is precisely what she did.

THERE IS NO doubt that Emily Warren Roebling was an extraordinary woman; and yet she recedes from us. Her vanished correspondence with Washington during the war leaves a silence that cannot be filled, except by imagination; the scrapbooks of clippings she made during the construction of the Brooklyn Bridge can never supply her private thoughts, whatever we might read into them. She was a woman of her time, too—that "masculine intellect" trapped in the nineteenth century.

She was never, however, an engineer, though some have claimed that for her. She did not claim it for herself; to do so diminishes both what she did achieve, and also the women who would follow her into a world where women had greater opportunity than she did. When she and Washington traveled to Chicago to the World's Columbian Exposition, she wrote to John of encountering a French duchess; they spoke together in French. "She received me most graciously and complimented me on my skill as an engineer! Wonder what she thought of me as a linguist?" Yet well into the twenty-first century the rumor persists that she was really behind the Brooklyn Bridge. Up until very recently, the American Society of Civil Engineers claimed on its website that a speech she gave to that society in 1882 kept her husband in his post; there is no record of such a speech. American National Biography Online, a generally reputable site, claimed that Abram Hewitt once spoke of her being the real brains behind the bridge, but again, there is no record of Hewitt making any such statement. Nevertheless, her work on behalf of her husband, and the Brooklyn Bridge, was invaluable. Not long before the bridge was finished, a report ran in the *Engineering and Mining Journal* of a dinner for Rensselaer alumni, at which one R. W. Raymond raised his glass at the toast "sweethearts and wives." "Gentlemen, I know that the name of a woman should not be lightly spoken in a public place; I am aware that such speech is specially audacious from the mouth of a stranger; but I believe you will acquit me of any lack of decency or of reverence when I utter what lies at this moment half-articulate upon all your lips, the name of Mrs. Washington Roebling!"

But it was hard for a woman—especially one of Emily Warren Roebling's class—to find useful work in the world. The Nineteenth Amendment to the Constitution, extending the right of suffrage to women, would not be ratified until 1920, many years after Emily's death. Finding occupation meant clubs, committees, volunteering for good causes. One such cause was the brief Spanish–American War, sparked by the sinking of the USS *Maine* in Havana Harbor in the first months of 1898; war was declared on April 25. John, still struggling to find a purpose in life, thought he might find one in battle, and wrote dramatically to his mother on April 21 that "war seems so certain that I think I may safely go ahead and make any final arrangements before going to the front." He made his will, leaving everything "as ship-shape as possible." "I will not attempt to make this an elaborate goodbye letter; for, not only do I agree with the Colonel that goodbyes are to be avoided in general principles, but have found further that after the most elaborate farewell one usually has to come back, or write again . . ." He wrote to the Navy Department offering his services, but despite his strenuous assurance of the value he might to be the war effort—"I am 30 years of age, a civil engineer by profession . . . In addition to the general knowledge of machinery etc incident to my profession, I have a fair theoretical grounding in the art of naval war"—he heard nothing. He did eventually serve with the Corps of Engineers, but his poor health meant that he never actually went to the front.

His mother had better luck in finding ways to serve the troops. The Spanish–American War seems a skirmish now, though well over 250,000 men served during the conflict, and its legacy is still felt: as a result Spain lost control over what remained of its empire—not only Cuba, but also Puerto Rico, the Philippines, Guam, and other islands. This war, too, brought future president Theodore Roosevelt to prominence: his volunteer cavalry, the "Rough Riders," became darlings of the press.

But the war was not fought by soldiers alone. To prevent the spread of yellow fever and other tropical diseases during and after the war, a quarantine camp—Camp Wikoff—was built on Long Island's Montauk Point. Thousands of men would pass through, but when they began to arrive, in the summer of 1898, work on the place had barely begun, and conditions were appalling. In August 1898 Emily visited the camp on behalf of the Women's National War Relief Association, and she was not impressed with what she saw. Nursing care for the men was minimal; and the camp's administrators refused to allow female nurses into the place at all. So Emily arranged for male nurses to tend to the men, and paid their salaries—$210—out of her own pocket. She arranged a cook, too, for the camp's kitchen, and sent supplies: 800 bottles of beef extract, 50 pounds of cocoa, 3 cases of jam, 75 pairs of socks, 24 pairs of shoes, 50 pillow cases, and 30 nightshirts. Early in 1899—the Treaty of Paris, ending the war, had been signed on December 10, 1898—the New Jersey legislature passed a special resolution thanking her for her efforts.

In the years to come—the excitement of the war over—her life in Trenton was confining. "I feel I am being buried alive," she had written to her son some years earlier. And so she began to travel—at first in the company of her husband, but later, on her own. Together they had taken a trip to Europe in 1892; four years later he backed out of another trip at the last minute and she set off without him. "I wonder if you are thinking that I went to Europe with your mother," Washington wrote to John in the spring of 1896. "I am sitting quietly in Trenton and she is squirming around England gratifying the ruling passion of all women—vanity." She went to London, Paris, Vienna, Warsaw, Moscow; she was presented at the English court. Washington's ironical tone indicates that he didn't regret his absence. "Your mother's presentation at the English court is already a thing of the past," he wrote to John in May. "Had I been in London I would have been obliged to go too, dressed in knee breeches with a small sword & cocked hat—Just think what I missed." In Russia she arrived in time to attend the coronation of Nicholas II and Alexandra, czar and czarina of Russia—Russia's last imperial rulers. To Washington she sent her impressions of these absolute, and doomed, monarchs—the wife who would fall under the deadly spell of Rasputin, the husband whose rule would end with the Russian Revolution and his family's obliteration. "He was as pale as marble," Emily wrote to her husband of the czar. Her impressions seem very striking now. He "looked sad and earnest with an expression on his face I shall never forget . . . While everyone is raving over the beauty and grace of the Czarina my heart has completely gone over to the Czar. He looks so gentle and loveable with the most beautiful blue eyes I have ever looked into. He looks very frail, is thin . . . and wears a sad thoughtful expression as though the weight of his responsibilities are greater than he can endure. The people seem to love him very much." She would not live to see the revolution that would bring down the Russian regime; she was simply dazzled by the grandeur before her eyes—although she also told Washington that she had seen the royal couple riding through the streets in "a quick and simple light carriage without a single attendant and not a footman."

Emily had a hankering for the old hierarchies of Europe. Before Moscow she had traveled to Vienna on "the famous Orient Express"; once there, "we drove in the Prater and saw the Crown Princess Stephen out riding on horseback. She looked very pretty and rode well—she had with her an officer and a little boy riding by her side on a pony." There was an error in her travel arrangements as she headed east; she didn't make it on the diplomatic train traveling to Moscow on which she was expected, "so we missed the pleasure of landing in the station on a red velvet carpet and walking under a silken canopy." To their son Washington wrote that she "goes everywhere & sees everything and don't stop day or night." And when she returned, late that summer, she needed time to come back down to earth. "Your mother still talks of Kings, Queens, Czars etc. in her sleep," he told John. "In another week she will get down to commoners and ultimately I hope to our low level."

And yet Emily was torn between her love of European finery and grandeur and a genuine wish to participate in something more democratic: the growth of a movement for women's rights. She wrote an anxious note to John from the Hotel Bristol in London, just before she was presented to Queen Victoria. "I am now beginning to wonder many things—How I will look with my nodding plumes and bridal veil! Whether I shall be able to curtsy and get up again all right! Whether I shall fall over my train! Whether I shall be self-possessed enough to take a good look at royalty—at close range! Or whether I shall be as charming with them as I have always been when I have charmed. I was visiting Queen Victoria!. . . I shall stay in London for a week after the Drawing Room so as to give the Royal Family a chance to return my call if they wish to. I do not want them to think me stuck up, even if I do belong to Sorosis."

Sorosis was an organization of professional women founded in New York in 1868 by Jane Cunningham Croly, a journalist. One of the first women in the United States to write a syndicated column, she had been outraged when the New York Press Club decided to bar women from a dinner honoring Charles Dickens on his American tour. In the following years women's clubs began to spread across the country. Emily was elected two years before her grand European trip; it was the beginning of her participation, in the last decade of her life, in the fight for equality. "What a queer medley of women's rights meetings at present!" Walt Whitman had written a few years before. "Women in breeches and men in petticoats—white, black and cream-colored—atheists and free-lovers, vegetarians and Heaven knows what—all mixed together, 'thick and slab,' until the mixture gets a little too strong, we should think, even for metropolitan stomachs." In her way, Emily Roebling was something of a pioneer in the cause. She was, of course, in a position of great privilege compared to many women in the United States, but she felt, nonetheless, at a disadvantage. She was dependent on her husband's fortune—a fortune about which he was always anxious—and knew her son was, too. His income, like hers, came from Washington's wealth, good fortune that still bore its own price. "People like you and I who live on what someone chooses to give us, are always poor! I have tried to impress on you the fact that no one can be entirely independent of those they are entirely dependent upon."

Her husband was content to stay at home, with his growing mineral collection, keeping his eye on the doings of the family firm—but she disapproved of his involvement. Toward the end of 1899 the company was contracted to make the cables for the Williamsburg Bridge, designed by Leffert L. Buck; another graduate of RPI, he was the same age as Washington Roebling, but had graduated from Rensselaer more than a decade later. Brother Charles was keen for the valuable contract—out of a wish to compete with his older brother, Washington thought. At the opportunity, "Charles's ambition was fired; he was determined to build the cables. They were more than twice the size of the Brooklyn Bridge cables; consequently the honor would be twice as great." But Washington kept his hand in, too, corresponding with Hildenbrand about the contract—to Emily's dismay. "I am broken hearted," she wrote to John. "The John A. Roebling's Sons Co has got the contract for the cables of Buck's bridge, and already your father is eagerly at work . . . your father is deeper in the mire than all the others. Already he is earnestly at work devising schemes to surmount the engineering difficulties. I totally disapprove of his whole connection with the contract, and the work, but I have requested him not to talk to me on the subject, and I will keep my opinions to myself—as far as I can and try to forget it. Of course I shall now feel free to do many things I have long wanted to do, as your father's business plans will take him quite out of my world."

After her return from her royal European tour, she organized a women's club in Trenton. "We are going to conduct the whole affair just like men do at their large public dinners, and I even have two very good slightly 'off color' historical anecdotes of Washington and Lincoln which I am going to tell." There were luncheons at the Roebling mansion; she traveled across the

country to meet other women who ran organizations like her own. Very occasionally her husband would go along: he didn't enjoy himself much. "The
dinner at Sorosis nearly finished me," he wrote to their son, "as the window
at my back was open and my dress suit very thin—Of course the female
rhinoceri (of whom a large herd surrounded me) did not mind in the least,
they being protected by a thick coating of vanity." No wonder she preferred
to leave him at home. "I am really discouraged with your father. I have made
up my mind to let him alone for the future, and not bother trying to get him
out of Trenton, or out of bed."

Now in her mid-fifties, she determined to study the law. New York University offered a course open to women—to a certain extent, at least. The
university's chancellor, Ohio-born Henry Mitchell MacCracken, understood that wealthy women, not only wealthy men, were potential donors to
the school. Accordingly, he admitted women as students of law and peda-
gogy to the university's campus at Washington Square. A women's law class
was started under the tutelage of Emily Kempin—a doctor of law who had
gained her degree at a university in her native Switzerland, but who had been
barred from practicing there. But the NYU course wasn't designed to produce working lawyers, either. The class did not "aim to prepare students for
the practice of law," as one of its attendees later recalled, "but to give to
women who are likely to have responsibility for the care of property or who . . .
desire to have fuller knowledge of the laws . . . the opportunity to study the
fundamentals of Modern American law."

The fundamentals were covered in a semester—not the full two years of
the men's degree. All the same, she studied contracts, property, domestic
relations; NYU's professor of law Isaac Franklin Russell wrote that "the
most wifely and motherly of women meet here with the most advanced and
aggressive of those who assert woman's demand for rights is still denied."
Emily Roebling, graduating with honors in the spring of 1899, understood
perfectly well that the legal system as it stood was of little benefit to her sex,
despite the argument of William Blackstone, whose *Commentaries on the
Laws of England*, published in the eighteenth century, was one of the founding texts of English—and American—common law. The idea that a wife's
lack of rights was "for her protection and benefit," as Blackstone had written,
"fades from her mind as she reads these laws, and finds that the favoritism

of women was a pretty compliment which had little foundation on facts," as she wrote in "A Wife's Disabilities," her essay published in the *Albany Law Journal* in 1899.

She argued forcefully for equality within marriage and before the law; her words cannot fail to conjure her own work during the years of the construction of the New York and Brooklyn Bridge. "No one denies that marriage is a contract, and that a woman gives all that she has to give to the man she takes for her husband," she wrote. Under the law a wife was only legally entitled to one third of her husband's estate, and "should he so choose in exercising his legal rights, he is not compelled to leave her one cent of his personal property." The statutes should be changed, she argued, "so that property, real and personal, belongs equally to man and wife, and can only be distributed and divided on the death of the last of the parties in the marriage contract." As she had written to her son: no one can be entirely independent of those they are entirely dependent upon.

She read her essay aloud at the graduation ceremony, which was held in the grand concert hall of the old Madison Square Garden. Her class was the ninth to graduate since the course's inception, and she was one of forty-eight women. "I am 'terrible' proud of myself," she told John, "and strange to say, [the] Colonel is both pleased and amused, and has accepted Professor Russell's invitation to sit on the platform with the Faculty of the Law School when they give us our diplomas." Miss M. Carey Thomas, president of Bryn Mawr College, gave the address: she spoke of the way in which the world was changing at the turn of the century; how attitudes to the rights of women were causing turbulence in every social class. "The higher education of women is sometimes blamed for this discontent. But discontent is the unavoidable accompaniment of every transition period of society," she said.

Washington's amusement, however, did not seem to stretch to actually agreeing with his wife. Emily told John that after she had read her paper he went up to Professor Russell and told him that "'I never heard her essay until tonight and I do not agree with one word she has said.'" Apparently, Russell agreed. "'Our pupils are allowed to advance any theory they like, if their legal reasoning is all right. We leave them the entire responsibility of their essays,'" he told the Colonel. A few months after her graduation, Washington wrote to John himself about Emily's enthusiasms. "Your mother goes to Rochester to speak on law, divorce & nurses—Knows nothing about either."

In the 1860s, in the days before they married, Washington had assured Emily that she would never end up like his own mother had, slaving over a stove in the backcountry while her husband traveled the world—and she didn't. A sophisticated, elegant woman, she played a crucial role in her husband's life in a way that was striking, to say the least, in her day—but they were both people of their time. Her campaigns were not always on behalf of those truly in need. She was, for instance, very interested in an appeal lodged by oil baron John D. Rockefeller against property taxes levied against the great estate he had built in Pocantico Hills, in Westchester County: Rockefeller believed that he should be taxed at the rate applied before the land had been developed, given that all further developments had taken place since he had begun building. The judge, she said with some satisfaction, "wound up by saying the rich have rights—as well as the poor and they should be respected by the law of the land."

She liked grandeur and privilege. Thanks to her descent from John Barrett, of Dutchess County, New York, who had been active during the Revolutionary War, she was enrolled in the National Society Daughters of the American Revolution, serving as vice president general. A biography in the NSDAR's papers gives a clear idea of her character: "It is unusual to find such executive ability so well developed in a woman who has not acquired them in the effort to support herself. She is firm and decided, with opinions on almost every subject, which opinions she expresses with great frankness." Her tact, energy, and "usefulness" were to be admired; furthermore, her "large, beautiful home on the Delaware River, at Trenton, built under her personal supervision, is well ordered and kept almost with a precision of a military post"—a sentiment with which her husband agreed. Her good nature, her skills in organization and socializing—what would today be called networking—made her "popular with the thousands of women who have met her in the many societies with which she is associated."

But her health, also, was beginning to suffer; she completed her degree despite real problems with her eyesight. They necessitated trips into New York with Washington where—as stationery letterhead shows—they usually stayed at the Waldorf Astoria. And there were other sources of difficulty, too; Washington's relationship with his younger brother Ferdinand was poisoned by their relative positions in the Roebling family hierarchy—a hierarchy set firmly in place by their father, his presence felt decades after his

death. At the beginning of 1898 Washington wrote to his son, John, of a trip
into the city to see a Dr. Knapp for Emily's eyes, "which are very bad, the
gout having struck in them . . . I am now the amanuensis and shift for myself
as much as my own numerous troubles will let me . . . I hope 1898 will
be more auspicious than '97—My relations at the Mill are becoming more
strained—owing to natural causes of the Cain & Abel variety."

The dispute with Ferdinand was, finally, over money: "My salary on the
bridge was $10,000 per annum, my expenses $20,000—I would have been
$100,000 better off if I had never seen it," Washington said. Ferdinand had
had an easy ride, as his brother saw it; as vice president of John A. Roebling's
Sons, Washington took no salary, whereas Ferdinand, as secretary and treasurer, did. Matters were exacerbated when in February 1899 the American
Steel and Wire Company made an eight-million-dollar offer for the company.
AS&W had been founded just a year before, by John W. Gates, who had
made his fortune selling barbed wire in Texas; he was known as "Bet-a-
Million Gates" thanks to his love of gambling. His plan was to create a "wire
trust"—along the lines of Rockefeller's oil trust—to buy up not only wire
manufacturers but steel mills and furnaces, and so lessen the cost of production. Here was the beginning of the globalized age. Sure enough, three weeks
after fourteen companies had agreed to join the "trust," wages were reduced
across the board by up to 13 percent; in August 1898 more than two thousand men went on strike.

Washington was far keener—at the time—to sell the mill than either Ferdinand or Charles was. Ferdinand thought the price was too low; Charles, in
charge of all the technical operations at the mill, didn't like the idea: "He
says he has been bossing his own place for so long that he could not stand to
boss it for some one else," Washington would write to his lawyer. Washington appears to have believed that selling the mill was the only way he would
get his just deserts: because Ferdinand had been reinvesting the mill's earn-
ings back into the company, Washington's dividends had gone down—a
5 percent return versus the 30 percent he had been getting in the 1880s. Ferdinand wouldn't countenance any offer less than twelve million dollars; Washington's willingness to accept a lower offer infuriated him. "Both Charles and
I are willing to sell at a fair price but not to give the property away," Ferdinand wrote furiously to Washington in February 1899. "I must say that you

are a bigger ass than I have considered you to be for a long time back and I don't think you ought to have control of your own property; it's not safe."

But was it Washington or Emily who wished to sell the mill to Gates and his wire trust? An undated letter from Emily to John seems to indicate that it was the wife, rather than the husband, who was keenest to take the deal. "In your wildest delirium you never [imagined] anything so complete as my triumph! The American Steel and Wire Company have *entirely* bought out the John A. Roebling's Sons Company—Peace to their ashes! They pay a good price all in solid cash—not a share of preferred or common stock, though of course you can buy it, the same as anyone else does in the stock exchange." She was, she told her son, "smiling my head open"—except when she spoke about it with "the Colonel." "He is broken-hearted. I have had some difficulty in getting him to sign the agreement which he did yesterday. After signing, he went to bed, and [later on he said] he had signed away his birthright. He says he feels he has nothing to live for."

But her delight was premature: the sale fell through in the spring of 1899 when Gates brought his offer back down to nine million, which none of the brothers would accept. Years later, Washington underscored his unwillingness to turn his father's business over to a conglomerate. "Ferdinand insisted on taking it; even Charles favored it. But I am free to say that I violently opposed it, and the deal did not go through," he wrote in a letter to his lawyer, even though he had, reluctantly, signed the agreement—his whole recollection a tidy shading of history. And the papers relating to the sale in the archive at Rutgers University are only there by happy accident. Washington, in the course of going through his records, describes the contents of the folders. "Correspondence in relation to selling the business of the John A. Roebling's Sons Co to the U.S. Steel Co* for $8,000,000 (their offer) which fortunately fell through. Today—1912—book value is over $30,000,000." But there is also a note appended, written on Waldorf Astoria stationery. "This correspondence should be destroyed—It never amounted to anything—We fortunately did not sell our Mill."

---

*In 1912, when Washington was writing, the American Steel and Wire Company had become part of U.S. Steel.

No marriage is an ideal marriage. No couple can really imagine the road ahead when they set out on the journey: what would Emily's life have been like if John Roebling had not died in 1869? What if her husband had not suffered so badly during the construction of the East River Bridge? There was stoicism and courage in her nature, as there was in her husband's; but her resolve must often have been badly tested, and perhaps it's no wonder that she expressed a wish for her husband to take the cash and so to free himself—to free them both—from the shadow of John A. Roebling's Sons Company. It was not to be. They quarreled; most of those quarrels vanished into air or were pulled carefully out of the family's papers. Washington's consolation was the quiet company of his mineral collection.

A SPECIMEN OF apatite: a rich purple color, deeper at the bottom, paler at the top where it is cut in a clean hexagonal; found in Auburn, Maine, it would fit in the palm of a hand. Beryls, aquamarine, green, heliodor—a glowing yellow—in rectangular cuts, shining like sea shifting to sunlight. A scatter of topaz, orange, yellow, blue, bright pink. A yellow diamond, 17.9 carats, found in Arkansas by a mine manager called Lee Wagner. A black diamond of 64 carats, one of the most extraordinary ever found, and an opal—only one of many—of 2,585 carats, uncut, found in Virgin Valley, Nevada, where it was formed from silica-rich water in the voids that remained after buried tree limbs had rotted away. The Roebling Opal is very dark, with flames of green and blue darting through it; it's so large it looks as if it must be made of glass. Another opal, cut as an oval cabochon, shimmers with the red and green remains of fossilized dinosaur bone. Two glowing cylinders of tourmaline, shimmering from red to pink to green, their stacked shapes settled in a quartz base, give rise to the specimen's name: the Steamboat Tourmaline, the whole thing nearly a foot tall. It was discovered in a mine in San Diego County, California, and bought by Washington all the way across the country.

In the photograph of Washington's "museum" in the grand Trenton house on West State Street, his minerals are ranged all around. The larger specimens are perched on the mantelpiece; smaller treasures are in glass cabinets along the wall. The room had a big partner desk, and a mandolin rests on a chair in this photograph, its case—or a violin case—stowed below the desk. Washington was always said to be a fine musician, as his father had been before his hand was so badly injured in the 1850s. The days when he could hunt for minerals himself were long gone; now he worked by correspondence, sending out all over the country for the specimens he wanted. He was always keen for "mineral gossip . . . every fellow wants what the other has," he wrote to another collector at the end of his life. His passion for the geological secrets of the earth, which had possessed him since he was a young man and which had served him so well in the dramatic sinking of the caissons of the Brooklyn Bridge, never dimmed. In his darkest hours, it was this that sustained him: "I am an invalid confined to the house and minerals are the only things that do not tire or excite me," was the note enclosed with a check for some new specimens in the early 1870s—once again in Emily's hand, because his sight was so bad in those days.

Minerals could give him pleasure then, even if he could not see them clearly. Weight, shape, density, texture are even more available to the hand than they are to the eye. Washington never considered himself an expert: but *amateur* has its roots in love, and his connection to the Earth that supported the structures he built was sustenance and inspiration. As with everything else he did in his life, his collecting had a purpose: his aim, he decided, was to gather into his collection a representative of every known mineral. He wasn't only after diamonds and opals and tourmalines: he wanted beaverite, graftonite, dehrnite; paradocrasite, koechlinite, rossite. A mineral might look unglamorous or insignificant, but he wanted them all, because he knew that, collectively, they were what made the world. He wouldn't achieve his aim, but it was not for want of trying.

In 1917 he was elected an honorary member of the New York Mineralogical Society; its President was George Frederick Kunz, a self-taught mineralogist whose father, like Washington's, had arrived in the United States from Germany. Born in 1856, by 1889 Kunz had joined Tiffany and Company, and he became the man responsible for sourcing its extraordinary gems. In that year the collection of stones he had prepared for display at the Exposition Universelle in Paris won a gold medal; tycoon J. P. Morgan stumped up fifteen thousand dollars to buy them all for the American Museum of Natural History, where they are still known as the Tiffany Morgan Collection. Washington wrote with humble gratitude to the younger man when he received news of his appointment to the society. "Although I have been collecting for 40 years I feel that I am still a neophyte in the science and that there is more than ever for us to learn in the future," he told Kunz. "To acquire at least one spec[imen] of every known mineral proves to be an impossible task for an individual," he added, but allowed that he had, with his ample resources, done his best. "When I began there were only a few hundred of the principal minerals—now there are thousands, with annual accretions of new varieties and species of fifty to a hundred more—My specialty has been to collect rare minerals—and in that way I have been able to be of a little assistance to the many masters of the Science in this Country." Some of the labels on his collection reveal how closely the minerals were linked to his life. A tiny nugget of gold, sent to him in December 1924, came from a mine on the Rappahannock, "in which I slept the night before the battle of the Wilderness."

He was generous with his collection in his lifetime, lending out specimens to scientists and mineralogists who wrote requesting his help. He knew that the work he had done was significant—perhaps as significant as anything he had ever built—and his son, John, understood this too. "Only museums live forever," Washington had written to George Kunz—and the year after Washington died, in 1927, John donated his mineral collection to the Smithsonian Institution in Washington, D.C., where it remains to this day. The collection, comprising sixteen thousand specimens in all, was given along with an endowment of $150,000 for its preservation and further development. The gift placed the national collections "on a par with those of the great museums of the world, and students and workers in mineralogy the country over are to be congratulated that, instead of being sold and distributed, they are to be retained where they will be cared for and accessible to all properly accredited workers as well as the visitor interested only in display," the *Scientific Monthly* reported.*

The minerals are dispersed among the museum's great holdings, some—like that spectacular tourmaline—on display, some deep in the bowels of the place, foundation stones of the country's store of knowledge, as their owner would have wished. But the whole collection, and Washington's fascination

---

*John's scientific inclinations—and his involvement with the Smithsonian—were wide-ranging. In 1920 he had paid for the construction of a solar observatory on Harquahala Mountain in Arizona; in 1925 he made another donation to erect an improved observatory on Table Mountain in California.

with geology, the science that leads us back to our planet's very beginnings and dwarfs any human achievement, is a further expression of his ironic stoicism. The acerbic writer who made a bone-dry comparison between the building of the Brooklyn Bridge and the construction of the Great Pyramid at Giza knew well that the thousands of years that had elapsed between the building of those two great structures was but the blink of an eye, geologically speaking.

# "The image of his wife floats before him"

NEITHER WASHINGTON NOR EMILY was in good health early in the new century, and toward the end of 1902 Washington would turn again to the water cure, journeying to Sharon Springs, New York, 250 miles north of Trenton and some fifty miles west of Troy. A thriving resort town, in those days the place was renowned for its mineral springs. This "resort of odoriferous waters" was as fashionable as Newport in its day, famous for the cures effected there by the miraculous powers of what flowed from the ground; its white sulfur springs were the only American mineral water awarded a diploma and medal at the Universal Exposition in Paris in 1900. Emily's ailments—digestive ailments, weakness, the failure of her vision—seemed to be as mysterious as Washington's. They crept on her, it seems, gradually debilitating her and making it impossible for her to travel or organize committees as once she had.

In a reversal of their usual habits—where she traveled the country and the world, and he stayed put—she remained in Trenton to find health. She was very frail, and now it was she who was nearly blind. And yet in scrawled handwriting she wrote to her husband in November 1902:

Dear Wash—

Each time I attempt to write I think it is for the last time my hand is so feeble. I have your letter of yesterday. Johnson goes to you tomorrow and will take you the slippers and tea. John is a perfect sunbeam in the house the sight of him does me more good than any of several nerve tonics. I am so glad you are sitting up & can't tell you how glad I shall be to have you home. I have so feared I should not see you again.

It now takes two people to hold me up . . . I live mostly on milk but I am further nourished by having my legs and arms rubbed with olive oil. They all say I am much better but I cannot see where. I certainly grow weaker each day . . .

Yours aff, Emily

He returned to her, but she was no better. Washington wrote to their son that "your mother reports herself just the same—no improvement—nurse says she is better—whom shall I believe."

Despite her failing health, however, she had been able to work on a book—no revelatory account of her life, but a transcription of the journal of the Reverend Silas Constant, an eighteenth-century ancestor who had served as a minister in the "prosperous pastorate" of Yorktown, New York. The diary—an often bald record of speeches given, marriages and funerals performed, farming tasks accomplished—has been annotated by Emily with notes on the local families Constant encountered. It makes for dry reading; but it demonstrates the careful industry that served her husband so well in her years as his amanuensis. Unlike John Roebling's account of his journey from the old world to the new, it is hard to argue that this is a volume ripe for rediscovery. Yet among the brief accounts of preaching and studying and farming are notes that chime against the image of a woman in failing health, but still at work on the task she has set herself. In the autumn of 1784 the Reverend's wife was very ill indeed. "Mrs. Constant more sick," is the simple entry of October 15. "Studying, &c," he notes the next day. But she is again "more sick" on the eighteenth; the following day "the Doctor came to see her; staid all night"; and then the next day "she remained very bad." Constant's descriptions sound not unlike the comings and goings in the Roeblings' own house.

The *Journal of the Reverend Silas Constant* was published in 1903—the year Emily died. She grew weaker and weaker in the first weeks of the year; and then, on February 28, with her son, her husband, and her sister at her bedside, she died. Most likely she succumbed to cancer, but the *Brooklyn*

*Daily Eagle* noted that the specialists called in by Clark, the family physician, "diagnosed the case as progressive muscular weakness." She had been struck by a bicycle—then a newfangled contraption—in Atlanta a few years earlier; her husband later believed that had contributed to her debility. But Clark's verdict was that her demise had been hastened by her years of hard work— a compliment to his patient. "Dr. Clark said that the complaint was not uncommon among persons who have displayed unusual mental or physical activity. He attributed Mrs. Roebling's breakdown to exhausted vitality and overwork." And while she was remembered for her role at her husband's side, the *Eagle*'s story celebrated her in her own right, as one of the most notable "club women" in the country. "In social life Mrs. Roebling occupied an important place. She was a charming hostess and her beautiful home a little beyond the City of Trenton was the scene of many a delightful gathering."

The paper, however, in noting that her son was by her side, named him as "Washington Roebling." The confusion regarding her widower's identity stretched even into the pages of the *Eagle*, which had once so faithfully reported his work at the water's edge.

However difficult their partnership had sometimes been, Washington, now alone in the ornate mansion that was Emily's monument, felt lost. "When real trouble comes I think you would miss your mother," he wrote to their son a little less than a year after her death. "I miss her all the time." And in a folder marked "Personal miscellany" in the archive at Rutgers University is an undated scrap of paper on which Washington has written out the words Mark Twain had carved upon the headstone of his beloved daughter Susy, who had died in 1896 at the age of twenty-four.

*Warm summer sun shine kindly here*
*Warm summer wind blow softly here*
*Green sod above, lie light, lie light*
*Good night, dear heart, good night, good night*

Her loss haunted him. Whatever strains there had been between them in all the years of their marriage faded in the shadow of her absence. "This is the anniversary of your dear mother's death," he wrote to John on the last day of February 1904. "The year that has passed has been to me one of anxiety, unhappiness, personal misery and gloomy forebodings—The year to come I hope will find me I hope more content and serene." A few months later, on his birthday, he looked at his life in a manner that eerily recalls the otherworldly beliefs of his long-dead father, making notes in a scrapbook, his words addressed to himself and no other. "Washington A. Roebling is 67 years old today. He feels more contented than he did a year ago—the contentment of old age is stealing over him and he is becoming resigned to the inevitable—The image of his wife floats before him as a fading image of the past—still acting as a monitor and guide—a spiritual vision which may again become a reality to me in the distant future—W. A. R."

IN DEATH EMILY would not be overshadowed by the Roeblings; she was buried in Cold Spring, the ancestral home of the Warrens, in the lush green cemetery that overlooks the town. Washington told John that her wishes on that point were clear: "You know she had a good deal of feminine spite in her make up and wanted to be buried up there because she hated Trenton, which was a sentiment she had worked up of later years—" She left John nearly five hundred thousand dollars in her will, much of it in stocks; the amount was then to be divided between his children after his death. There was some wrangling between Washington and John regarding the latter's sense of responsibility; Washington instructed his son—now in his mid-thirties—that it was time he learned to manage his affairs "in a common-sense and business-like manner." John wished to ensure that when Washington died, he would at least inherit his father's books, sword, and the family papers, which were "the things I care most about."

Money, as ever, plagued Washington's thoughts, even in that sad year after Emily's death. He endowed a bed at New York's Roosevelt Hospital in Emily's name, though not entirely willingly, to hear him tell it. Dr. Robert F. Weir, one of the leading figures at the hospital (and a pioneer of antiseptic surgery in the United States), "got me in a corner," he told John, and before he knew it he had handed over five thousand dollars. Then in the summer of 1904 two fires almost completely destroyed the Rensselaer campus; Washington was dragooned in to help raise money to rebuild his alma mater. He agreed, despite his bad memories of the place; John was ordered to contribute—with money provided by his father. "I am appointed chairman of the begging committee to raise $500,000 to rebuild the RPI—(They have selected the poorest man in the US for the job!) You must give $1,000 which I will send you when I feel like it—Please suggest victims to me." He signed off: "(Poor beggar) WAR." And not a month had passed before another needy specter loomed out from the past. "Yesterday an old-fashioned Mühlhausen beggar struck me—they are the worst kind," he grumbled to his son. "The tears roll down in torrents—they get on their knees and clasp your knees in their arms and don't let go until they get what they want . . . This man was specially manufactured by God Almighty—he was too good to live—he had never swore, drank, smoked or done bad in his life—He had certificates— first, that he was born, 2d that he was baptized . . . 6 he had a mother, 7 he could fry eggs . . . Each certificate was 2 feet square and stained with German tears and I had to read it."

The years passed in the big house; he kept an image of Emily by his bedside, always. In February 1908 he lost a good friend when Wilhelm Hildenbrand died at sixty-three; "careful and conscientious in his work," the New York Times wrote, an engineer who, like his long-ago master Washington Roebling, "preferred to err on the safe side." Washington told his son, John, the news in a letter that, however, brought brighter—if surprising— news, too.

"Would you fall off your chair in astonishment if I were to marry a 'merry widow' of 40?" he asked his son in a letter dated March 7, 1908. Washington and Cornelia Witsell Farrow had met at the home of a mutual friend earlier in the year, and she had been a frequent visitor to Trenton since. She was a southerner, from Charleston, South Carolina; her first husband had died in 1896, and at the time she met Washington she had a sixteen-year-old son, also called John, who was about to head off to Princeton. She was, the Trenton Evening Times reported, "a woman of charming personality and many accomplishments." Concerned that John might think he was being taken advantage of, Washington assured him that an "ante-nuptial agreement" could be made protecting the Roebling assets; but John (in words that recall old John Roebling's words to Washington when the latter hesitantly announced his engagement to Emily) wrote back immediately with reassurance. "I am very glad to hear that you contemplate remarrying," he wrote by return. "I have thought for several years that you ought to marry, both for your own sake and that of posterity—but I had a natural delicacy about suggesting it. Now that you are about to do it, I am delighted. You omitted to tell me the lady's name. Is it anyone I know? Tell me all about her. I feel quite sure I will like her." He was not concerned for his fortune. "You speak of prospective 'ante-nuptial agreements'! Why have any? A woman worthy to be your wife is entitled to all the rights of her high position: anything less would be unjust."

Washington seemed surprised at his own vitality. A couple of weeks later he wrote to John again, describing his bride-to-be and considering his situation just past his seventieth year.

"It sounds queer to talk about my wedding; the wedding of an old man who ought to be thinking about his grave rather than the vanities of life.

"But these relationships are those of the heart, not governed by reason or judgment (fortunately so perhaps)—A second marriage late in life cannot be judged by the standard of the first because its motives are usually quite different . . . Mrs. Farrow's grandfather was a Connecticut Yankee who came South after the war and entered business. She is thin, slender, brown-haired, of my height, with much personality and extremely amiable, speaking with a strong Southern accent. You will like her like a sister presently—

"As regards our mutual relations you know that I am just and no wrong will come to you or yours—How these things come about is always a mystery, and I feel somewhat guilty in inflicting myself upon Cornelia.

"At any rate I invite you most cordially to come up and attend the simple ceremony—I am not strong and feel like breaking down without some support—I have no one to help me and must do everything myself." He offered John in closing some photographs of his Emily; "her mobile face would never photograph well," he said.

Colonel Roebling and Mrs. Farrow were married, on April 21, 1908, in Boston, near the home of her "friends and protectors," as Washington called them, Mr. and Mrs. Frederick G. Crane. ("The Cranes make all the bank notes of the U.S.," Washington told John—as they do to this day: Crane Currency supplies paper money for banks all over the world.) The church was decorated with Easter lilies and American Beauty roses; a string orchestra played the wedding music, and "bride and groom were unattended," the papers reported. Her gown was of blue satin, trimmed with old Irish lace; "her only jewel was a diamond pendant, a gift of the bridegroom." At a celebratory lunch a few days before the ceremony, the happy couple's table had been decorated with "a remarkably realistic miniature representation of the

Brooklyn Bridge" as a centerpiece. They headed down south for a honeymoon, and then returned to Trenton, where she settled in happily—John
Roebling, as he predicted, seems to have delighted in his stepmother. They
wrote to each other frequently; she called him "My Dear John" and signed
herself off "Corrie." Letter after letter thanked him for sending flowers,
roses, mainly, American Beauties, the same as she had at her wedding. She
was a keen gardener. He also sent her books—including a copy of *The History of Butler County*, about Saxonburg and its environs, which she said she
was very glad to have.

Washington had someone to stand by his side, then, when a statue of
his father was unveiled, with great ceremony, in Trenton's Cadwalader Park
on June 30, 1908. The bronze figure—as imposing now as ever it was—com-
mands the heights of the park, standing "at the apex of a lovely green
mound . . . a place where the sun will always smile from the time of its rising
until it sets." It had been commissioned, on behalf of the Roebling Memorial
Association, from the noted sculptor William Couper—who had also im-
mortalized Abram Hewitt in marble for the city of New York; Hewitt had
died a few weeks before Emily Roebling. The day of the statue's dedication
in full bright summer was, by all accounts, a day like no other in Trenton,
New Jersey—unless you looked back to the patriarch's funeral. The papers
reckoned that nearly fifteen thousand people—a full one sixth of the city's
population—turned out to commemorate Trenton's first citizen. Trenton's
businesses were shut so that people could join the parade, which began at
eleven A.M. from the Roebling plant on South Broad Street and made its way,
a winding two miles, to the park, passing by Washington's mansion on West
State Street. The parade was comprised of "5,000 persons who gain their
livelihood in the works which were founded by John A. Roebling and every
man and boy of the great marching column was filled with enthusiasm."
Nine marching bands played "lively martial music"; when they passed Washington's house—and the grand houses of Ferdinand and Charles—they
played with "renewed vim" in honor of the Roebling progeny. At two P.M. the
statue—which up until that point had been hidden from view—was unveiled
by Charles's daughter Emily, "who pulled a silken cord which released the
United States flags which covered the magnificent bronze workmanship"; as
she did so the band played the bridal chorus from *Lohengrin*.

An oration was given by Henry D. Estabrook, general counsel of the Western Union Telegraph company; he was also the father of Ferdinand's daughter-in-law, Blanche. His speech—"spellbinding," the Trenton paper said—gave a full account of John A. Roebling's life, and was, for the most part, what would have been expected on such an occasion, a paean to this "man of iron," as Estabrook called him, a man who had stood "head bared to the blows of Fortune or the storms of Heaven; eyes fixed unwaveringly on whatever object he had in hand; poised, confident, unyielding, imperious and proud." His accomplishments might be called too numerous to mention— but Estabrook, before the assembled citizens of Trenton, mentioned them all.

But he mentioned something else, too. The plan for the statue had turned Washington's mind to his father's biography again; around this time he had picked up his pen once more, and he had shared what he had written with Henry Estabrook as the latter prepared his address. Estabrook described the rule of law in John A. Roebling's household plainly as "Draconian." "John Roebling sometimes punished in anger, which is not punishment but truculence . . . It may be parental neglect or maudlin selfishness to spare the rod and spoil the child, but on the whole I had rather spare the child and spoil the rod . . . I am sorry this faulty thread should be traced in the seamless shroud I would fain weave for so great a man. It is a defect exaggerated, perhaps, by that very greatness, like a pinch too much of carbon in a mass of metal." No wonder Estabrook also said that the son's account of the father's life "might never be published."

Estabrook's remarks on John Roebling's "truculence" passed unnoticed in the celebrations of the day; they were a few sentences within a very long speech. And the monument itself ("a true and faithful likeness," the audience was told) looks rather more sympathetic than Estabrook's words might suggest. The reason for that was not simply a wish on the sculptor's part to idealize his subject. The statue in Cadwalader Park is as much a portrait of Washington as it is of his father.

Washington had been in correspondence with William Couper in 1907— initially regarding the text of the memorial plaques on the plinth. His tone is blunt in wishing to ensure that the father did not get credit for what the son had, in fact, accomplished. "You will please put 1852 in front of 'Niagara Bridge' and 1855 behind it. In regard to the East River Bridge you will put 1869 in front of the name and nothing behind it. This is a monument to John A. Roebling who neither built or finished it," Washington wrote. But then his involvement with Couper went one step farther. John Roebling was hardly photographed in his lifetime; in the one good image that survives he stares at the camera with characteristic fierce intensity: this could hardly have served as a model for his memorial. Only one solution arose, as Washington told a correspondent a few years later. In 1916 he wrote to General James R. Rusling about the statue: "The features of the statue resemble me to a certain extent for the reason that I sat for it several times—The photographs were not satisfactory to Mr. Couper and he had never seen the original."

Once again, it seems, Washington had no choice. There was to be no escape from the blurring of his identity with his father's. But more than once he would stress that the lending of his features did not mean the erasing of his work. "Forty years had elapsed since Mr. Roebling's death," he wrote to Henry Jacoby, professor of engineering at Cornell University—"I was one of the few persons living who remembered his features and appearance sufficiently to enable the sculptor Mr. Couper to reproduce a life-like image." As to the bridge over the East River: "Mr. Roebling died four months before any actual work was done on the structure—For the building of it, which took fourteen years, I claim the full credit—other structures have been built since, costing two and three times as much, but falling far short of the amount of traffic they carry or in convenience to the public, for the simple reason that the first Brooklyn Bridge was built in the right place."

Life with Cornelia put a spring in his step. He went with her to New York for the first time in two years. They saw *The Spendthrift* at the Hudson Theater (about "a spendthrift wife and a hardworking husband, driven to bankruptcy through her extravagance"—all the rage in the city that spring) and generally took in the sights. "Went to Brooklyn over the Br—looked at the Manhattan Br," he told John. There were tailors to visit, the hatter, the shoemaker: "Cornelia is exhausted trying to keep up with me." Despite his own resistance to the automobile, he bought his wife a car—a Peerless, one of the best of the early American makes. He was influenced, perhaps, by his nephew and namesake, Washington A. Roebling II, Charles's son, who was becoming quite a figure in the new world of motoring; after the inevitable spell at John A. Roebling's Sons Company, he bought up the Walter Automobile Company and hired the Frenchman Etienne Planche to design him a race car. "It goes like hell—spits flames in all directions," Washington told John.

The Walter Automobile Company eventually filed for bankruptcy, however; and young Washington's life was cut short. In early 1912 "Washy"—as he was known to all the family—traveled with his friend Stephen Blackwell, son of a former U.S. senator, to Europe, to drive across the Continent: a fine diversion for a couple of wealthy young men. But the ship they boarded to take them home was the *Titanic;* they were among the fifteen hundred people who perished in the freezing sea when she sank. In the weeks to come there were those who sent flowers and condolences to Cornelia Roebling, thinking it was her husband who had died in the disaster.

The sinking of the White Star liner—hastened, it's thought now, by steel hull plates that became brittle in the cold—presaged a darker time. When the First World War broke out, Washington had no sympathy for the country from which his family came. Germany invaded Belgium at the beginning of August 1914; German troops moved across the country, slaughtering civilians—men, women, and children—as they went. "I am sorry to see Germany overwhelm the poor Belgians who never did them any harm—It has come to this pass, that for an extra German to live, he must kill somebody else to make room for him—We can all play at that game—It means perpetual universal war," Washington wrote gloomily to John that summer. By the spring of the next year he had acquainted himself with the report prepared by Viscount Bryce, former British ambassador to the United States, on the "alleged German outrages," which had been published in London that year. "You must read Mr. Bryce's report on German brutality," he urged his son. "At no time in the world's history has it been exceeded—I feel like publishing a letter of my father written in 1830 when he immigrated to this country, stating that he came in order to escape the intolerable slavery existing throughout the land, and crushing all possible progress in a man's life." Little had changed, he reckoned, in the country of John Roebling's birth. "From the cradle to the grave you are the victim of a murderous aristocracy."

Yet the war was good for business. Washington kept John up to date with the success of John A. Roebling's Sons. There was a forty-thousand-dollar order from Russia for drilling lines to be sent to Baku, plus cables for a suspension bridge in the Urals. The Russians wanted rope wire, telegraph wire, barbed wire; and now orders for wire-wheel automobile tires were "immense." "They come in 100 million feet at once—This year's produce will be 550 million feet—we are buying machines outside to make it and are now short of wire!" There were thousands of feet of rope to be sent to the newly opened Panama Canal; and while they had lost orders from both England and Germany, Brazil and Argentina stepped in to fill the gap. The U.S. Navy ordered torpedo netting and submarine netting; the company was swamped with orders. "Every harbor is being mined—the mines are anchored with durable rope spun over with tarred hemp marline—We are making 800,000 feet and more coming—Also a million feet of phosphor bronze rope for salt water work, endless quantities of antennae wire and aviation wire." A little more than a dozen years earlier there had been no such thing as "aviation wire"—or indeed, wire for automobile tires. Just about every new technology, every new invention required the products of the Roebling mill.

The United States entered the war on April 6, 1917—and in the closing months of that year the mill began to work even harder on a government order for millions and millions of feet of rope—an order that Washington was not supposed to know about at all. The mill, he told John, "is overwhelmed with an order for untold millions of feet of rope, 50 to 100, for netting and mining purposes—destination a secret—it may be for closing the North End of the North Sea—If you hear, don't tell me, I am not supposed to know." Washington was right about the use to which the rope was being put. In the final year of the war a stretch of sea 230 miles long between Orkney and Norway would be closed off by a wire screen rigged with seventy thousand mines—"the barrage that stopped the U-Boat," as publication released by John A. Roebling's Sons in 1920 called it. The idea for the barrage echoed the one General Meigs had proposed John Roebling string across the Potomac during the Civil War; now a far grander scheme was realized. Nearly eighty million feet of rope was produced by 13 manufacturers, and by far the largest share, more than 27,000,000 feet, came from the mill at Trenton. According to the pamphlet, "the fate of the world hung by a wire rope." This is arguable: the effectiveness of the barrage has been disputed, considering the immense effort (and danger incurred) required to lay seventy thousand mines, each loaded with three hundred pounds of TNT.

Washington knew of the work going on in secret—but he didn't know everything. At the front of his copy of the pamphlet there is an emphatic note in his neat hand, written in the last year of his life: "I never saw this book until 1926! W. A. R."

But all those orders during the war didn't mean that it was smooth sailing at the mill. There were fires, and strikes—and Washington's ongoing tussles with Ferdinand, and to a lesser to degree, Charles. But soon enough neither of his younger brothers would be around for Washington to chafe against. Who, then, could take over the running of the vast enterprise that John A. Roebling's Sons Company had become? "My brothers are getting old," Washington wrote to John in 1917, as if he were the younger man. Ferdy—as Washington had once called him fondly—was seventy-five; Charles nearly

Participating in impressive flag-raising ceremonies at Roebling, N. J. in June, 1917, the group above includes, from left to right, the late Charles G. Roebling and Colonel Washington A. Roebling, both former Presidents of the Roebling Company; the late Edmund Riedel, William Gummere and Walter Murphy. Mr. Gummere is now Assistant to the Manager of Industrial Relations and Mr. Murphy has retired from the Company's employ.

A flag-raising ceremony at the Roebling Kinkora works—the 1918 ceremony was meant to inspire patriotism in the workers. Charles Roebling is on the left, Washington in the middle seventy. The last years of the war brought change to the whole country—and, not so long after, change for Washington, too.

THE CLOSENESS THAT Washington and Ferdinand had shared as young men was long gone. Ferdinand had been a confidant during the Civil War, his ally in his unsuccessful attempt to make a business from Langen's gas engine, his supporter in the worst days of the East River Bridge—and Washington had been, in those days, supportive of Ferdinand's efforts to build the business. Washington was the working engineer, Ferdy the businessman, his spectacles perched on his nose. "I think with you," Washington had written to Ferdy in 1878 when his brother had proposed expanding the capacity of the company. "It is only by constant improvement and extension that we are able to keep up the rate of dividend . . . One thing leads to another. When we make 6,000 tons of wire a year we will want another rolling mill. I even expect to see the day when the whole block between Hudson and Clinton will be full of wire cloth looms."

Another matter drew them together then: the question of what was to become of their youngest brother, Edmund. Their mother's death had affected Edmund deeply; the death of the "brutal father" who had beaten him so badly when the boy had run away from boarding school hadn't improved matters. Washington's guardianship during that time had not been a success. "I had no time to look after him, being overwhelmed by the Brooklyn Bridge," he would write in later years. He had corresponded with Ferdinand at the time, trying to figure out what would be best for their brother, especially after he had vanished from Washington and Emily's side on the trip they all had taken to Europe in 1873, when Edmund was nineteen; he had disappeared for six weeks, and no one had had any idea where he had gone. It would be a repeated pattern in Edmund's life. Washington tried to work out with Ferdy what to do: "He has a strong aversion to going to school again when he returns to America and I wish you would write to me whether you think there will be any thing to do for him in the mill next fall in the way of hard work—He can loaf splendidly and is already acquiring a European polish which has to be taken out of him." Edmund, however, didn't want to be in Trenton; New York was more appealing, and he thought he might clerk in the Brooklyn office of the bridge. But such a position "would only result in his becoming a New York street loafer. He can't be trusted alone in a large city."

By the end of 1874, Eddy was approaching his majority; and Washington would no longer be his guardian. He shared his anxiety with Ferdinand, asking how best to keep the securities Eddy had inherited safe. "He is very careless himself," Washington said.

There was no such worry over Charles, brother number three, who joined the mill in 1871, at the age of twenty-one, just as soon as he had graduated from RPI. "He was his father over again, to a far greater degree than any of the other children," Washington wrote, "[with] the concentrated energy which drives one to work and be doing something all the time." He took charge of production and modernization of the company, while Ferdinand held the financial reins; Charles oversaw the extraordinary growth and development of the company in the years to come, and would expand its capacity to produce wire for uses John A. Roebling had not dreamed of. Like his father, he wanted to figure everything out for himself. "This tendency of self-reliance was one of Charles's chief strengths. It strengthened his capacity for successful work and enabled him, in later years, when he was in his prime, to undertake great projects where he had no precedents and was forced to rely on himself alone." This extreme self-reliance, however, had its disadvantages, according to his elder brother: "Sometimes it is much cheaper to buy what you want than to waste thousands in experimenting." But Charles also admired the inventions of others, and knew a good thing when he saw one. "When Mr. Bell first developed the telephone he experimented in Princeton and came to Trenton to get copper wire, where Charles became acquainted with him, and I have understood he put in a little money on the ground floor, which has been returned to him a thousand fold," Washington wrote to John.

The battle over the proposed sale of the company in 1898 ended any semblance of cordiality among the brothers and allowed Washington to vent the resentment he had harbored over the years. The terms of his father's will, which left the company, rather than cash, to the brothers, worked very much to the advantage of the younger siblings, Washington wrote to his lawyer, R. V. Lindabury, in a long, discursive letter penned during the wrangling over the fate of John A. Roebling's Sons. At the age of thirty-two—in 1869, the year of his father's death—Washington had managed to scrape together four thousand dollars in savings, he noted; when Charles reached the age of thirty-two he was worth four hundred thousand dollars. During the war, "Ferdinand stayed at home, bought a substitute, and made himself solid in the mill office during that time"; when Washington came home, a newly married man in 1865, he was "without a cent and broken in health."

Even long after his death, John Roebling continued to exert control over his sons. They had inherited great wealth, certainly: but the manner in which that wealth was distributed—or, crucially, not distributed—drove a wedge between them, one that became apparent even before the East River Bridge was finished. From the beginning both Ferdinand and Charles were in favor of expansion—rightly so, for as the twentieth century loomed, oil companies, steel companies, were growing ever larger, bigger firms swallowing up smaller ones, and a family company like John A. Roebling's Sons had to work hard to compete. But as the years passed, Washington became more and more anxious about spending money, and by March 1880—not long after the four cables of the East River bridge had been wrapped—his opinions on expansion had shifted dramatically. Ferdinand wanted to spend one hundred thousand dollars improving the mill; Washington was exasperated. "What all these extensive improvements are to end in except ruin and bankruptcy for all interested I cannot see. First of all the wire mill had to be built to use up the rods, now the rod mill must be enlarged to correspond with the wire mill. That once done another wire mill will have to be built and so on indefinitely." In his "humble opinion," Washington said, it would be better to buy in steel rods when they were needed. "Of course you and Charles will in this as in everything arrange between yourselves what you will do—but I send you my opinions for what they are worth."

They were not worth much to Ferdinand, at least not as Washington saw it. By 1898 he was furious at his brother, convinced that Ferdinand's desire to continue to grow the company was purely selfish. "Ferdinand is a pessimist," he wrote to Lindabury, his lawyer. "He keeps Charles constantly frightened as to the terrible things that are going to happen . . . His own services he values so highly that it would be difficult to name a figure which would be an adequate recompense." He claimed his brother was full of ill will toward him; but the sentiment he expressed was no finer. "He is waiting for my death or a miracle to make it all his own," he told Lindabury bluntly. "In the meantime he gratifies his vindictiveness by preventing his brothers from getting what they have a right to." His work on the Brooklyn Bridge not only cost him his health, he averred, but his fortune. Once again he affirmed: "I would have been $100,000 better off if I had never seen the bridge."

This grasping fury came from a man possessed of so much more than so many of the citizens of his country, of the world, would ever have. His anger was, at least, not only on his own behalf: Edmund, too, was owed dividends from his shares in the company—but Ferdinand would not pay them; Edmund was thought to be keeping a mistress who would squander anything he got. Edmund, in 1897, took enough note of Ferdinand's worry to visit a lawyer, who wrote to Washington concerning Edmund's "relations with a woman"; Edmund had assured the lawyer that "he [Ferdinand] need fear no trouble from her," and had already made his will—a fact that would have some bearing on Washington's actions in trying to ensure that all of Edmund's nieces and nephews benefited from his legacy. The way Edmund had lived his life did not mean, thought Washington, that he or his heirs should be deprived of what was rightfully his. "Edmund has been much sinned against," he told Lindabury. "Maltreated at home when young and then driven out to be the prey of gamblers, prostitutes and drunkards—He was born with as good a hand as any of us—Now I want F. W. to put Edmund's dividends into a Trust Co. for the benefit of his legal heirs—but he won't do it."

Those dividends would have been generous. In the last quarter of the nineteenth century, and the early years of the twentieth, John A. Roebling's Sons went from strength to strength. When the company had been incorporated in 1876 its total assets had been less than nine hundred thousand dollars and by 1899 (after a boost from the Spanish—American War, which had stimulated production) revenues were up past teen million dollars; 50 percent of all the wire rope made in the United States was made by John A. Roebling's Sons Company. But in 1901, U.S. Steel, the first billion-dollar corporation, came into being, created from the united power of J. P. Morgan, the Carnegie Steel Company, the American Steel and Wire Company, National Steel and others: in a very short time U.S. Steel was 120 times as large as John A. Roebling's Sons Company—and Charles, especially, reckoned it was time to make their own steel. Now even Washington could not disagree: "*Toujours l'audace*, the Frenchman says. You should never wait for something to turn up. Things should be made to turn up by impulses emanating from your own volition, then you are the master, otherwise the servant."

This was the beginning of Roebling, New Jersey, a brand-new town. It was built on a parcel of land then called Kinkora, some ten miles south of

Trenton, on the bank of the Delaware River. Washington made his first visit to the 150-acre site in the summer of 1904 and invited his son to visit "our new principality—the Duchy of Kinkora." It was, he said, "rather interesting. Plenty of wildflowers, Jersey scenery—which means sand and clay, nice farms, orchards, fields, etc. When you appear again with your family we will make a trip down there via trolley. It takes fully an hour to 1½ hours and costs 45 cents per person." They could have driven, of course, if Washington had been willing to climb into an automobile. And Washington could now only admire the determination of his brother Charles—another demonstration of how he was "his father over again." "Charles found it necessary to become an expert metallurgist. He knew all about the physical properties of steel and its chemical constitution; he kept pace with the progress and microscopic examination of steel," his older brother wrote.

Roebling, New Jersey, became a model of its kind. While the Kinkora works are now relegated to the role of museum, the streets and houses built for the workers there are as sturdy as they ever were, the straight line of Main Street echoing the Main Street John Roebling had laid out in Saxonburg more than seventy years before. The railroad magnate George Pullman had offered a model for Roebling, in the company town he had built for his workers outside of Chicago; Saltaire, in Yorkshire, had been built by Sir Titus Salt for the workers in his woolen mills. Ground was broken in Roebling in 1905; rents went from $2.50 per week in the boardinghouses for unskilled laborers, to $25 a month for a supervisor's detached house. There was a school, stores, a volunteer fire department, a baseball diamond with a fifteen-hundred-seat grandstand. It became a place where people loved to live, and loved to work. The food in the dining halls was all homemade: "They had the best cherry pie in the world," recalled Ruth Egan, who joined the company out of high school in 1944 and stayed for thirty years. Her nostalgia for the company's canteen recalls Johanna Roebling, toiling in the kitchen to feed the men in Saxonburg in the company's earliest years; and a connection to the company's founder was never far from anyone's mind. In the inaugural issue of the company's magazine, *Blue Center*, published in 1925, an interview appeared with its longest-serving employee, Ernest C. Ermeling, who had worked for more than a half century in the rope department. His father had been a "school chum" of John Roebling's in Germany; and the father, too, had worked for his old friend, serving as a steam engineer for forty-five years.

"Mr. Ermeling recalls many interesting incidents when he was a boy and when the Roebling plant had only a few hundred employees. One concerns the Sunday mornings in the old apple orchard, where Charles G. and Ferdinand W., then boys, indulged in the manly art of self-defense."

Now Charles and Ferdinand—observed closely by Washington—kept pace with the twentieth century's developments. Even before 1900 the rise of the telephone and electrical power had meant a huge increase in demand for wire—particularly copper wire, as copper is highly conductive. "The whole manufacturing world was going mad about electricity," Washington wrote. "We had never drawn or rolled copper wire: it was a new problem which Charles of course had to solve, and as usual he succeeded." Charles was an innovator in the early days of the cable car, designing a rope machine sixty feet tall, to produce one-and-a-half-inch ropes, each more than thirty thousand feet long and weighing up to eight tons. "For a while we had all the cable road business to ourselves," Washington noted. And as the demand for elevator rope began to increase, the company was there to meet it. "The Otis Elevator Co owns or controls ⅔ of the elevator business in the U.S. We furnish almost all of their rope—and our firm owns a large block of their stock—but not quite enough—I have just bought $45,000 worth," Washington informed his son. When the Wright brothers built their Flyer, Roebling wire was used for the trusses between the wings; fifty contracts were fulfilled for the Panama Canal alone—including wire screens to protect the workers from mosquitos. After the Williamsburg Bridge would come the Manhattan Bridge, and there were orders from all over the world.

But there were the beginnings of industrial unrest, too. In 1908 a fire destroyed the rope shop on Clark Street, part of the Trenton works; "the work of several years swept away in a few hours," Washington said. It was a premonition of a sequence of blazes that would affect the company seven years later—the worst of which came about, Washington was sure, as a result of broad industrial unrest in 1914, when the International Association of Machinists demanded a standardized wage across two dozen plants in Trenton. As a result, the Roeblings had blacklisted some machinists—and Washington thought the blame lay with the disaffected workers. The plant was "undoubtedly set on fire by an incendiary . . . the fire alarm was systematically cut so that fifteen minutes were lost waiting for fire engines," Washington said. "We live in a reign of terror. Everyone expects the whole mill to go in which case we will be ruined. Further testimony about the [Elmer Street] fire confirms the incendiary plot—a ball of fire was seen coming through the ceiling under the jute pile directly after the fire started." Some, however, thought the blaze, which did at least one million dollars' worth of damage, and was the worst fire in the city's history, was the result of German sabotage—despite (or perhaps because of) the German origins of the company, which had stepped up production to supply the Allied armies.

Charles was as tireless and innovative as his father had been, with some of his reticence and sternness, too. But after the fire of 1915 his older brother had noticed a change in him; there was some hesitation, some weariness, when it came to rebuilding the plant. "Charles is getting to feel his years," Washington said to his son, "and this may be the last big job he undertakes . . . Is there ever a time in life when you can sit down in peace and contentment?" By then, they were all feeling their years. In the spring of 1917, in the midst of the war, Ferdinand—now seventy-five—became ill, "dangerously so," Washington told John; and within a month, on March 16, he was dead. His fortune he left to his sons, Ferdinand Jr., who took up his father's post of secretary and treasurer of the company, and Karl, who became second vice president. In death, all animosity was forgotten. In a tribute presented to the board of directors, Washington wrote that he was "sincerely grateful" to his brother for having "carefully and honorably" looked after his interests in the business in the years when he had been incapacitated by the strains and illness of bridge work; "this meant a great deal to me and mine." He went on, in a vein that recalls something of his father's mysticism—if not with his father's belief in a life beyond the grave. "As time goes by, our memories retain only the good. The other things fade away. What death really means I do not understand, and do not want to. We enter life without knowing it, and leave unconscious. Cut off thus at either end, all we can do is to obey the dictates of the infallible conscience with which we are endowed by nature.

"But who succeeds in doing that?"

A little more than eighteen months later, Charles was gone, too. Huge pressure had been placed on the company by the war; supplies of raw materials were low and demand was high—and once again, workers were after higher wages. The strain told on Charles, who suffered from kidney disease as well as high blood pressure, hardened arteries, and a "tired heart," his brother said. He had been deeply affected by the death of his son, lost with the *Titanic*, and he never remarried after his wife, Sarah, died in 1887. By the autumn of 1918 Washington was very worried indeed about his brother, and about the prospects for the future. Charles's illness, he wrote his son, was a matter of "supreme consequence to me . . . He was the directing head, the man who looked ahead, planned and worked and designed and executed with tireless energy year in and year out, and usually successfully—He never copied, was always original even when at times it might have been wise to attain those results in some other way, but it all helps to strengthen one for future efforts. How to fill the place of Charles is a serious problem."

He died on October 5, 1918. Washington's letter to Charles's physician, Dr. Evans, offers an extraordinary image of the passage from birth to death. "One winter evening, December 9, 1849, a woman in labor next to the room I was in gave birth to a child," Washington wrote to Evans, five days after his brother had died. "I heard its first faint cry. On Saturday last I heard the last sigh of the same child. His death was peaceful. The eyelids suddenly fell, a faint expiring breath, and it was all over. But what a life of endless activity great deeds, hard work and indomitable courage, has been embraced between those two sighs!" To another correspondent—a Mrs. Stockton, who had written to Washington when she had heard of Charles's death—he expressed the depth of his loss and the extent of his anxiety. "By the death of Charles I not only lose a friend and brother, but Trenton loses its greatest citizen and the country at large its foremost mechanical engineer," he told her, thanking her for the kind note which had touched his heart, he said. Charles, he went on, "was a paragon of industry. We all relied upon him. With his wise counsel, his instant decision and effective action we all felt we had a strong tower to lean upon. We already miss him and look around in vain for someone to take his place." This was the heart of the matter. "His death seems to be the turning point in the fortune of the Roebling Company. Time, which usually heals all things, can not replace him. Coming so soon after the death of F. W. it seems like a double strike of fate," he wrote.

When Charles's estate was finally settled a few years after his death, it was worth well over fifteen million dollars; John A. Roebling's Sons Company was worth 240 times what it had been when the brothers had inherited the firm in 1869. For all his complaining, Washington admired the financial know-how of both his brothers. "C. G. made his money by not losing

any—that is the great secret . . . C. G. was endowed with the faculty of saying no! Nobody could wheedle him into an investment," Washington wrote
to John. But now the line of succession was less clear. Karl, Ferdinand's son,
took on the presidency of the company following his uncle's death. War had
brought enormous prosperity—an increase in production of 75 percent—
but also enormous strain. Ferdinand Jr. continued as secretary and treasurer but neither man—though Karl was only forty-five and his brother five
years younger—was in good health. Ferdinand Jr. suffered a stroke at the
end of the war; Karl had heart trouble, among other ailments. It was an
anxious time for Washington, and for the whole family. Cornelia's son, John
Farrow, had joined the army; in November 1918, just after the Armistice, she
had not heard from him in over a month. "She scans the casualty lists with
nervous dread," Washington wrote to John Roebling. Not long after his
own grandson, John's son Paul, perished in the influenza pandemic that
swept the world in 1918. John Farrow, Cornelia's son, had been gassed, it
turned out; but he did recover, and come home.

Ferdinand's sons were struggling with their responsibilities, according to
their uncle. Ferdinand Jr. had gone to White Sulphur Springs, in West Virginia, for his health—the very resort that Washington had seen in ruins in
1862. After all, the Roeblings still held to the water cure. Karl was doing his
best, but "he will soon break down too," Washington wrote to his son. He
described his nephew's attacks of angina. "I wish you could see your way
towards taking an interest in our business—Scammel [one of the company's
lawyers] thinks our outlook is dubious—nobody to run it—which I knew
long ago . . ."

On May 29, 1921, Karl collapsed at his summer home in Spring Lake,
New Jersey, and died. He was forty-eight years old. According to the *New
York Times*, the state of his health, both physical and mental, meant that
Washington's concern for the future was not misplaced. For several years, the
paper said, Karl's health had been precarious, "following a nervous breakdown at the time of his father's death. Although titular head of the company, he has not since that time engaged actively in the business. Nor did he
take an active part in the social life of Trenton, as it was thought he would
do when his father's house was remodeled." He had stayed, instead, in either
Spring Lake or his house in Philadelphia; for some years, then, he had not
really been able to run the company at all.

There is a hint in a note sent to John's wife, Reta, in the autumn of 1920, nine months before Karl's death, as to who was in charge of the family business. "I am running the 'mill' very satisfactorily," eighty-three-year-old Washington wrote to his daughter-in-law. This is just what he would continue to do.

# "You can't desert your job"

IN 1912 WASHINGTON WROTE to John because he was concerned about his grandson, Siegfried, then twenty-two and shortly to graduate from Princeton. Washington had arranged a job at the mill for him as a wire inspector. "I wish you would impress on Siegfried the necessity of sticking to this job and paying some attention to it by being regular in attendance . . . It is absolutely necessary that the young man has some definite employment, otherwise his whole life will be ruined!" Most of the young men at the plant started there at twelve or fourteen, Washington said, and knew the work inside out: "Siegfried has none of these great advantages—The little college experience he has had has only been a detriment in my opinion." Although he was a young man of expectation, Washington said sternly to John, "all this amounts to nothing unless he familiarizes himself with the way in which affairs are conducted in this world, comes in contact with actual workers, and realizes that nothing is easy and that nothing does itself and that wealth flies away before you know it . . . Please read this to him, and get him ready."*

During the years of his marriage to Emily, after the Brooklyn Bridge was built, Washington clearly wanted to be more occupied with work than she

---

*Siegfried would eventually be second vice president of the company, but—like his cousin Karl—died suddenly at the age of forty-five; he had a heart attack in his sleep. It was his wife, Mary, whose achievements were remarkable: after her husband's death in 1936 she took his seat on the board of the Trenton Trust Company, and the following year was elected president—at thirty-seven she became the first woman to head a major U.S. bank. She continued to be influential in the world of finance throughout her long life.

liked him to be. Before the end of the century she had expressed her anger to their son regarding Washington's involvement with the new Williamsburg Bridge; in early December 1899, she had written to John that his father "is more than half sick with his preliminary work on the cables. I think he will last about ten days when he gets finally in the field and takes to his bed as he did in Brooklyn!" She called him "a broken reed," and yet at the end of that month, in aid of the work at Williamsburg, he produced the strikingly beautiful drawing that is still the best illustration of what a cross-section of a Roebling cable looks like.

When the preliminary cables—the ones that would enable the working platforms to be made—were drawn across the river in April 1901, Washington, not Buck, the bridge's engineer, was described as being on the scene. "Now lying on the bottom of East River is the first tangible connection between Brooklyn and Manhattan of the many that must pass from borough to borough before there is the sound of traffic on the new East River Bridge," the *New York Times* wrote. The piece recalls the stories written in the summer of 1876, when Brooklyn and New York were joined by a slender thread, when both shores were packed full of cheering crowds and the air rang with the whistles and sirens of every ship in the harbor. Now the excitement wasn't as great—but there was excitement all the same when the *Champion*, captained by George Earl, pulled the cable strands across the river at slack tide, just after noon on April 9. "From the heights of the tower, the decks of tugs about, and even from the eastern shore, there came a cheering and then the snap of cameras. Every whistle within penetrating distance shrieked," the paper's reporter wrote.

It took twelve minutes to pull the line across the water: three wire ropes two and a half inches in diameter, running from the anchorage at Delancey Street in Manhattan, six hundred feet from the saddle at the top of tower, and then down, three hundred feet, to the reels on the float on the river below. And on the float, in "general charge," was Colonel Washington Roebling—keen, as ever, to draw attention to the innovation on display, and the economy achieved thereby.

"The method of carrying the strands across the river adopted yesterday is declared by Col. Roebling to be something new here. When the Brooklyn Bridge was built a small line was dragged across, then a larger and a larger, most of the exchange being done in the air. But in this case strands of the temporary cable are uncoiled along the bottom of the river, then raised, and later combined. Then the permanent cables are manufactured on the structure built in midair upon the temporary cables thus obtained. The saving in time by the new method is twelve days in all, or three days for each of the temporary cables."

Two decades later his memories of work on the Williamsburg Bridge cables—for which he gave full acknowledgment to Charles, saying that "the credit for building them belongs to my late brother Charles, assisted by Mr. Hildenbrand, with an occasional suggestion on my part"—were not entirely happy ones. There had been a lawsuit over payment for the work on the cables, and Buck's use of ungalvanized wire meant the cables were discovered to have rusted less than a decade after the bridge opened. "It became necessary to rip off the flimsy sheet iron covering and padding, and replace it by proper wire wrapping with caulked joints next to the cable bands," Washington wrote.

And now, after the death of his two brothers, after the death of his nephew, he was called on again as the one man who could run John A. Roebling's Sons Company.

In June 1921 the *New York Evening World* sent a correspondent to interview the new president of the company. The reporter was a remarkable woman in her own right. Marguerite Mooers Marshall was thirty-four years old, a graduate of Tufts University and a widely read journalist at the time she went to Trenton to meet Washington Roebling. She was the author of a regular column—syndicated across the land—called "The Woman of It," which would not be out of place in a twenty-first-century paper: her pieces had titles such as "The Woman Who Had Everything" and "Emotional Love." She was a novelist of romance tales—*Wilderness Nurse*, *Arms and the Girl*, *Not in Our Stars*, to name just a few of the fourteen she published before her death in 1964. And there is something of romance in her portrait of Washington, who had no love for journalists—but spoke openly to the young woman who arrived on his doorstep from New York that summer.

"A little old soldier of eighty-four, Col. Washington A. Roebling, the man who built Brooklyn Bridge and the son of the man who planned it, is fighting today his last fight," her piece began. "His fight TO GET HIS WORK DONE in spite of all his enemies—illness, debility, pain, loneliness, bereavement, the terrible depression of a man who has outlived his generation." A man born in the 1830s, who had fought in the Civil War, whose father had come to America in a sailing ship and who had lived to see the first airplane take flight—an airplane held together with the wire his company had made. The year before Col. Roebling met Miss Marshall, a bomb set off at the corner of Wall Street and Broad Street had killed thirty-nine people and wounded hundreds more—the first terrorist attack in downtown Manhattan. These days women cut their hair and wore their skirts shockingly short; the Volstead Act's puritanical ban on liquor only made the stuff more desirable. The nineteenth century was long gone. But little of this mattered to Washington Roebling: he had, he told Marshall, a job to do, working nine A.M. to five P.M., and taking his lunch in the office. "'My appetite is pretty good; I still eat plenty of pie. But I haven't eaten a raw apple in 35 years. I'm no Prohibitionist,' added Col. Roebling, with candor. 'I never could see why a gentleman shouldn't have a glass of wine with his dinner.'"

She was charmed by him. He was, she wrote, "a slight, wiry man, whose iron-gray hair is still plentiful, whose shoulders are only a little bent, whose blue eyes look keen, who appears not a day over seventy." But this young woman was puzzled as to why a man of eighty-four wouldn't give up the hours of his desk, relax and enjoy his life. "'How can you give up what's a part of you?' he asked, slowly. 'In Germany, men retire at fifty; you never see that in America. I guess it's a good thing; as long as you keep on with your work you have an object in life.'"

His object now was the mill, and he threw himself into the task. He was, he would write, "still in possession of all my faculties, although impaired in bodily health." "During the past 30 years I have lived in Trenton and have visited the works almost daily. There is no branch of our extensive business with which I am not thoroughly familiar," he said. He rode the trolley to work every morning—no automobile for him, of course—accompanied by his faithful companion, Billy Sunday, an Airedale terrier to whom the colonel was devoted. Named for a wildly popular baseball player turned evangelist—Washington maintained his scorn for organized religion—Billy had special dispensation to curl beneath the feet of his master as the pair journeyed to work. Billy's doings often feature in Washington's correspondence. He would run off; John would console his father; the cycle would begin again. In photographs he may often be seen standing at his master's side.

In his conversation with Marshall, Washington was as forthright as ever about his ailments. His youthful aspect, he told her, was an illusion. " 'I haven't kept young and fit,' he told me sadly. 'I can't hear out of this ear'— touching the right one—'I can't see out of this eye'—laying a finger bedside the left eye—'my teeth aren't right, my chest hurts me when I talk; it takes me ten minutes to go up and down stairs. If I could only feel well I could stand anything else. But I don't even for a few minutes at a time.' " But somehow the bridge builder in him was still energetic, and when the prospect of bidding to make the cables for a great bridge between Philadelphia and Camden, New Jersey (now called the Benjamin Franklin Bridge), arose, he could not resist. Clarence Case, the Roebling estate's executor, put it perfectly when he went through Washington's exchanges from the time. "Reading between the lines we may gather that there was a strong urge within the veteran bridge builder to add this structure to the long list of notable bridges with which the name 'Roebling' had been connected, particularly as to the making of the wire rope for the cables, and, if possible, the construction and laying of the cables."

The Camden bridge had been formerly proposed in 1919, and it was confirmed in the summer of 1921. It was to cost twenty-nine million dollars and would be, when finished, the longest suspension bridge in the world, with a span of 1,750 feet. The chief engineer was the Polish-born Ralph Modjeski, who had worked on the reconstruction of the massive Quebec Bridge over the Saint Lawrence River after the first bridge had collapsed in 1907. The cables for this bridge would be enormous—each with a thirty-inch diameter, twice as big as anything that had been made up until that point; each cable would consist of twenty thousand wires. "We need the work badly," Washington wrote to John in April 1921, though he expressed his doubts about taking the project on. "We would only lose on it as we did in NY. There is no one to look after it & it is hard and risky work." No longer were engineering projects the concern of one man, one visionary, as they had been in the days of Telford, Brunel, Eads, and Roebling. "The modern engineer only makes plans—very seldom does he execute them himself. He looks around for an abler contractor to execute them for him. Formerly this was not the case," he would reflect around the time the Camden bridge was finished.

But the pull was irresistible. Yes, he wanted to supply the wire—but he wanted to make the cables, too. He wrote to John in September 1921, once again hoping to draw him into the business. "Please brush up on your mathematics and help me make a few cable calculations," he wrote. "I do not know how much mathematics our cable expert, Holton Robinson, knows nor how much calculating the Bridge engineer, Modjeski, proposes to indulge in." Robinson had been the cable engineer on the Williamsburg Bridge. Washington was concerned about the height of the cables above low water— taking into account the saddle movements, the temperature, and the deflection caused by the load on the bridge. His son's reply, however, sent at the beginning of October, was something of a reprimand. "While your health is very good now, you are occupying too responsible a position to run any risk of a breakdown through undertaking such an uncertain and nerve-racking piece of work as the erection of the unprecedented cables of the Camden Bridge," he wrote. "Bid on making the wire; but do not bid on making the cables."

But in the spring of the following year, Washington told his son, the engineers of the Camden bridge came to see him, entreating him to bid on the cables. "I do not see very well how to get out of it," he wrote to John. The son's reply to the father was swift, reiterating his earlier point—and suggesting that Washington should watch his back. "You need every ounce of health you possess in order to keep the mill going. You cannot afford to take on the heavy responsibility of the cable making contract. Both the cables and the towers are unprecedented. It is unknown ground with an unknown risk. No one has a clearer understanding of that than the engineers of the Camden Bridge, or they would not be so persistently anxious to unload it on your heavily burdened shoulders . . . It is Modjeski's bridge built on his own new lines. If it is a success he will get the glory. If anything goes wrong, the blame will probably be laid on your cables." (And he commiserated, too, regarding Billy Sunday's latest exploits: "Am very sorry about your dog Billy. We have a dog, 'Nick,' who runs away regularly and is gone for weeks at a time but always turns up finally in a half-starved condition.")

The American Cable Company got the contract for the Camden bridge, in the end; by the late summer of 1922 Washington recognized that it wouldn't be wise to take the project on. "I regret to state that owing to death, to old age and other disabilities, the firm of JAR's Sons no longer have anyone at their disposal who could take entire charge of such a work," he told a correspondent, somewhat despondently. "A three million dollar contract, with a time limit, with heavy responsibilities and the hardest kind of work, is a burden that a broken down old man of 85 should seek to be relieved from, especially as the prospects of any profit are remote. Competition is inevitable. With Modjeski, I have never had any talk or correspondence about the bridge." But he remained regretful that John A. Roebling's Sons Company didn't get the contract to supply the wire, as he wrote to Clarence Case in September 1923. "I regret that so much time & thought have been wasted by us on this work in the last two years, all to no purpose—We have knowledge & practical processes to make such cable wire exactly right, which no one else knows—The cables can only suffer in consequence."

He complained of his frailty to Marshall when she came to interview him—but he acknowledged his resilience, too. "I've buried eighty doctors," he told her. "There were ten years of my life, the time I was 40 till about 50, that I never stirred out of one room. There was a time when I thought I'd be blind." Now he read and wrote avidly. He loved novels, he told her, they were "a mild form of mental intoxication"; they distracted him from the burden of his work, he said, but the energy with which he approached it makes it seem far from a burden. As his son John said, he was never "a mere figure head" at the company; "his executive handling of the business was an example of the best modern practice." Thanks to him the main plant's power supply was switched from steam to electricity, an enormous undertaking. He developed an entirely new department for the electrolytic galvanizing of wire. And the loss of the contract for the cables of the Camden bridge perhaps still stinging, he took and completed the contract for the cables of the Bear Mountain Bridge over the Hudson River. That bridge—still going strong—was the first vehicular bridge crossing south of Albany; when it opened in 1924, its span of 1,632 feet made it thirty-two feet longer than the Williamsburg Bridge, the longest span in the world at the time. By the close of 1923, Washington wrote to John, work on the bridge was progressing well—the towers were nearly completed, and the anchorages too. By the spring of 1924 work had begun on the cables, and Washington's words convey the enthusiasm of a much younger man—but one keen to see a new generation follow in his footsteps. "The technical difficulties of cable-making at

Bear Mt. are being rapidly overcome," he told his son. "[Six] strands already done in each cable—The driving machinery was designed in our office—Some of our men are up there helping—In addition there are 12 R.P.I men who want to learn the art of cable-making—Likewise 12 West Pointers do." Almost out of instinct, the letter closed: "Have indigestion from eating a blackberry and am going blind."

The *New York Times* ran a brief item on May 27, 1923, marking the occasion of Colonel Roebling's eighty-sixth "birthday anniversary," noting that the presence of Billy Sunday on the previous day had been "a factor in the joy of the celebrations". The summer of the following year, just after he had turned eighty-seven, the Trenton Chamber of Commerce honored him with a "loving cup" at a "Faithful Service" dinner—in the presence of more than two thousand diners in the Trenton Armory—to recognize that his was the longest service in any Trenton business: seventy-five years in the saddle. Washington sat among the five hundred employees of John A. Roebling's Sons Company who were present; Herbert Hoover, then secretary of the U.S. Department of Commerce, was the distinguished guest. "I went home before his speech," Washington wrote on the margin of a clipping recording the event. "Having become deaf I scarcely heard what anyone said except myself." But there was melancholy in the day; next to the Trenton paper's report is another little note in Washington's hand, as neat as it ever was. "I never expected to outlive my brothers—My brother Charles should be alive today . . . An occasion like this would have delighted him. We are passing away so rapidly that in a very few years the present Co. will have gone the way of all flesh, with me among them—Good bye—"

Despair never quashed his interest in the wider world. His curiosity could be sparked in a moment, by a chance conversation. An undated scrap of a note to John tells of a conversation with a waiter at the Waldorf one night, a young man from the Middle East, who showed him how to write his name in Arabic: Washington sent the calligraphy to his son. "My new waiter was raised in Egypt but finds no use in the Waldorf for his philological accomplishments," he said, clearly impressed with the bright young man. "At 88 one is apt to think he has been forgotten," he wrote in July 1925. "But I never allow myself to think about it—I am still active and on my feet in spite of many afflictions and carry more responsibilities on my shoulders than I ever did, with waning strength to meet them." Now he really was a man who had outlived his generation. "In a sense I stand alone—and have no one to fall back on."

Four years earlier, Miss Marshall had asked him what advice he would give to those who wished to live as long as he had. He had laughed at the question. "I wouldn't advise them to do it!" he said. "I know that I'd be glad to go anytime. But you've got to take the days that are sent you." But as to longevity, common sense was the key. "We should have common sense in eating, drinking, working, playing, choosing a wife. A wife is a great help in living many years." But still, Marshall was amazed at his endurance. "How do you do it?" she asked him simply, and he answered with a creed that's good enough for anyone to live by.

"Because it's all in my head," he answered quickly. "Sixty years I've known it and it's all there. It's my job to carry the responsibility. And you can't desert your job; you can't slink out of life, or out of the work life lays on you. I haven't any new business plans, but I've lived through hard times before and I can do it again."

FINALLY, HOWEVER, HE really was the last of them all. "I was very much shocked to read in the morning's paper that McNulty is dead," Washington wrote to John, on April 21, 1924. McNulty, who had named one of his sons after his old boss, was seventy-three. "He came to me very young, only about 20 y old," his friend mused, thinking of those days when he himself had been a young man too. "This death leaves me as the last survivor of the original engineering force of the Brooklyn Bridge—a circumstance never even dreamed of by me 50 years ago." Reminders of the past—of his extraordinary survival—were everywhere. A month after McNulty's death a woman in Seattle sent him her poem in praise of the builders of the Brooklyn Bridge. "Life is so brief at best/ All time conserved is gain;/ A vision vast and clear/ God gave to Brain!" one of the verses runs. The engineer sent a kindly reply, the draft of the letter written in pencil on the back of her envelope. "Poems—allusions—anniversary remarks on the opening of Brooklyn bridge in 1883 still make their appearance from time to time," he said. "Yours is one of the best and I congratulate you on your effort—I am now almost 88 years of age, have survived all my associates on that work—that phase of my varied life is so far back in the dim past that sometimes I wonder if it really happened." The poem, he noticed, was simply dedicated to "John A. Roebling & Son," with no mention of his own name. "Long ago I ceased to endeavor to clear up the respective identities of myself and my father — Many people think I died in 1869." But, once again, he explained to Mrs. Costigan, a stranger on the other side of the continent, the events of those long-gone days. "The simple fact is that my father prepared the original plans and promoted the project—through an accident he lost his life in 1869 before actual work on the bridge was begun—as his associate it devolved upon me to take charge and build the bridge." As if it had been as simple as that. "I am still active in business," he closed, "and take interest in the life of the day—But the infirmities of old age are closing in around me, and the end cannot be far off." *Yours truly*, he signed himself, *W. A. R.*

The confusion between himself and his father would never disappear. It had been there since, on the day of the opening of the Brooklyn Bridge, the *Brooklyn Daily Eagle*—the paper that had taken the greatest interest in the construction over the fourteen years it had taken to rise over the river—referred to "the death of Colonel Roebling" in writing of John Roebling's death.

He still took his role as head of the family seriously, still trying to ensure that his youngest brother, Edmund, would help provide for those more distant members of the clan who were in distress. Now that Ferdinand was gone, he wrote directly to Edmund; by then the pair had not met for a quarter of a century. "Death has been busy thinning the ranks of the Roebling family," he told Edmund. He made a plea to Edmund to make gifts from his vast—and mainly untouched—fortune to the sons of their sister Josephine, and the five children of their sister Laura. This gift, he said, is "simply meant to relieve the distress of a few of your relatives. They will thank you for it as long as they live. They will appreciate it much more now than later on after their lives have been embittered by dire poverty." He was concerned that his brother was too far gone, perhaps, to comprehend this request. "If this letter is too long for you to understand, let Mrs. Carrington explain it to you, point by point. I am sure she will see the propriety of it." Mrs. Carrington appears to have been Edmund's companion. A doctor had declined to say whether Edmund was compos mentis when he had examined him; a few months later Washington explained the situation to Clarence Case. The present sad situation arose "from his surroundings from boyhood—No real

home, no friends, no ties of relationship, no wife, no occupation, not sufficient force of character to rise above the circumstances and perhaps too
much money when young." He would disappear "off the face of the earth"
for years at a time, he told Case, vanishing to China, to Colorado, to
Nebraska—where he had bought a huge farm which he then had struggled
to sell. "Today Edmund is a harmless bedridden white haired old man of
over 70." He was gratified to report to Case that Edmund had given his consent that three dollars a day be paid to his seven needy nephews and nieces.

A new century was a quarter gone. The Great War was in the past, and the
future was arriving. In 1925 *The Great Gatsby* was published, and John Logie
Baird demonstrated his ability to transmit moving pictures by wireless—
television—in London. And in March, beloved Billy Sunday was found dead
in his bed, Washington wrote sadly to John, equating Billy's situation, in
some ways, with his own: "About 6 years ago he was run over by an automobile, receiving an injury in the spine from which he never fully recovered,"
Washington said. "I have had many dogs in my day, but never was I attached
to one like Billy—in a sense we were both invalids—he was my constant
companion all day long and we understood each other perfectly . . . His intelligence was remarkable," he wrote. Although he was denied the gift of
speech, with "the wag of his Tail [he] conveyed many meanings." He connected Billy's afflictions with Emily's, too, the bicycle that had struck her
eight years before she died having affected her, as he saw it, as Billy Sunday
had been by that car, he told his daughter-in-law, Reta. Billy Sunday was
buried beneath Washington's window in the Trenton house.

There was a "modest repast" in May, to celebrate his eighty-eighth birthday; he didn't feel up to more. But he was happy to have his son and his son's
wife by him: "Cornelia thinks he grows handsomer every year," Washington
had said of John to Reta not long before, "with his curly gray hair, which he
gets from his mother." Street lighting was being installed along West State—
"the Great White Way in Trenton," Washington called it, certain that no one
in the house would ever sleep again. But some lights were welcome: "From
the top of the Trust Co building I had the pleasure of seeing last evening the
3 triple stars Venus, Mars & Mercury close together," he wrote to John in
July. The planets would have been visible around eight P.M. that night, about
seven degrees above the horizon in Trenton. "This is only the 2d time in my
life that I have seen Mercury—First time in 1882 at Mt. Desert. Few people

ever see him—It will be about 80 years before they are in such close proximity again, I mean for all three—" His assessment was accurate, though he had no background in astronomy. Conjunctions of the three planets are rare, and a conjunction in which the three could be seen so close together in the sky very rare indeed. Washington's eighty years is, in fact, a conservative estimate: there would in fact be only three such close events in the period between 3000 B.C. and A.D. 3000. The planets looked favorably down on Washington in the last full year of his life.

His eyes, however, were not solely on the heavens. "Our business averages the same," the letter to John went on, "but there is an absence of creative energy." That creative energy was what had once driven him to send company circulars all around the country, asking Ferdinand to provide him with state directories so he could target his mailings. But now that energy could not come from him. "This summer I feel the heat, that and other troubles pull me down so that I scarcely get around, and suffer much—have not been to the mill for 4 days—Have become very blind and am looking around what to do about all my complicated & extensive financial matters, checks, correspondence, investments, etc."

Ever the interested engineer, he cut out and kept an article from the *New York Times*, from March 13, 1926, announcing that the New York and New Jersey Hudson River Bridge Advisory Committee had approved a fifty-million-dollar bridge between Fort Washington, in Manhattan, and Fort Lee in New Jersey. "Towers to be higher than the Washington Monument," the headline read. "Span the greatest in the world." The engineer who had prepared the report was named as O. H. Ammann; he announced that the bridge could be finished as early as 1933. Washington's penciled notes of criticism were written in the margins of the piece. "Project too gigantic . . . Equal to all 5 Brooklyn bridges combined in one place . . . He must come down out of the clouds . . . Strains are not certain." Othmar Ammann's bridge across the Hudson, the George Washington Bridge, was opened in 1931, an astonishing feat of engineering completed during the Great Depression. The wire for its enormous cables was supplied by John A. Roebling's Sons Company.

Washington had shrunk to ninety pounds in the spring of 1926. "Have broken down completely—losing strength daily—cannot eat, sleep walk or stand . . . Have not blood enough in my body to live on," he wrote to

R. B. Gage, a fellow Trenton mineralogist in May 1926; but he went on to describe a correspondence he had with Paul Kerr, professor of mineralogy at Columbia University. Kerr "has asked me for 'Newtonite' from Arkansas. I sent him 3 pieces . . . Have a beautiful small boulder of aqua marine. Absolutely clear, polished on 2 faces. When you want to see anything, come in and look at it, without me. I can't come down." And so it seems, by this point, he was confined to his bedroom, unable to visit even his own "museum" downstairs. "Think not that I am improving," he wrote to Reta the next day. "Growing weaker daily—body racks with pain—Head bowed down in sheer apathy—Bones crack when I am rolled over—Fall down when I try to stand." Yet, though bedridden, he was still able to find wonder in the world. His letter to his daughter-in-law continues: "A surprise: For several years—10—a night blooming cereus stalk has been knocked about in the greenhouse—last night it suddenly bloomed—was brought to my bedside at 10 P.M. A delicate odor filled the room—a wonderful flower—much larger than a rose—a calyx filled with snow white petals curved outward and oval pointed—This morning it is gone—to sleep the sleep of ages again."

Washington Roebling died on the afternoon of July 21, 1926. In the last day or so of his life he had lapsed into unconsciousness; his end, it was reported, was peaceful. His wife, Cornelia, was at his bedside, as were his son and daughter-in-law, John and Reta; his grandson Siegfried was there, too, along with Cornelia's son John and his wife, Adelaide. Faithful William Clark, Washington's doctor, was there as well. He had outlived so many—not only his brothers Ferdinand and Charles, and McNulty, the penultimate survivor of the work over the East River, but also James Buchanan Eads, whose lawsuit over the construction of the caissons had given him so much grief, dead in Nassau, in the Bahamas, in 1887; and Seth Low, the upstart mayor who had been so strongly in favor of his removal as Chief Engineer, had died a decade earlier, aged just sixty-six. William Paine had died in 1890. Eugen Langen, whose engine had nearly diverted the course of Washington's life, was gone by 1895, though he was only four years older than Roebling. William Kingsley had died ten years before that, at the age of only fifty-two; his grave in Brooklyn's Green-Wood Cemetery is marked with a monumental block of stone that "formerly was a part of the East River Bridge."

Three hundred people attended the funeral, held at the house on West State Street. The mill had closed at noon that day so the men and women who worked there could pay their respects, just as they had done for his father. A special train took his body to Cold Spring, New York, where he would be buried beside Emily, in a grave on the heights of the verdant cemetery there. For a while he had not been sure where he ought to be buried. "The grave situation worries me very much," he had written to John just before Emily's death. He said he did not care where he was buried; but burial in Trenton had a singular disadvantage—a familiar one. "There is room for me in my father's lot, where I would be completely overshadowed by his big monument and name," he said. A plot in Cold Spring seemed best; his and Emily's graves were each marked with "a runic cross," his the same as hers.

Cornelia was not forgotten. He left one third of his estate—worth twenty-nine million dollars—to her, and the other two thirds to his son. While nearly all of his vast mineral collection went to the Smithsonian, "three small cases of cut stones" were reserved for his second wife, as well as "a set of Nankin china," which must have had special meaning to her. After his death, John wrote a sketch of his father's life that was used as a basis for accounts published in both local and national papers; Cornelia wrote to John on black-bordered paper about what he had written. "Your sketch of your father's life is just as he would have wished it. I thank you more than I can say for the tribute you paid me; nothing could be said of me that would mean as much to me." Among the condolence letters she received was one from George H. Brown, with whom Washington had served during the war; they had shared a tent at Budds Ferry, in Maryland. "Nearly everyone could, or thought he could, play some kind of musical instrument, but none of them could come within a mile of your late husband, who could make a violin talk."

Cornelia died in 1942; she is buried beside her first husband, in Live Oak Cemetery, in Walterboro, South Carolina. The mansion that Emily built, and where Washington and Cornelia had lived, was demolished in 1946 to make room for the New Jersey State Library. Donald Roebling, John's youngest son, removed the stained-glass window of the Brooklyn Bridge from the top of the staircase in 1936, and moved it to his estate in Florida; the place was restored in the early years of the twenty-first century, but the window, alas, seems to have disappeared.

The year after Washington's death—just a few days short of what would have been his ninetieth birthday—Charles Lindbergh's *Spirit of St. Louis* landed at Le Bourget, in Paris, completing the world's first solo transatlantic flight. The airplane's control cables had been manufactured by John A. Roebling's Sons Company; the plane was braced with Roebling wire, and its lighting and ignition cables had been made by the company, too: it constituted a new kind of bridge, thousands of miles long, stretched across a great ocean. "In the death of Colonel Roebling," said Governor Arthur Harry Moore at the time of Washington's death, "New Jersey has lost one of, if not its most distinguished citizen. His accomplishments during the seventy-eight years he was a resident of Trenton have reared a monument to his memory, the equal of which his most ardent admirers could not have devised." The *Bernardsville News* put it plainly: "The story of Colonel Roebling's life is deeply interesting."

# "Time & age cures all this"

THREE YEARS AFTER WASHINGTON Roebling's death, an article about him appeared in the *New York Times Magazine*; it is preserved in the Rutgers archive. "His friends knew him as more than an engineer," wrote the author, identified only as "R.L.D." "He was a close student of the classics, a linguist, a mineralogist and an accomplished musician. Perhaps this versatility is not as surprising as it seems. Something of the poet's imagination as well as the soldier's gallantry and the engineer's accuracy was needed to build the Brooklyn Bridge. The structure, for Roebling, was more than a mass of stone, cement and steel, put together by applying brute force to inanimate nature; it was a realized dream, the result of a mastery of methods and materials as perfect as that possessed by any artist."

Artists converse with each other over years, decades, centuries: the Brooklyn Bridge has been a potent subject of conversation. In the early years of the twentieth century photographers like Eugene de Salignac—who took the famous 1914 photograph of a group of men perched nonchalantly among the bridge's stays and suspenders—Berenice Abbott, Walker Evans, and Karl Strauss played with the light and shadow of the structure, its solids and voids. Closer to the present day, Richard Estes's vibrant realism, or Barbara Mensch's atmospheric prints keep that conversation up to date. The paintings of Childe Hassam, Frank Stella, Georgia O'Keefe, and Ellsworth Kelly carry the bridge from realism to abstraction, from a path over the river to an exercise in pure form. The Brooklyn Bridge, in the twenty-first century, continues to be an inspiration for artists: from the Berlin-based duo Mischa Leinkauf and Matthias Wermke, who hoisted white flags atop the towers of the Brooklyn Bridge on the anniversary of John Roebling's death in 2014, to the "human harp" devised by London artist Di Mainstone, which uses contact microphones and "musical prosthetics" to elicit extraordinary sounds, chords of resonance and vibration, from the cables and steel of the bridge.

In his obsessive epic *The Bridge*, Hart Crane wrote:

*Through the bound cable strands, the arching path*
*Upward, veering with light, the flight of strings,—*
*Taut miles of shuttling moonlight syncopate*
*The whispered rush, telepathy of wires.*
*Up the index of night, granite and steel—*
*Transparent meshes—fleckless the gleaming staves—*
*Sibylline voices flicker, waveringly stream*
*As though a god were issue of the strings. . . .*

Marianne Moore in "Granite and Steel," hymned the "enfranchising cable, silvered by the sea,/ of woven wire"—and yet to her this "climactic ornament" was "John Roebling's monument," not his son's.

But the realized dream was Washington's. "Whenever a Roebling dies in Trenton he is immediately credited with having built the Brooklyn Bridge," he wrote to the editor of the *New York Sun* in 1918, after the death of his brother Charles. "I furnished the information contained in the obituary published in a Trenton paper. The reporter thought it would enhance the article by adding the Brooklyn Bridge to the other really great achievements of Charles G. Roebling. My late brother was still at school when I took charge of that bridge." The paper also reported that the original plan for the bridge followed one by Isambard Kingdom Brunel: "the height of nonsense," Washington wrote. "This simply shows how history can be distorted as the years go by." The *Christian Science Monitor* picked up the same distortion, and in a piece on "The Roeblings, Bridge Builders," mentioned Charles's part "in the building of the Brooklyn Bridge." Washington wrote emphatically in the margin: "Charles Roebling was not an engineer of the Brooklyn Bridge!"

In each situation confronting him, each time a practical decision had to be made, he would make it, and trust his own ability as one founded on hard-won knowledge. He was able, in the closing years of the nineteenth century, to take the kinds of risks that are not available to his modern counterparts, now that government regulations limit the freedom of individual engineers—a wise precaution against bad decision making, though not one that always averts bad decisions. Washington Roebling had a complete and extraordinary overview of the projects he was undertaking; such an overview is almost impossible now. Henry Perahia, who for fifteen years was Chief Engineer and Chief Bridge Officer for the New York City Department of Transportation, says that "now you have whole companies with dozens of people doing what one man did." As Washington himself said, "it's all in my head." Perahia notes: "You had someone who says: This is my bridge. And this is one of the few bridges where there were major changes to the design as it was being built—much of the engineering was done on the fly." Changes to caissons, cables, roadway, trusses were Washington's, and Washington's alone.

He had been forged in a war, a soldier learning that in the heat of battle he had no one to rely on but himself. And he had learned that, too, in Saxonburg and Troy, Pittsburgh and Cincinnati. What he built—over the Ohio, over the East River—has endured. The Covington and Cincinnati bridge, now called the John A. Roebling Bridge (arguably the Roebling name is far better known in Cincinnati and its environs than it is in Brooklyn) and a national Historic Civil Engineering Landmark, has undergone several reconstructions and rehabilitations, the latest and most extensive during the final years of the twentieth century and the early years of the twenty-first. The Brooklyn Bridge had its first major renovation beginning in 1948, overseen by David. B. Steinman; the work, which considerably strengthened the roadway—unfortunately altering its graceful appearance—was finished in 1954.* All of the trolley and railway tracks were removed, the roadway was widened to three lanes in each direction, and the truss work was strengthened; the flooring was completely replaced—as it would be again in 1999.

In 1983, on its centenary, a thorough survey of the Brooklyn Bridge was undertaken, a test of Washington's conjecture that it would last for another couple of centuries. Blair Birdsall, an engineer who had started his career at John A. Roebling's Sons and gone on to work with David Steinman, recalled his first investigation into the bridge's condition, in 1945, when a diver

---

*Steinman, one of the leading engineers of his day, was a poet, too: "The Harp" plays both with the idea of the bridge as a harp, and with the idea of the author as a biblical David, who plays on its strings.

brought Birdsall chips from the wood of the caissons under the river "still redolent with pine pitch." At the time, while repairs were certainly necessary (there was "savage" corrosion where the diagonal stays were supported at the tower tops, for instance), the bridge was in fine shape. The masonry of the towers was in "perfect condition": "the mortar is particularly notable—it is just as sound today as it was when it was installed." The main cables were found to be in "excellent condition"; and while all the suspenders and diagonal stays would need to be replaced, and the roadway and truss work given a good going-over, Birdsall had no cause to doubt the old chief engineer's conjecture. "The question has been asked how long the bridge will last. My answer to that question is that the bridge will go on into the future as long as we are smart enough to keep the materials in the condition in which Roebling left them." The four main cables of the Brooklyn Bridge went into service in the spring of 1883—more than 130 years ago. They remain the primary support holding up the suspended structure; and although the burden carried by the cables has been reconfigured several times, the cables themselves remain essentially as they were when the bridge opened for business. J. Lloyd Haigh's brittle wire remains within them. And while it may be true, when speaking of steel cables, that "if a weak wire is spliced to a strong one and stretched from tower to tower, the strength of the weak wire is the measure of the whole length," compensation can be made for that weakness. We all have flawed wire in our structures. And yet we are able to bear the weight that we must bear, because we find a way to make repairs. Allowances must be made for imperfection, as Washington wrote to the bridge trustees in 1878.

The latest major renovation, to repaint the structure and improve roadway access, began in 2010—the original estimate was $550 million; and complaints about delays and cost overrun sound uncannily like the very same kinds of complaints that were made during the bridge's construction.

But Washington Roebling's legacy, and his achievement, is far greater than can be embodied in a single structure, however useful and beautiful that structure may be, and deserves to be understood far beyond the world of engineering. His voice speaks clearly from the past. Great bridges are still built—and some of them, despite vast leaps in technology in the past decades, still fall. When a section of the I-35W over the Mississippi River in Minneapolis collapsed at the height of rush hour on August 1, 2007, killing thirteen people and injuring another 145, it was no accident. Not only was there a lack of robustness in the original design; four of the bridge's gusset plates (the thick sheets of steel used to connect beams and girders to columns, or to connect truss members) were thinner than all the others. The contractor had stored construction materials on the bridge; loads had increased and maintenance was poor. All these factors virtually ensured the disaster would happen. As Washington wrote to Henry Murphy when a brick arch collapsed during the building of the Brooklyn approach to the Brooklyn Bridge—killing one of the workers—"the brick arch fell because it had a right to fall." Its central support had been removed before it should have been; "the real accident was not so much that this arch fell, as that the other one stood." It was another instance of his precept that "not one accident in a hundred deserves the name."

His assistant engineer George McNulty, who was overseeing that part of the work, was responsible. But Washington's account of his conversation with the younger man sounds as if it would have been very different from any such conversation initiated by John Roebling. "The precise reason why Mr. McNulty removed the arch centering I do not know, in fact he does not know himself. It probably never occurred to him the pier could turn, if he had thought about it a very slight investigation would have shown him . . . Ambitious natures are apt to be overconfident and shrink from asking counsel from more experienced persons for fear their infallibility might be impugned. Time & age cures all this."


JOHN A. ROEBLING'S SONS was sold to the Colorado Fuel and Iron Corporation in 1952, the year Washington's son John died—"a scholar, a philanthropist, a scientist and a good citizen interested in all matters of public welfare," said the *Bernardsville News*, in marking his passing. The sale was a watershed in the decline of Trenton's industry that had begun as early as 1904, when U.S. Steel had taken over the Trenton Iron Works; in the 1920s, local pottery and rubber firms had been swallowed up by the big nationals, too. John A. Roebling's Sons had been a holdout, one of the last bastions of a family-run industrial company.

But the years before that had been good for the firm, even in the hard times of the Depression. Ferdinand Jr. took over as president after Washington's death, and not very long after the company got the contract for the cables of the George Washington Bridge over the Hudson. Undoubtedly Washington would have changed his mind regarding the sheer impossibility of building such a bridge: but still, the scale of the thing was awesome at the time. With a span of a mile from anchorage to anchorage, it took 106,000 miles of wire to make up the four cables—the wire in just one could be looped right around the Equator. All in all, the cables weighed 28,500 tons. If the cables of the Brooklyn Bridge, the Manhattan Bridge, the Williamsburg Bridge, the Philadelphia–Camden Bridge, and the Bear Mountain Bridge together were piled in a heap, they would weigh a mere 22,400 tons. Speaking to the novelty and danger of the construction, the *New York Times* assured its readers it was safe; they had spoken to the chief engineer. "The whole undertaking involves no untried operations, Mr. Ammann assures us, this type of steel wire cable having been used in all large cable bridges since John A. Roebling first spun it over the Brooklyn Bridge."

No correction by Washington could be made in the margin of that newspaper clipping—but John A. Roebling of course had had nothing to do with spinning the cables of the Brooklyn Bridge, for he had been dead seven years before the first wire went across.

When the Empire State Building was completed in 1931, its 120 miles of elevator rope were supplied by John A. Roebling's Sons Company, along with 1,550 miles of electrical wire. The next year the company won the contract to make the cables of the Golden Gate Bridge in San Francisco, which would eclipse the George Washington Bridge as the longest suspension bridge in the world when it was completed in 1937. "Trenton Makes, the World Takes" says the illuminated sign that still graces the Lower Trenton Bridge over the Delaware River. But the words are less true than once they were; John A. Roebling's Sons Company closed its doors for good in 1974. In Trenton the Roebling works will be converted into fashionable, loft-style apartments; in Roebling, what remains of the plant is now a fine little museum, and the whole two-hundred-acre site is undergoing a gradual cleanup and decontamination by the Environmental Protection Agency. John A. Roebling's Sons Company is a ghostly presence in this once-vibrant industrial city.

John A. Roebling believed in ghosts; indeed, he could speak to them, over a wooden table. His son had no time for such things. In 1922 he was asked whether he might consider writing an autobiography. He dismissed the idea out of hand. "For me to write an autobiography when I am approaching 86 and can scarcely find time to attend to my own affairs is out of the question," he said. In any case, he felt such an undertaking would be doomed. "I doubt if any man's true and inner life was ever portrayed in a biography." He knew that too well: he had tried to capture his father's life and must have felt he had not succeeded, however vivid the portrait seems now. He focused instead on running John A. Roebling's Sons Company, and bringing it into the modern age. "In all business, some one must say the potent words, yes or no! And in order to say them right, he must keep in intimate touch with all details. He must be able to gauge the future—be a judge of human nature, on which all depends, and act at once . . . one has to be a financier, a tax expert, a lawyer and a technical expert. With an endless string of untoward happenings cropping up almost daily, a man must have stability enough to take them at their face value, and not lose his balance."

There was one subject, however, he really didn't wish to cover. "You must excuse me from writing about the Brooklyn Bridge, which took up twelve years of my active life. But I assure you that the periodic outbursts about its downfall are all fakes, inspired by other would-be bridge builders at new locations, who think that by decrying the old structure they can promote the prospects of their enterprises."

# Cold Spring

THE LITTLE TOURIST INFORMATION booth is down by the old train station in Cold Spring, New York. The line runs right by the river, West Point just to the south, Albany to the north. Tucked back behind the trees are the remains of the foundry that ensured that the town had a vital role to play in the Civil War; when Emily Warren was growing up here, the boom of cannon would have echoed overhead, over the water, as new and deadly weapons were tested on the grounds. Now the ironworks—founded by Gouverneur Kemble, for whom Emily's brother G. K. was named—is a rolling, wooded park. The three of us had arrived just before lunch—my husband, my son, and I, setting off early to head north on the Palisades Parkway, through Bear Mountain Park, where the bridge comes as a surprise, appearing suddenly as you curve toward the river. My husband, Francis, was at the wheel, our son, Theo, in the back. I remember driving up the Palisades with my own parents—another group of three, like Washington and Emily and John— when I was small, when we still had a car, my dad's ancient VW Beetle, and I remember, or at least I think I do, the sweep of silver blue river beneath the bridge, the graceful slope of its cables spun by John A. Roebling's Sons.

An older gentleman is in the booth by the train depot, neatly dressed, wearing a tie even on a summer day. His silver hair is swept back from his forehead, and there are spectacles on his nose. I ask him where the cemetery is, and how best to get there. He laughs. "I don't know that!" he says. "I haven't been there—and I hope I don't have to go there too soon!" Well, fair enough, we all think. My phone tells us where to go. It's pleasant, in any case, to wander past the shops and restaurants along Main Street, which feels a little like Brooklyn-on-Hudson, with plenty of artisanal ice cream and lifestyle accessories on offer.

The streets slope upward, away from the river. The burial ground is perched above the town. It's a half-hour's steep walk, and no one else is on the pavements. The houses look a little less prosperous the farther we move from the river, but they are still fine American homes set in their yards, American flags waving before them, or appearing on the bumper stickers of the cars parked out in front.

We take a wrong turn, my phone notwithstanding, and now we are a little hot and bothered. Pine Street, Division Street, Bank Street—but we finally get to the cemetery gates. It is a grand place, well-kept, with a gothic-style building near the entrance. The lawns stretching away are close-clipped, evenly watered, warm green stretching off through branching paths and between stone monuments of every size and description. Mature trees cast welcome shade; we stand under a big old oak and drink from our plastic water bottles.

"Where's he buried?" my son asks.

I don't know. This has been a family holiday, not a research trip. The only notion I have to guide me is Washington's mention of a runic cross—but there are plenty of them here. This is a big cemetery. Husband and son are regarding me evenly, waiting to see what I come up with. So I stride purposefully over to the little gothic building; it is closed and locked. A piece of paper taped inside a window lists prices for burial here; there is a phone number to call for assistance. I don't have much hope of an answer, but it's picked up on the first ring.

Doug Loden drives through the gates in the next five minutes; he lives across the street. He's happy to help: certainly he knows where the Roebling plots are. He opens the doors to let us in to his tidy sedan. Uphill again, just for a moment or two, curving right, and now we really are high above the town, the Hudson River in the distance, behind the screen of trees. "Here you are," Doug says, and nothing else. There are two runic crosses just across the dappled grass.

How long should I stay? What should happen here? Who comes here? There are matching plastic buckets filled with wilting sunflowers in front of each grave; who has left them, I don't know. The crosses are precisely the

same height, though the ground is not quite level, sloping back down towards the gate. Pink granite makes a coverlet for each resting place, side by
side. My husband and son keep back as I step forward silently to stand by
Washington's memorial, which has no words carved on it but his name, and
the dates of his birth and death.

&#x2720; *GIFTED* &#x2720;
&#x2720; *NOBLE* &#x2720;
&#x2720; *TRUE* &#x2720;

are the words carved into the body of the cross on Emily's grave, where I
sit a while, too, tracing my fingers on the letters, resting my palm on the
stone, on both stones, before I get up and once more take my phone out of
my pocket, this time to take pictures, because that is the thing to do, to
make a memory into a possession. Without some record of events, a record
I have been tracking since I first slipped Washington's photograph into my
wallet all those years ago, what is left? Memory itself, Washington wrote,
becomes "pictures of the imagination"; reality fades away. Here, says Theo,
let me—and he takes the phone from my hands as I pose here, smiling. But
after a little while there are no more pictures to take, nothing else I believe I
can bring here.

Come on, I say. Let's go.

# Acknowledgments

Like a bridge, a book is not in fact built by one person alone; there have been many along the way who have helped shepherd *Chief Engineer* to publication. I have been working on this project—or a version of this project—for longer than I sometimes care to remember: I still have a black box of index cards covered with notes taken in the New York Public Library when I was still at university and supposedly doing an English degree. There are quite a few people to thank.

Of course I am indebted to the work of scholars who have gone before me. The work of David McCullough, Clifford Zink, and Henry Petroski has been especially valuable. Of incalculable value has been the work—and friendship—of Donald Sayenga, who first brought Washington Roebling's memoir to light. My correspondence with Don, my long lost pal, means more to me than I can say. He has been a valued reader, too: one who has saved me from error on more than one occasion. I cannot thank him enough.

It was my time in the archives at Rutgers University and Rensselaer Polytechnic Institute that truly made Washington come to life for me. Everyone I encountered in those places was welcoming and tirelessly helpful, not least because I was often working from an ocean away. At Rutgers: Fernanda Perrone and Albert C. King set me on the right path when I was heading to New Jersey and first delving to the collections in New Brunswick; Ronald Becker, David Kuzma, and Linda Zuckerman produced dozens of grey boxes filled with wonder; Christine Lutz and Erika Gorder helped me with further research and images. Meghan Rinn was of great assistance when I was back in London, as was Stephanie Crawford.

When I made my first visit to RPI, I was ably assisted by Amy Rupert; more recently the terrific Tammy Gobert, Automation Archivist for the Institute Archives and Special Collections, unlocked the treasures of the Fixman

Room for me. Jenifer Monger, Assistant Institute Archivist, sat with me every day as I pored over letters, notebooks, and drawings; she was a sounding board and a good friend, and has remained so, going above and beyond the call of duty when it comes to tracking down queries sent all the way from London. Washington Roebling may have been miserable at RPI: I never, ever was.

Also of great help were Chela Scott Weber of the Brooklyn Historical Society; Marcia Kirk, now Director of Public Programming at the New York City Department of Records, and Barbara Hibbert at the city's Municipal Archives led me to a trove of working drawings and designs for the construction of the Brooklyn Bridge. At the Smithsonian Institution, I'd like to thank Kirk R. Johnson, Director of the National Museum of Natural History, and Paul W. Pohwat, Collection Manager (Minerals), who shared stories and images of the handwritten labels on Washington Roebling's mineral collection. Thanks also to Tammy Kiter in the Manuscript Department of the New-York Historical Society, and to Kris Eckelhoff, who did some work for me there. Huge thanks to my new friend Fran Rosenfeld at the Museum of the City of New York, as well as Alex Yankovich, and Lauren Robinson. At the Brooklyn Museum, thanks to Monica Park. Jee Leong Koh and Olivia Harris helped with some research in the New York Public Library.

In Trenton, Don Jones opened the Roebling Museum especially for me on a beautiful spring day; Erica Harvey put me in touch with him. Since then the museum's executive director, Varissa McMickens Blair, has been on my side. In Saxonburg I had a lovely base at the Mainstay B&B from which to explore; Bob Kaltenhauser shared gin and Roebling lore with me. John Fleischman and Drenda Gostkowski pitched in, too. When I got to Cincinnati, I was lucky enough to encounter Don Heinrich Tolzmann, whose passion for the Roebling Bridge across the Ohio was inspiring; he also sent Don Sayenga my way, and I'm enormously grateful for that. And in Cold Spring, Doug Loden took us to the graves of Washington and Emily Roebling. Yvonne Skala and Patrick Walz in Wiesbaden introduced me to the true delights of the water cure. Susan Nalezyty at Georgetown Preparatory School was most helpful.

This book would never have been written if it were not for the shelter, assistance, and financial support offered to me by the British Library. I was honored to be one of the winners of the 2014 Eccles British Library Writer's Award. My hours spent in Rare Books and Music, and in the British Newspaper Archive, were of incalculable value, as was access to the staff canteen!

Professor Philip Davies, Director of the Eccles Centre for American Studies, has always been enormously supportive of this project, as have Dr. Matthew Shaw, Cara Rodway, and Jean Petrovic. My friend Rebecca Carter pointed me in the direction of the Eccles Centre in the first place. When I was not working in St. Pancras I could be found in St. James's Square, in the London Library. Thanks to librarian Inez Lynn and everyone there for my berth.

In the final stages of this book I had a wonderful journey to Doncaster and Tyneside to see the Bridon-Bekaert rope works. Dave Hewitt gave up two days of his valuable time to instruct me in the making of wire rope in the twenty-first century; Dave Thompson, Joe Inkskip, and Steve Heyland were patient in answering my questions.

I've had more than one stab at writing this book. Ian Jack, then at Granta, saw a proposal in the 1990s and had faith in the project then; my friendship with Barbara Mensch—and her beautiful photographs of the bridge—was an inspiration, too. Historian, playwright, actor, and firm friend Ian Kelly forged a link with George Gibson and Bloomsbury: the right connection at the right time. Henry Perahia, a valued early reader, took me for a walk over the Brooklyn Bridge quite a few years ago, and subsequently offered excellent advice on bridge engineering. Nele Guentheroth at the Stadtmuseum in Berlin provided information on John Roebling's beginnings as an engineer; Philip Oltermann helped with some German translation. Martin Rees, Astronomer Royal, was of enormous assistance when it came to checking Washington's remarks regarding the conjunction of Mars, Mercury, and Venus in 1925; thanks to him Dr. Robin Catchpole at the Institute of Astronomy in Cambridge and Dr. Steve Bell, head of HM Nautical Almanac Office in the United Kingdom Hydrographic Office, also came to my aid. This was an especially satisfying puzzle to solve.

I am lucky to have Bloomsbury as my publisher. George Gibson, remarkable both as an editor and as a friend, believed that Washington's story should be told, and waited ever so patiently for me to finish it; when at last I did finish it, his work made this a much, much better book. Cindy Loh, Nancy Miller, Laura Phillips, Callie Garnett, Grace McNamee, Sara Kitchen, Janet McDonald, Laura Keefe, Sara Mercurio, and Marie Coolman were a stellar team at Bloomsbury USA. In the UK, my editor Michael Fishwick, Alexandra Pringle, Anna Simpson, Marigold Atkey, and Jude Drake could not have been more supportive, as was Bill Swainson. And of course: Alexis

Kirschbaum, now part of the Bloomsbury Band! Ant Harwood has been a wonderful agent; his good counsel and good company are unequaled. Thanks, too, to James MacDonald Lockhart.

Suzanne Gluck offered an Upper West Side welcome and the support of her friendship and wisdom; she also introduced me to Min Jin Lee, author, heroine, Wonder Twin. Tom Gatti, a great editor and a great friend, read a draft of this book and offered sound advice, never mind all those years of moral and practical support. Peter Stothard, Keith Blackmore, Richard Whitehead, Alex O'Connell, and Megan Walsh: you will know why your names are here. My days in Heyden White Rostow's classroom live on in my head and in my heart; she offered me a bed, too, when I was working in New York. Sylvia Kahan, my dear friend and cheerleader. Jen Doman, my bosom buddy in Brooklyn. Sarah Maslin Nir, the Brearley girl and stalwart pal I met in London's East End. Deborah Broide for sage advice and theatrical tips, too. Simon Winchester, Ken Burns, Andrew Solomon, and Andrea Wulf—so many thanks for going to bat for Washington Roebling.

My adored parents, Ellen and Arthur Wagner, are gone now, but their names must appear in this book. Gail and Donny Abrams revealed New Jersey's wonders to me and sent me off to New Brunswick every day in the company of Ms. Garmin; you can never get lost when you have the love of your family to guide you. Stephanie Guest and Richard Ellis, without whom nothing would be possible. Dear Alan and Griselda Garner have given me a seat by their fire—and an extraordinary model of dedication to craft. Kate Bassett, Ruth Scurr, Jill Waters, Melissa Katsoulis: I am lucky to have such friends.

As for my husband, Francis, and my son, Theo: no words will ever do. You said I could do it. And I did. You're the best.

"The experience gained by one failure is often of the greatest value in pointing the way to the right path," Washington wrote in his memoir. Any errors in *Chief Engineer* are mine and mine alone.

# Notes

In these notes, a simple "WAR" followed by a page number always refers to Washington's memoir, held in the archive of Rutgers University, which includes material about his service in the Civil War. Page references are to the holograph manuscript, as the memoir has also been transcribed. References to letters, etc., give the sender, addressee and the date, and the archive in which the material is held.

In the Rutgers archive are four volumes of Roebling letters, selected and arranged by Clarence E. Case, the family's executor. Where I have referred to these volumes, rather than to individual letters found in file boxes, they are listed as Case I or Case II, along with the sender, addressee and date.

Names provided in brief form, such as "Morris," refer to authors listed in the Bibliography.

## Abbreviations used in endnotes:

EWR = Emily Warren Roebling (also used for Emily before her marriage, for clarity's sake)
FWR = Ferdinand W. Roebling
JAR = John A. Roebling
JARII = John A. Roebling II
LER = *East River Bridge, Laws and Engineer's Reports*, 1868-1884, Brooklyn: 1885
PTF = Washington Augustus Roebling, *Pneumatic Tower Foundations of the East River Suspension Bridge* (New York: Averell & Peckett, 1873)
RUL = Rutgers University Library
RPI = Rensselaer Polytechnic Institute
WAR = Washington A. Roebling.

## Foreword

**knew the Roebling family well once wrote:** RUL, Introduction by Clarence E. Case to Transcript book, I.
**The admiral is 69:** WAR to JARII, November 24, 1910, John A. Roebling II Papers, RUL.

head of the whole Royal Navy: Morris, 12.

"genius for enjoyment": Morris, 209.

artists or even old friends: Morris, 8–9.

confess to inglorious failure: RUL, Address Delivered by Abram S. Hewitt on the Opening of the New York and Brooklyn Bridge (New York: John Polhemus, May 1883).

known the Chief Engineer of the bridge; McCullough, 12.

Arcadian and original: John A. Roebling: An Account of the Ceremonies at the Unveiling of a Monument to His Memory, 23.

walk over the Brooklyn Bridge: Mumford, 225.

the identities of most engineers: Mumford, 61.

That he never did: New York Evening World, June 11, 1921.

precaution must be taken: Blockley, 240.

the monstrous organism: James, "New York Revisited," The American Scene, 75.

banklike blandess and solidity: Malcolm, 8–9.

in the case of others: Schuyler, 155.

where he had left off: Roebling family papers finding aid, 33, RUL.

his own exclusive property: RUL, WAR to Henry S. Jacoby, July 24, 1910.

might consider negligible: John A. Roebling: An Account of the Ceremonies at the Unveiling of a Monument to His Memory, 23.

scraps: RUL, WAR to JARII, Case, III, December 1893.

almost masculine intellect: Schuyler, 233.

she sent him before their marriage, WAR to EWR, July 7, 1864, Emily Warren Roebling papers, RUL.

and effective working: Blockley, xv.

## Chapter One: "No one ever does just the right thing in great emergencies"

Nothing happened to me: WAR, 265.

will not impede navigation: New-York Daily Tribune, March 27, 1857.

through the state legislature at Albany: LER, 3, "An Act to incorporate the New York Bridge Company for the purpose of constructing and maintaining a bridge over the East River, between the cities of New York and Brooklyn," April 16, 1867.

"Greatest Possible Load Upon It": Brooklyn Daily Eagle, June 25, 1869.

at first what had happened: WAR, 265.

big shears would do: RUL, WAR to FWR, June 28, 1869.

foundations of the bridge: McCullough, 166.

a few blocks away: WAR, 258.

in great emergencies: Ibid., 265 ff.

monster in human form: Ibid., 44.

in a violent manner: Ibid., 266.

per million population per year: Diego F. Wyszynski and Mariana Kechichian, "Outbreak of Tetanus Among Elderly Women Treated with Sheep Cell Therapy." Clinical Infectious Diseases 24 (April 1997).

respiratory arrest and death: J. J. Farrar, et al., "Tetanus," Journal of Neurololgy, Neurosurgery & Psychiatry, 3 (September 2000): 292–301; doi: 10.1136/jnnp.69.3.292.

write the death certificate: WAR, 267.

future has in store for us: Ibid.

criminations [sic] are futile: Ibid.

dated September 1, 1867: JAR to New York Bridge Company, September 1, 1867, Brooklyn Bridge, Reports, RPI.

to do his business: *Brooklyn Daily Eagle*, September 10, 1867.

continuous village by 1860: Jackson, 284.

turned inside out: Ibid., 269.

third city of Christendom: Stone, 243.

855 piano makers; Ibid., 245–46.

thousands were ruined: "Black Friday, September 24, 1869," *American Experience,* www.pbs.org.

Metropolitan Museum of New York: Brown, *Fifth Avenue Old & New*, 120.

Parisians were worse off: *Catholic World* 9, no. 53 (August 1869): 553–66.

other results of neglect: *Catholic World* 9, no. 53 (August 1869): 553–66.

rates were one in twenty-five: Ibid., 553–66.

outside of the metropolis: Ibid.

rivals to their own: Stiles, vol. 2, 248.

of the city of Brooklyn: Stiles, vol. 1, 23–24.

of their boyhood: Stiles, vol. 2, 35.

neighbor, New York: Ibid., 420–21.

Dutch by 1642: Stiles, vol. 1, 25.

No one seems very sure: Leonard Bernardo and Jennifer Weiss, *Brooklyn By Name: How the Neighborhoods, Streets, Parks, Bridges, and More Got Their Names* (New York: New York University Press, 2006), 57.

wealthiest ferry company in the world: *New York Times*, September 4, 1870.

hundreds that cross, returning home: Whitman, 129.

in 1852 and 1856: Stiles, vol. 2, 489.

differing only in degree: *Brooklyn Daily Eagle*, January 23, 1867.

across the East River: Stiles, vol. 2, 496.

navigation of the river: *The Statues at Large, Treaties and Proclamations of the United States of America, from December 1867 to March 186,9* vol. 15 (Boston: Little, Brown, 1869). www.archive.org.

no less than President Lincoln: McCullough, 73.

on June 7th 1869: WAR, Engineering notebook 5, 2, Washington A. Roebling Papers, RUL.

Cornford & Collins: Ibid., 14.

marked by a crowfoot: Ibid., 26–27.

to July 22/69: WAR, Engineering notebook 1, 93, Washington A. Roebling Papers, RUL.

in some shape: WAR, 88.

died in infancy in 1838: Sayenga, 89.

before the Civil War began: Sayenga, 145.

weeks before John Roebling's death: McCullough, 95–97.

ended with his life: WAR, 89.

full partner in the business: McCullough, 96.

against them in my private ledger: JAR, Last Will and Testament, Personal Miscellany, John A. Roebling Papers, RPI.

no one else was: WAR to James Rusling, January 23, 1916, Washington A. Roebling Papers, RUL

before the letter above: Sayenga, xviii.

desired to attend: *New York Times*, July 26, 1869.
at the sides and ends: *Brooklyn Daily Eagle*, July 26, 1869.
what it all implies: WAR, 268.
valued at $1,250,000: *Brooklyn Daily Eagle*, August 4, 1869.
guardian of his youngest brother: WAR, 220.
At first I thought: McCullough, 411.
for an East River crossing: WAR, 244½.
would have to compete: McCullough, 119.
to their last resting place: *Brooklyn Daily Eagle*, July 26, 1869.
not wholly irreparable: *Brooklyn Daily Eagle*, July 22, 1869.
like a vision of heaven: *Brooklyn Daily Eagle*, February 11, 1884.
calculated and completed: *Brooklyn Daily Eagle*, July 22, 1869.

## Chapter Two: "The finest place in the world"

into old age: WAR, 4.
progress of a century has produced: Brown, *History of Butler County*, 14–18.
escape with her baby: Ibid., 35–36.
Main Street—he got away: WAR, *Early History of Saxonburg*, in Schuyler, 38–39.
the gospel of the New World: Brown, *History of Butler County*, 466.
One of the first attempts: R. C. Brown, *History of Butler County*, 1895 (Tucson: Americana Unlimited, 1974) 26–27.
brought them to the United States: Sayenga, 13.
are well set forth: WAR, 16.
Socrates as director: Ibid., 27.
$1.00 per acre: Ibid., 17.
what would become the center of town: Ibid., 60; author's correspondence with Drenda Gostkowski.
a mistake after all: Ibid., 24.
a Dutch shepherd: Ibid., 21.
sprouts came up for years afterwards: Ibid., 22.
worked to get the beast out: Ibid., 49.
an immense mystery to me: Ibid., 49.
more educated people: Ibid., 57.
for miles around: Ibid., 72.
routes of his boyhood: Sayenga, 53.
the flute and clavier: WAR, *Early History of Saxonburg*, in Schuyler, 163–64.
the playwright said of the drama: Hempel, 154.
think what I would say next: WAR to JARII, April 15, 1894, Case, III, RUL.
that boys delight in: WAR, 70.
the early impressions disturbed: RUL, WAR to JARII, January 5, 1926.
never left Mühlhausen: WAR, 1.
should become a great man: Ibid., 2–3.
Richard Gottheil and Isaac Broydé, "Erfurt." *The Jewish Encyclopedia* (1906). www.jewishencyclopedia.com.
included the study of suspension bridge: Guentheroth, 8.
and I second it: WAR, 5.

be asked to do so: Schuyler, 13.

master passion: WAR, 7.

forbid the newspaper craze: Ibid., 7.

And it was a craze: Margaret Latimer, et al., eds, *Bridge to the Future* (New York: New York Academy of Sciences, 1984), 81.

convey to him our encouragement: Kahlow, 1757.

such turmoil brings with it: Schuyler, 17–20.

then as well as now: WAR, 10.

diligence of his studies: Guentheroth, 10.

master builder: Ibid., 13.

which seawater has: JAR, Travel Journal, 35, John A. Roebling Papers, RUL.

water into the air: Ibid., 59.

induce a good movement: Ibid., 44.

as soon as possible: Ibid., 68–70.

all-disturbing European: Ibid., 84.

marriage to Johanna, in 1836: WAR, 40ff.

fell by 90 percent: Erie Canal website, www.eriecanalway.org, History and Culture.

water—and that is all: Harriet Beecher Stowe, "The Canal Boat," in Hecht, 97.

four dollars a day: Schuyler, 46.

after the father of his country: WAR, 54.

an American citizen, 1837: Schuyler, 46.

into the mountain depths below: Dickens, *American Notes*, 164–65.

could cost three thousand dollars apiece: Schuyler, 49.

Supt. of the mines in 1832: WAR, 73.

for making ropes of hemp: Donald Sayenga, "Modern History of Wire Rope," www.atlantic-cable.com.

whole rope went to pieces: WAR, 73.

pointing the way to the right path: Ibid., 73.

Method of and machine for making wire ropes: United States Patent and Trademark Office, Patent #US000002720.

the sum of seventeen thousand dollars a year: JAR, "Wire Rope Proposal—Allegheny Portage Railroad, 1844," John A. Roebling Papers, RPI.

fulfillment of the contract on my part: Ibid.

until this essay has been made: Thomas P. Jones, U.S. Patent Office, to JAR, March 24, 1841, John A. Roebling Papers, RUL.

foundation for his fortune: WAR, 78.

Germans work in Germany: Ibid., 77.

Mühlhausen savagery: Ibid., 37.

commerce to a shuddering halt: "1837: The Hard Times." Harvard Business School, Historical Collections, www.library.hbs.edu.

the Opposition for two years past: The *Times* (London), June 13, 1837.

they are no longer necessary: WAR, 81.

## Chapter Three: "Something of the tiger in him"

the Pittsburgh infirmary: Schuyler, 146.

a half million dollars today: www.measuringworth.com.

employer and the employed: Stuart, 325–26.

Charles Sumner Lobingier, "Personal Reminiscences of Henry Dodge Estabrook," *St. Louis Law Review* 9 (1924): 245.

iron to his account: Schuyler, 144.

trembled in his presence: WAR, 42–43.

great use in later life: Ibid., 35.

found room for these people: Ibid., 37.

ever dared enter it: Ibid., 37.

but it was impossible: Ibid., 42.

blow of the fist was nothing uncommon: Ibid., 42.

Heale wrote, to do so: Foyster, 64.

against such behavior on their books: Sommers, 205–6.

brought up a set of sneaks: WAR, 42.

life or death presented to it at once: Ibid., 43.

inexpressible grief: Ibid., 62.

A curious world indeed: Ibid., 43

it grows on me: Ibid., 45.

appointed him hospital chaplain: *New York Daily News*, Tuesday, May 28, 2013.

separates man from God: James, *Substance and Shadow*, 219.

university life in Berlin: WAR, 45.

the poor can have them: Ibid., 44.

not long after the birth of Jesus: Ralph Jackson, "Waters and Spas in the Classical World," Porter, 1.

amniotic fluids: Porter, viii.

England's Malvern baths: Janet Browne, "Spas and Sensibilities: Darwin at Malvern," Porter, 104.

cure is no quackery: Ibid., 109.

faith in the water cure: Ibid., 111.

as early as 1760: Weiss and Kemble, 169.

the idol of the water fiends: WAR, 94.

his occupation in the fields: Priessnitz, 6.

cold water is a certain remedy: Ibid., 38.

the answer to everything: Ibid., 9–16, 38–44.

were also crucial: Ibid., 30.

it is always much increased: JAR, water cure, John A. Roebling Papers, RPI.

that is worth everything: WAR, 95–96.

symbolic cultural meanings: Porter, xii.

back to her true meaning: JAR, "The Truth of Nature," "Life and Creation," John A. Roebling Papers, RUL.

happiness or misery results: JAR, "Man. Conscience." February–April 1863, John A. Roebling Papers, RUL.

"What hath God wrought?": "Today in History, May 24, 1844." Library of Congress, www.memory.loc.gov.

the greatest feat of his life: WAR, 82.

with wire for his ropes: Sayenga, 79.

executing the work: WAR, 82.

aqueduct 2,200 tons: Schuyler, 60–61.

at the ends into which they were laid: WAR, 84.

could not have made the cables in time: Ibid., 87.

black-haired Charles Swan: Ibid., 89.

everything was drenched: Ibid., 88.

unhealthy precocity can be engendered: Ibid., 89.

had to fly for his life: Ibid., 34.

rather than the English: Schuyler, 169.

the three R's in German: WAR, *Early History of Saxonburg*, in Schuyler, 159.

occupation of Paris, too: WAR, 91.

farewell to poor Riedel: WAR, *Early History of Saxonburg*, in Schuyler, 160.

happened to me in my life: WAR, 92.

affected my whole life: Ibid., 93.

like those of her eldest son: Zink, 26.

amounts to nothing in comparison: WAR, 107.

the first complete rolling mill for iron: "Chronology by Decade," Historic Pittsburgh, www
.digital.library.pitt.edu.

built over the Monongahela River: Tarr, 32.

half-frozen and hungry: WAR, *Early History of Saxonburg*, in Schuyler, 162.

days for a month in advance: WAR, 111.

I went to an English school: Ibid., 108.

into my dutch ears: Ibid., 111.

a rice pudding beggar: Ibid., 109.

dragoons at the battle of Ligny: Ibid., 110.

the door in the dark: Ibid., 109.

best of clothes himself: WAR, 112.

and is entirely consumed: *Pittsburgh Daily Gazette and Advertiser*, April 11, 1845.

a little money was made: WAR, 98.

"Historic American Engineering Record, Smithfield Street Bridge, Pittsburgh," Bridges and
Tunnels of the Alleghenies, www.pghbridges.com.

evil results: WAR, 115.

remained generous and kindly: Schuyler, 167.

the markets in New York City: Booth, 12.

Rondout at High Falls: WAR, 118.

was also consumed: Ibid., 119.

piled into four wagons: Ibid., 125.

the hungry army invade her premises: Ibid., 127.

persecution in Britain: Richman, 87.

by every method and in all directions: WAR, 119.

He knew exactly what he wanted: Ibid., 121.

apparently lifeless: Ibid., 130.

his greatest engineering works: Ibid., 131.

latest fads of Priessnitz: Ibid., 132.

heard howling at night: Ibid., 132.

never shall I forget that ride: Ibid., 134.

spinal concussion: Ibid., 136.

just my luck to strike it: Ibid., 144.

load of the water itself: Vogel, 11.

weather for seventy years: Ibid., 16.

20 or 30 of them: WAR, 137.
a small competency: Ibid., 56.
God of water cure: Ibid., 149.
enslaved to their machinations: Bourne, viii.
out of the swill barrel: WAR, 150.
another man's wife: Ibid., 151.
in the register during 1851–52: Sayenga, 137.
happy and prosperous: Kaestle, 82.

## Chapter Four: "I was not a chip off the old block"

4,588 patents: Hindle and Lubar, 79.
before railroads were ever dreamed of: Hawthorne, May 5, 1850.
only in the mind of the Engineer: Stuart, 260.
at Fairmount in Philadelphia, Pennsylvania: Ibid., 268.
of their equilibrium: Ibid., 270–71.
supporting the bridge: Ibid., 307–8.
and killed upon the spot: *Illustrated London News*, May 29, 1847.
revolutionized the industry in 1829: "Rainhill Trials," www.rainhilltrials.com.
because of metal fatigue in a girder: Peter R. Lewis and Colin Gagg, "Aesthetics Versus Function: The Fall of the Dee Bridge, 1847," *Interdisciplinary Science Reviews* 29, no. 2 (2004).
"Homan J. Walsh and the Kite that Helped Build a Bridge": Nebraska State Historical Society website, www.nebraskahistory.org, May 2010.
money Ellet proposed to keep: Stuart, 275.
a regiment to dislodge him: WAR, 156.
after 55 years becomes truth: Ibid., 156.
keep the directors of the bridge company off the structure: JAR, Niagara Falls, October 17, 1848, John A. Roebling Papers, Bridge Projects, RPI.
Ellet was a failure: WAR, 101.
50 plans that were never executed: Ibid., 153.
from Manchester, in England: Sayenga, 148.
gave rise to this magnificent structure: *New York Times*, March 21, 1855.
Sic Transit Gloria Mundi: WAR, 162.
aggregate to positive unhappiness: Twain, 16.
sometimes there is a little too much of it: Schuyler, 173.
wasn't a very good trade: Rittner, 23–24.
the Silicon Valley of the 19th century: P. Thomas Carroll, http://www.rpi.edu/dept/NewsComm /Magazine/March99/designing_america.html. "Designing Modern America: In the Silicon Valley of the 19th century," *Rensselaer Magazine* (March 1999), www.rpi.edu.
president of the state's first board of agriculture: Ricketts, 12–14.
within the reach of all: Ibid., 5–7.
learned their trade at West Point: Sayenga, 159.
established at Glasgow: Ricketts, 69–71.
and a zoologist too: Ibid., 22–26.
informal examinations for women: Ibid., 61.
excused as aforesaid: Ibid., 37–38.
performed by the student himself: Ibid., 64.

sixty dollars a year for the privilege: Ibid., 123.

Mrs. Griffin, in Niagara Falls: *New York Times*, December 30, 1884.

crushing institution in the whole world: WAR, 166.

left the school as mental wrecks: Schuyler, 174.

which he accused the Institute of producing: Ibid., 174.

squares of distances: Ricketts, 82–83.

not a chip off the old block: WAR, 166.

& then he is wrong: Ibid., 170.

lectures of the day before: Ibid., 167.

this was successfully accomplished: WAR, Personal Miscellany, Washington A. Roebling Papers, RPI.

weakening self-reliance: WAR, 167.

of the terrible grind: Schuyler, 174.

make allowances: unnamed friend to WAR, October 8, 1856, Washington A. Roebling Papers, RUL.

God knows what I shall do: Ibid.

Keep of my things whatever you like: unnamed friend to WAR, Thanksgiving day, 1856, Washington A. Roebling Papers, RUL.

sense enough to understand his love: WAR to EWR, April 1864, Washington A. Roebling Papers, RUL.

sixty-two pages in tiny, sepia-inked script: WAR, "Design for a Suspension Aqueduct—Washington A. Roebling," 1857, John A. Roebling Papers, RPI.

## Chapter Five: "It is curious how persons lose their heads in times of excitement"

$120 million: *New York Times*, "On This Day," October 24, 1857. www.learning.blogs.nytimes.com.

'this is the land': Cole, 55

take his rope orders away: WAR, 173.

Some of the land was bought: Iowa Land Grants, 1855–1861, John A. Roebling Papers, RPI.

till everyone got tired: Ibid., 174.

profitably sold later on: Ibid., 176.

their heads in time of excitement: Ibid., 179.

received us with great derision: Ibid., 179.

Civil War was nearly over: Ibid., 178.

producing $26,563,000 worth of goods: Lorent, 93.

a permanent domination: Thurston, 6, 20.

cables being made in place: WAR, 189.

Something of that kind occurs every day: Schuyler, 177.

passionately fond: WAR, 190.

ten times as many acquaintances as in Trenton: Schuyler, 179–80.

glad to escape with my life: WAR, 188.

I still live: Ibid., 190.

a huge Trussed structure: WAR, 190.

first mixed with cold water: WAR, "List of Persons . . ." Pittsburgh, PA, Notebook, Washington A. Roebling Papers, RPI.

This is a fixed rule: RPI, JAR, "Water Cure."

well-known local musician around that time: WAR, 197.

dodging his father: Ibid., 197.

little sympathy with the South: Ibid., 191.

the clearest light that shines on this land: Henry David Thoreau, "The Last Days of John Brown," www.walden.org.

clumsy jests and clownish grimace: May 24–28, 1860, in Gillette, 79.

electoral vote to Lincoln: Ibid., 3.

they are not responsible: "Lincoln and New Jersey: A Bicentennial Tribute by the New Jersey State Archives," www.nj.gov.

he warned: Ibid.

the American Civil War had begun: Fergus M. Bordewich, "Fort Sumter: The Civil War Begins," *Smithsonian Magazine* (April 2011), www.smithsonianmag.com.

Company A of the National Guard of Trenton: New Jersey Civil War Record, page 1427, Company A, National Guard, of Trenton, New Jersey State Library, www.njstatelib.org.

He did not buy a new hat: WAR, 182.

unless sooner discharged: "Civil War Enlistment," National Park Service, www.nps.gov.

514 K Street, Washington: Schuyler, 199–201.

pieces of the gallows: Miers, 9–10.

from entering the body: WAR, 199.

2.5 per cent of the country's whole population: "The Civil War by the Numbers," *American Experience*, www.pbs.org.

replies the laconic one: Schuyler, 195–96.

## Chapter Six: "The urgency of the moment overpowers everything"

rank of sergeant: Miers, 5.

neither does it kill you: WAR, 202.

through Charles County: Waite Rawls, "The Potomac Blockade," in Washington Post, *Civil War Stories*.

and hit my house: WAR, 203.

we would all go to the devil: WAR to FWR, December 19, 1861, Case, I, RUL.

tremendous explosions: Schuyler, 209.

harder times coming: Schuyler, 211–12.

for the enemy to hit: McPherson, 374–75;

also: "USS Monitor: Preserving a Legacy," www.monitor.noaa.gov.

Pearl Harbor in 1941: McPherson, 376.

There is an end of wooden ships forever: *Times* (London), March 29, 1862.

very small as she really is: WAR to Charles Swan, April 10, 1862, Case, I, RUL.

refrained from sinking her: WAR, 204.

nearly a month was wasted: Ibid., 205.

McPherson, 447; also Cynthia G. Fox, "Income Tax Records of the Civil War Years," *National Archives* (Winter 1986), www.archives.gov.

conscription law in American history: McPherson, 430.

except by complete conquest: Ibid., 414.

put a bullet through his brain: WAR to JAR, April 16, 1862, John A. Roebling Papers, RUL.

arms taken off yesterday: Ibid., April 18, 1862.

and do it himself: McPherson, 426.

slang at the time for experiencing combat: Ibid., 409.

experience to put them up: WAR, 206–7.

which I prized highly: Ibid., 207.

according to Washington's account: Ibid., 208.

with military suspension bridge: Ibid., 208.

with the tape in my mouth: WAR to FWR, June 8, 1862, Case, I, RUL.

official red tape: WAR, 209.

just to get rid of me: Ibid., 210.

all the rest put together: Ibid., 210.

he had been promoted in February: WAR to JAR, February 5, 1862, John A. Roebling Papers, RUL.

in the jail for safekeeping: WAR, 211.

his back on the land span: WAR to Charles Swan, August 3, 1862, Roebling Family, Charles Swan, RUL.

I was concerned personally: WAR, 211.

almost a spy: Ibid., 213.

staff at breakfast: WAR to JAR, August 24, 1862, Case, I, RUL.

no forage for the animal at all: "Claim for indemnity for a horse lost by Lt. W. A. Roebling," Washington A. Roebling Papers, RUL.

your bitterest enemies: WAR to JAR, August 24, 1862, Case, I, RUL.

never combined in one man: "The Second Battle of Manassas," History E-Library, National Parks Service, www.nps.gov.

that is about the proportion: WAR to JAR, August 24, 1862, Case, I, RUL.

resist shedding a tear: WAR, 214½.

wind out of my mouth: WAR, 216.

would have won the battle in his position: Ibid., 218.

most people get who go to a war: Ibid., 218.

place where Hooker was shot in the foot: "Map of the Battlefield of Antietam," Washington A. Roebling Papers, RUL.

photographs of the battlefield: National Parks Service, Photo Gallery, Historic Photographs, Antietam National Battlefield, www.nps.gov.

lying where they fell: "Map of the Battlefield of Antietam," Washington A. Roebling Papers, RUL.

recrossed the Potomac unmolested: WAR, 218.

and forever free: McPherson, 557.

All diseases are unnecessary: WAR, 218.

a picnic and were glad to come: Ibid., 220.

assumed a new aspect directly: Miers, 19–20.

in keeping with the rest of the bridge: WAR to JAR, January 12, 1863, John A. Roebling Papers, RUL.

Ambrose Burnside as commander of the Army of the Potomac: McPherson, 569.

Hooker was put in command: Ibid., 585.

appetites of the cannons' mouths: Ibid., 601.

the worst in American history: Ibid., 609.

the remainder of the war: WAR, 221.

know how to swing an axe: Ibid., 268.

useless work is done in a war: Ibid., 269.

Chancellorsville was lost right here: Ibid., 272.

only half drunk: Ibid., 273.

a used-up man: WAR to Gamaliel Bradford, December 15, 1914, Case, III, RUL.

might still have been saved: WAR, 272.

privilege of the private: Ibid., 273.

his utmost to help him win: WAR to Gamaliel Bradford, December 15, 1914, Case, III, RUL.

first battle ever to be won thanks to control of the air: U.S. Centennial of Flight Commission, "Military Use of Balloons During the Napoleonic Era," www.centennialofflight.net.

stitching the pieces together: Lowe, 11.

ice through the nozzle: Ibid., 46.

chief of the Corps of Aeronautics of the United States Army: Ibid., 75.

almost nothing is heard: WAR, 276.

taking the war into the Union's home territory: McPherson, 647.

we commenced to follow Lee: WAR, 282.

my father could not comprehend: Ibid., 200.

never to recover: Ibid., 283.

I knew whole armies were near by: Ibid., 284.

showed in his every act: WAR to Gamaliel Bradford, December 15, 1914, Case, III, RUL.

I'm fighting because you're down here: Shelby Foote, *Paris Review*, "The Art of Fiction," no. 158 (Summer 1997).

vantage points for defense: McPherson, 653.

pushing them up by hand: WAR, 286–87.

must have killed him: Ibid., 288.

staying there without getting killed: WAR to James F. Rusling, February 18, 1916, Washington A. Roebling Papers, RUL.

one of the most famous assaults of the war: James R. Brann, "Defense of Little Round Top," July 2, 1863, Civil War Trust, http://www.civilwar.org/.

invincibility is broken: McPherson, 662.

slight rise were reasonably protected: WAR, 291.

starve or retreat: Ibid., 293.

much pity in the human breast: Ibid., 295.

and there it stays: Ibid., 294.

and they would not close it: McPherson, 667.

disappointment to Gen. Lee: WAR, 297.

our army crossed on it: Ibid., 298.

wooden trestle pier with axes: WAR, "Diary of the suspension bridge at Harpers Ferry, June 1862–January 1864," Washington A. Roebling Papers, RUL.

## Chapter Seven: "I am very much of the opinion that she has captured your brother Washy's heart at last"

sitting still with all our might and main: Schuyler, 224.

was his principal aide-de-camp: WAR to James F. Rusling, February 18, 1916, Washington A. Roebling Papers, RUL.

no matter who the generals are: WAR to Charles Swan, August 5, 1863, Roebling Family Papers, Charles Swan, RUL.

waited for nothing: WAR to James F. Rusling, February 18, 1916, Washington A. Roebling Papers, RUL.

Plantation of New Plymouth: Morton, 89.

as William Bradford recorded: William Bradford, ed. *Samuel Eliot Morison, of Plymouth Plantation*, 68–72.

Speaker of the House of Representatives: Emily Warren Roebling, *Richard Warren of the Mayflower and Some of His Descendants*, 26.

   This copy in the New York Public Library, number 19 of 50, had a card included, "With the compliments of Mrs. Washington Augustus Roebling."

was thirteen years her senior: Weigold, 4.

keep their feet dry: Taylor, 5.

choose a suitable husband: Ibid., 6.

needlework, painting, and music: Weigold, 4.

not so very small: WAR to EWR, April 21, 1864, Emily Warren Roebling Papers, RUL.

object of interest to contemplate: WAR to EWR, July 7, 1864, Emily Warren Roebling Papers, RUL.

a most lovely complexion: WAR to Elvira, February 26, 1864, Case, II, RUL.

various emotions: Ibid.

an old maid if she had waited for you: Johanna Roebling to WAR, July 12, 1858, translated from German, Washington A. Roebling Papers, RUL.

the shins: WAR to EWR, April 4, 1864, Emily Warren Roebling Papers, RUL.

proxy you know: Ibid., March 28, 1864.

Your adoring Washy: Ibid., undated.

with great gusto: Ibid., December 16, 1864.

much in that line as I do: Ibid., October 29, 1864.

don't obey willingly: Ibid., September 19, 1864.

making her personal acquaintance: JAR to WAR, March 30, 1864, Washington A. Roebling Papers, RUL.

before he was married: WAR to EWR, April 4, 1864, Emily Warren Roebling Papers, RUL.

respect and esteem naturally follow: Ibid., April 5, 1864.

long may he wave: Ibid., April 21, 1864.

her loving Wash: Ibid., June 8, 1864.

cashier him on the spot: WAR to Morris Schaff, May 18, 1909, Washington A. Roebling Papers, RUL.

to the rank of major: Schuyler, 189.

New York State Archive, Gouverneur Kemble Warren Papers, www.nysl.nysed.gov.

most Southern ladies are: Miers, 26.

their homes and their country: Ibid., 28.

stand up and get shot: Ibid., 27.

heard of his assassination: Schuyler, 196–197.

washed I lie abed: WAR to EWR, June 22, 1864, Emily Warren Roebling Papers, RUL.

hope for the best: Ibid., May 21, 1864.

tickled to death at it: Ibid., June 10, 1864.

she never had the opportunity: Ibid., June 19, 1864.

inducement for investment: *Cincinnati Enquirer*, May 6, 1864.

vastly improved thereby: Greve, 849.

formidable than ever: McPherson, 757.

walk in his ways: WAR to EWR, July 27, 1864, Emily Warren Roebling Papers, RUL.

another of the kind: Ibid., August 2, 1864.

the way to the assault: *New York Times*, August 2, 1864.

had been gained was quickly lost: McPherson, 760.

the flowing bowl: Miers, 29.

suppose is true of worms; Ibid., 29.

a barren waste: McPherson, 778.

all were dead within me: Masur, 256.

even at the President's levees: Kaplan, 71.

in case we ever have a fight: WAR to Charles Swan, September 1, 1864, Roebling Family Papers, Charles Swan, RUL.

I was not there to vote: Miers, 31.

entire recovery will take a year: JAR to WAR, November 17 1864, Washington A. Roebling Papers, RUL.

Mother died this P.M.: Charles Swan to WAR, November 22, 1864, Washington A. Roebling Papers, RUL.

greatest giver of us all: WAR to EWR, December 25, 1864, Emily Warren Roebling Papers, RUL.

it will all be forgotten: WAR, 229.

to right the wrong: WAR to James F. Rusling, February 18, 1916, Washington A. Roebling Papers, RUL.

promoted to colonel: Schuyler, 189–90.

service during the war: Schuyler, 189–90.

## Chapter Eight: "All beginnings are difficult, but don't give up"

young women in the front seats: *Trenton Evening Times*, March 31, 1899.

success and prosperity: EWR, "A Wife's Disabilities."

Have you got one: RUL, WAR to EWR, January 4, 1865.

married Cornelia Barrows: Weigold, 15–16.

White kids: WAR, notebook 1865, Washington A. Roebling Papers, Personal Miscellany, RPI.

looked upon like anything else: War to EWR, March 16, 1865, Emily Warren Roebling Papers, RUL.

a person's neck aches: WAR to Charles Swan, March 16, 1865, Roebling Family Papers, Charles Swan, RUL.

hoisting machines, etc: WAR to FWR, March 29, 1865, Ferdinand W. Roebling Papers, RUL.

will make for a beginning: Ibid., February 14, 1866.

so why should I: WAR, 226.

Sesesh house: WAR to FWR, April 17, 1865, Ferdinand W. Roebling Papers, RUL.

Washington and Emily eventually lodged: Tolzmann, "Roebling Heritage Tour."

sympathizers as they are called: WAR to FWR, May 12, 1865, Ferdinand W. Roebling Papers, RUL.

looked a year in advance: Ibid., July 1865.

one and a half million dollars: Annual Report of the President and Directors to the Stockholders of the Covington & Cincinnati Bridge Company, 1867, John A. Roebling Papers, RPI.

**Porkopolis:** Steve C. Gordon, "From Slaughterhouse to Soap-Boiler," *Journal of the Society for Industrial Archaeology*, 16, no. 1 (1990).

**200,000 souls:** Greve, 848–49.

**all the more felt:** 1867 C&C Annual Report, John A. Roebling Papers, RPI.

**a pontoon bridge out of old coal barges:** Geoffrey C. Walden, "Ohio in the War," Cincinnati Civil War Round Table, www.cincinnaticwrt.org.

**novelty of the thing:** Richard McCormick, "Bridging North and South: The Pontoon Bridge from Cincinnati to Kentucky," My Civil War Obsession, September 10, 2010, www.civilwar obsession.com.

**bridge offices were just around the corner:** Tolzmann, "Roebling Heritage Tour."

**while the iron is hot:** WAR, 223.

**begging off another:** June 2, 1865, Ferdinand W. Roebling Papers, RUL.

**Anton Methfessel:** "Charles Roebling," Roebling Museum, www.roeblingmuseum.org.

**fifty-one feet wide:** Farrington, 6.

**agitating my mind:** WAR to FWR, June 29. 1865, Ferdinand W. Roebling Papers, RUL.

**hot as you may imagine:** Ibid., June 21, 1865.

**chance on your hide:** WAR, 228.

**These days it looks:** Arthur C. Benke and Colbert E. Cushing, *Rivers of North America* (Burlington, MA: Elsevier Academic Press, 2005), 381.

**to receive the cables:** Farrington, 6–7.

**than the sound wire:** Ibid., 9.

**steam-powered engine:** Ibid., 7.

**middle of November:** WAR to FWR, October 5, 1865, Ferdinand W. Roebling Papers, RUL.

**when part way over:** Farrington, 8.

**operation of the workmen** *Cincinnati Enquirer*, October 9, 1865.

**hideous nightmare to me:** WAR, 234.

**before we come home:** WAR to FWR, October 14, 1865, Ferdinand W. Roebling Papers, RUL.

**from the Roebling mill in Trenton:** Farrington, 13.

**vastly for the better:** WAR to FWR, November 8, 1865, Ferdinand W. Roebling Papers, RUL.

**many gray hairs:** WAR, 234.

**damaged the iron:** Ibid., 223.

**all crushed alike:** Ibid., 236.

**but don't give up:** Ibid., 231.

**Michigan and Indiana:** WAR to FWR, April 27, 1866, Ferdinand W. Roebling Papers, RUL.

**young hurricane:** Ibid., June 2, 1866.

**wrapped by the end of July:** Ibid., July 22, 1866.

**a truss in name only:** WAR, 238.

**on site for only a few weeks:** Ibid., 239.

**deserve to succeed:** Ibid., 226.

**this great enterprise:** *Cincinnati Enquirer*, January 2, 1867.

**executed the design:** Farrington, 15, 16.

**people can buy on the spot:** WAR to FWR, April 21, 1867, Ferdinand W. Roebling Papers, RUL.

**completed by January 1st, 1870:** McCullough, 24.

**likely to accomplish this great enterprise:** New York and Brooklyn Bridge Proceedings, 320.

**whether that was in fact:** WAR to FWR, February 28, 1866, Ferdinand W. Roebling Papers, RUL.

**burgeoning since the 1840s:** Kate Scott, October 29, 2013, OUPblog, www.blog.oup.com.

**short trip to Europe:** WAR, 241.

Brooklyn Br of the future: Ibid., 242.

deserving of yours: McCullough, 55.

rope shop in Trenton: WAR, 66.

come back again in two weeks: FWR to WAR, November 12, 1867, Washington A. Roebling
      Papers, RUL.

A thing is and is not: JAR, "Truth of Nature," February 1862, John A. Roebling Papers, RUL.
All else refers to it: Ibid.

Little Willy: JAR, November 24, 1867, Spiritualism Meetings, Questions, and Notes, 1867–68,
      John A. Roebling Papers, RPI.

at the age of thirty-three: trentonhistory.org, index for 1859, retrieved April 17, 2016.

Which had left her deaf: WAR, 44.

that any death could affect me as much: JAR to WAR, October 14, 1861, Washington A. Roe-
      bling Papers, RPI.

no use for metaphysics: WAR, 118.

many and varied ramifications: Delp, 48.

spirit is dissolved: Ibid., 52.

Paradise, effectively: Ibid., 54.

the Spiritualist paradise: T. Peter Park, The Anomalist, www.anomalist.com.

which he conveys: *New York Times*, December 27, 1872.

also "in detail": JAR, November 24, 1867, John A. Roebling Papers, Spiritualism, RPI.

written beside his words: Ibid., December 29, 1867.

she answered *Yeas*: Ibid., November 24, 1867.

lime ores he was analyzing: WAR, 117.

Reichenbach correct in his views: JAR, March 22, 1868, John A. Roebling Papers, Spiritual-
      ism, RPI.

Willy are you grown any: Ibid., January 25, 1868.

never satisfied: WAR, 151.


## Chapter Nine: "I will have to go to work at something"

no end of trouble: WAR, 243.

traditional agricultural ones: Burrows and Wallace, xvii.

along the chief thoroughfare: Ibid., 653.

on the swift little steamers: WAR to JAR, July 14, 1867, Brooklyn Bridge Papers, RPI.

larger than our own mill horses: WAR to Charles Swan, July 17, 1867, Brooklyn Bridge Papers,
      RPI.

a half million barrels of beer: T. R. Gourvish and R.G. Wilson, *The British Brewing Industry,
      1830-1980*, Cambridge: Cambridge University Press, 2009, 610–11.

laid in July 1865: *Times* (London), November 4, 1869.

slightest camber to the floor: WAR to JAR, July 19, 1867, Brooklyn Bridge Papers, RPI.

useful insects: *Paris Universal Exhibition 1867, Complete Official Catalogue*, contents page.

metallic production of the world: Hewitt, 1–2.

to the age of steel: Ibid., 29.

in the way of traveling abroad: WAR to JAR, July 29, 1867, Brooklyn Bridge Papers, RPI.

St. Pancras station to his credit: History of St. Pancras Station, www.stpancras.com.

drawings of that bridge's caissons: WAR to JAR, August 3, 1867, Brooklyn Bridge Papers, RPI.

looks better than the reality: Ibid., August 16, 1867.

all I came to England for: Ibid., August 16, 1867.

wire-making considerably: Museum of Science and Industry, Manchester, England, Rolling Mill, item 1976.45.

deserved the highest prize: Hewitt, 5.

see it for himself: WAR to JAR, October 2, 1867, Brooklyn Bridge Papers, RPI.

the English Hof: WAR to Charles Swan, September 13, 1867, Brooklyn Bridge Papers, RPI.

whether you still have them: WAR to JAR, November 14, 1867, Brooklyn Bridge Papers, RPI.

Krupp factory in Essen: Ibid., August 23, 1867.

alone weighed forty-eight tons: Modern Industries, 100.

Bessemer steel is by far the best: WAR to JAR, January 1868, Brooklyn Bridge Papers, RPI.

with a pointed chin: WAR to Charles Swan: December 21, 1867, Brooklyn Bridge Papers, RPI.

Bach had once played the organ there: "Divi Blasii Church," Mühlhausen city website, www.muelhausen.de.

without which such ceremonies: WAR to JAR, January 6, 1868, Brooklyn Bridge Papers, RPI.

famous as you have done: EWR to JAR, January 6, 1868, John A. Roebling Papers, RUL.

was not lost on the father: McCullough, 167.

the steady use of one: WAR to JAR, August 25, 1867, Brooklyn Bridge Papers, RPI.

will take well in the U.S.: Ibid., August 25, 1867.

may as well do that: WAR to FWR, November 14, 1867, Brooklyn Bridge Papers, RPI.

make something of it: WAR to Charles Swan, September 13, 1867, Brooklyn Bridge Papers, RPI.

let me know what you think: WAR to FWR, November 12, 1867, Brooklyn Bridge Papers, RPI.

by Washington's account: WAR, 1.

following the Roeblings: Ibid., 145.

There is no money in it: JAR to WAR, November 10, 1867, Brooklyn Bridge Papers, RPI.

high-pressure steam: Thulesius, 71.

had sunk in a storm of New York: *New York Times*, April 28, 1854.

Wilhelm Maybach: Day and McNeil, 717.

anything on the matter myself: WAR to Charles Swan, September 13, 1867, Brooklyn Bridge Papers, RPI.

He refused ever to ride in one: McCullough, 558.

during the stormy winter months: WAR to JAR, undated 1867, Brooklyn Bridge Papers, RPI.

Signed yours &c: *Journal of Commerce*, March 25, 1857.

and a pension bill: *Brooklyn Daily Eagle*, January 29, 1869.

had come to blows: WAR, 162.

between the two men: Ibid., 163, 244, 245.

friendship with Adams; Ibid., 224½.

the whole time of the fair: *Engineering News-Record* 10 (January 20, 1883): 25.

where his great works stood: McCullough, 35-36.

bring it in with them: *Brooklyn Daily Eagle*, April 17, 1869.

$6,675,257: "Report of John A. Roebling, C.E., to the President and Directors of the New York Bridge Company of the Proposed East River Bridge," *Brooklyn Daily Eagle*, September 10, 1867.

twice that sum: McCullough, 506.

projector of the East River Bridge: *Brooklyn Daily Eagle*, May 24, 1883.

free and above board: WAR, 246.

promotion of the enterprise: Ibid.

Just like the East River Bridge: WAR to FWR, April 7, 1869, Ferdinand W. Roebling Papers, RUL.

full about the Bridge: Ibid., April, 1867.
roof of the Grand Central Depot: *New York Times*, January 21, 1910; "Report of the Chief Engineer," January 1, 1877, Brooklyn Bridge Papers, RPI.
in regard to lumber: WAR, East River Notebooks, book 2, 25–26, Washington A. Roebling Papers, RUL.
more money than is in the country: FWR to WAR, May 1869, Ferdinand W. Roebling Papers, RUL.
approval of the Executive Committee: McCullough, 144.
the one held accountable: Ibid., 147–48.

## Chapter Ten: *"Good enough to found upon"*

At the time of his death: WAR, 221.
its way toward completion: Ibid.
suitable shafts and air locks: Washington A. Roebling, PTF, 14.
high-pressure atmosphere: Schuyler, 241.
How to sink it: JAR, "Foundations and Caissons—Notes & Sketches, 2/4," Brooklyn Bridge Papers, RUL.
*nouvelles annals des constructions*: WAR, "Notes on European Bridges," Brooklyn Bridge Papers, RPI.
a caisson in Brest: WAR, "Description of a Caisson & Cofferdam erected at Brest in France—1867 & 1868, as described in the Government descriptions of designs and plans of such works sent to the Vienna exhibition," Brooklyn Bridge Papers, RPI.
picking up things: WAR, "First Annual Reports of the Chief Engineer and General Superintendent of the East River Bridge, June 12, 1870," Brooklyn Bridge Papers, RPI.
comparatively short distances: PTF, 14.
layers of earth and rock: New York City Municipal Archives, image 2028.
fancied contingencies: PTF, 18.
folly that was ever seen: *Mechanics' Magazine*, December 15, 1857.
promptly and instantaneously: *Brooklyn Daily Eagle*, March 21, 1870.
Dante's 'Inferno': *New York Times*, December 16, 1879.
digging down in the river: WAR, September 28, 1869, Municipal Archive, 656.
density of the atmosphere: *Brooklyn Daily Eagle*, June 29, 1870.
steel pick had no effect: PTF, 27.
northeast of Brooklyn: Ibid.
deep natural fissures: WAR, East River Notebooks, book 4, November 14, 1869, Washington A. Roebling Papers, RUL.
an earthquake had happened: WAR, "Calculations of Pressure—Brooklyn Caisson," Brooklyn Bridge Papers, RPI.
five tons per square foot: PFT, 43–45.
air should blow out: Ibid., 46.
sufficient presence of mind: Ibid., 47–49.
turning on the compressed air: WAR, "Blowing out of the air in Caisson through Supply Shaft," January 5, 1871, Brooklyn Bridge Papers, RPI.
beyond a scare: WAR to FWR, January 9, 1871, Ferdinand W. Roebling Papers, RUL.
The masonry: "Notes on the Masonry of the East River Bridge, A Paper by Francis Collingwood," presented to the American Society of Civil Engineers, November 1, 1876.
if it had been sound: WAR to FWR, October 23, 1870, Ferdinand W. Roebling Papers, RUL.

no air bubbles: WAR, Miscellaneous note, Notes and Designs 1856–1896, Brooklyn Bridge Papers, RPI.

has not been seen since: *Brooklyn Daily Eagle*, December 5, 1870.

fire extinguishers had no effect: McCullough, 233.

reinflation impossible: PTF, 52–53.

precedes paralysis: WAR, December 1, 1870, Brooklyn Bridge Papers, RPI.

eighty degrees: PTF, 56.

Caisson was burning yet: WAR, December 1, 1870, Brooklyn Bridge Papers, RPI.

where uncertainty is profound: Blockley, 240.

to improve if possible: PTF, 35.

four minutes with ease: Ibid., 37.

seven more courses: WAR, "Third Annual Report of the Chief Engineer, New York Bridge Company, June 1, 1872," Brooklyn Bridge Papers, RPI.

large scale in this country: Ibid., 71.

Washington's notebooks: WAR, East River Notebooks, book 1, Washington A. Roebling Papers, RUL.

by a large fragment: PTF, 82–84.

world above: Ibid., 89.

twenty-four pounds per square inch: McCullough, 298.

paralysis of arms and legs: PTF, 87.

to attend to the men: Smith, 17.

human interaction with developing technology: Phillips, 2–4.

an ocean of air: West, John B.

at sea level in Paris: Phillips, 17.

would effect a cure: Ibid., 17-54.

attending deep pneumatic foundations: WAR, "Third Annual Report of the Chief Engineer, New York Bridge Company, June 1, 1872," Brooklyn Bridge Papers, RPI.

torn from the bones: Smith, 41.

to which it may be necessary to go: WAR, "Caissons & Air-Shafts—Notes," Brooklyn Bridge Papers, RPI.

quite insensible: Smith, 63.

than upon any similar work: WAR, "Third Annual Report of the Chief Engineer, New York Bridge Company, June 1, 1872," Brooklyn Bridge Papers, RPI.

where the blockages occur: Phillips, 119–20.

Jaminet and Smith were considering: Ibid., 102.

the real cause of decompression sickness: Ibid., 118.

disagreed with Bert explicitly: Smith, 48.

decompress from such a depth: Phillips, 73.

depth of seventy eight feet: PTF, 85–87.

enterprise depended on their success: WAR, "Third Annual Report of the Chief Engineer, New York Bridge Company, June 1, 1872," Brooklyn Bridge Papers, RPI.

intense anxiety for Col. Roebling: Biography of WAR, Notes & Designs, Memoranda and Letter copies, Brooklyn Bridge Papers, RPI.

## Chapter Eleven: "I have been quite sick for some days"

Exposure to the compressed air: Smith, 32–37.

construction of the East River Suspension Bridge: WAR to John W. Glenn, April 5, 1871, Washington A. Roebling Papers, RPI.

concentration of the load on the bridge: WAR to Wilkins, May 15, 1872, Washington A. Roebling Papers, RPI.

enjoying life for the first time: WAR's "history" of Edmund Roebling and his financial affairs, March 19, 1922, Edmund Roebling Papers, RUL.

Wash's dole: lists of household expenses, 1871, Washington A. Roebling Papers, RUL.

his death was hourly expected: Schuyler, 243.

stupefied sufficiently: WAR to FWR, May 23, 1872, Ferdinand W. Roebling Papers, RUL.

a seat of extreme suffering: Smith, 43.

afforded by morphine: PTF, 87–88.

doubtful: Smith, 54.

even more vital now: Schuyler, 243–44.

may be able to get to Trenton: WAR to FWR, January, 1873, Ferdinand W. Roebling Papers, RUL.

requested a leave of absence: McCullough, 321.

Dumas and Manet: Roe, 198.

London's Millwall Docks: *Survey of London*, vol. 43 and 44, *Poplar, Blackwall and Isle of Dogs.* Originally published by London County Council, London, 1994. "The Millwall Docks: The Buildings," 356–74. www.british-history.ac.uk.

It makes a nice splice: WAR to FWR, June 1873, Ferdinand W. Roebling Papers, RUL.

Emily is spending more than I expected: Ibid., June 1873.

a power of ink: Ibid., August 1873.

resolving not to get it: Schuyler, 96.

until our suit is decided: WAR to FWR, June 4, 1873, Ferdinand W. Roebling Papers, RUL.

Washington's brother-in-law: McCullough, 346.

Washington handing Eads five thousand dollars: McCullough, 345.

could not be subject to a patent: Gandhi, "The St. Louis Bridge . . ."

more difficult one than the East River: McCullough, 345.

remote parts of the Caisson: WAR, East River Notebooks, book 1, March 19, 1870, Washington A. Roebling Papers, RUL.

previous to the launch: Lawsuit statement of James B. Eads v New York Bridge Company, Brooklyn Bridge Papers, RPI.

affected by his failing health: *Scientific American*, April 23, 1887.

New York anchorage: McCullough, 565.

## Chapter Twelve: "Now is the time to build the Bridge"

rougher sort of people: Jackson, 1205.

sainted him and keep his day: Golway, 4–5.

payment padding: Ibid., 91.

$73 million in 1871: Ibid., 92.

known to have cost $45,000: *New York Times*, June 1, 1871.

his arrest on October 26: Jackson, 1206.

he had been to the first meeting: McCullough, 127.

**This ring was overthrown in 1871:** WAR to William Couper, July 26, 1907, Washington A. Roebling Papers, RUL.

**honest judgment on this question:** McCullough, 273.

**$9.5 million:** WAR, "Report of the Chief Engineer of the East River Bridge on Prices and Materials and Estimated Cost of the Structure," June 28, 1872, Brooklyn Bridge Papers, RPI.

**largely financed the Union Army during the war:** Harvard Business School Historical Collections, "1873: Off the Rails," www.library.hbs.edu.

**even when offered below par:** WAR to FWR, October 28, 1873, Ferdinand W. Roebling Papers, RUL.

**To build now is to save money:** "Report of the Officers of the New York Bridge Co to the Board of Directors," February 1875, Brooklyn Bridge Papers, RPI.

**blind, deaf, and mute:** U.S. Department of Transportation, Federal Highway Administration, "Women in Transportation: Bridge Construction"; https://www.fhwa.dot.gov/wit/bridgec .htm, accessed October 26, 2016. *New York Daily News*, May 24, 2012.

**engineer and foremen:** Francis Collingwood, December 31, 1875, Brooklyn Bridge Papers, RPI.

**when actually present at it:** WAR to H. C. Murphy, December 6, 1875, Brooklyn Bridge Papers, RPI.

**support the weight of the bridge:** McCullough, 330.

**socially or in any other way:** WAR to H. C. Murphy, December 6, 1875, Brooklyn Bridge Papers, RPI.

**Brooklyn Boys' Preparatory School:** National Register of Historic Places Continuation Sheet, section 7, page 2, www.archbold-station.org, http://www.archbold-station.org/documents/ ABSatRedHill-NRHP-HistoricalNarrative-200609.pdf, accessed September 10, 2015.

**puzzling their brains:** *New York Times*, May 22, 1883.

**he gets all the credit:** *Boston Weekly Globe*, May 29, 1883.

**Mrs. W. A. Roebling:** EWR Brooklyn Bridge Scrapbook No. 2, Emily Roebling Papers, RPI.

**to give the work his personal supervision:** EWR Brooklyn Bridge Scrapbook No. 1, Sunday *Eagle*, Sunday morning, May 20, 1877, Emily Roebling Papers, RPI.

**345 feet above the foundation:** WAR, "Annual Report of the Chief Engineer," 1876, Brooklyn Bridge Papers, RPI.

**345 feet above the foundation . . . six hundred miles away:** Ibid.

**would be borne not by stockholders:** LER, 12, "An act providing that the bridge in the course of construction over the East River, between the cities of New York and Brooklyn, by the New York Bridge Company, shall be a public work of the cities of New York and Brooklyn and for the dissolution of said company, and the completion and management of the said bridge by the said cities," May 14, 1875.

**Closely written document:** WAR, "Cables—Note of manufacture and design," Brooklyn Bridge Papers, RPI.

**ultimate strength of 10,730 tons:** Hildenbrand, 10–11.

**will take 2 years in any event:** WAR, "Cables—Note of manufacture & design," 2, Brooklyn Bridge Papers, RPI.

**jump out of the sheaves:** Ibid., 11.

**would be wrapped with wire:** WAR, "Annual Report of the Chief Engineer," 1876, Brooklyn Bridge Papers, RPI.

**a thousand men employed for years:** WAR to Henry Cruse Murphy, November 21, 1876. Brooklyn Bridge Papers, RPI.

**beyond all contingencies:** WAR, "Annual Report of the Chief Engineer," 1876, Brooklyn Bridge Papers, RPI.

**laden with a picnic party:** *Brooklyn Daily Eagle*, Monday August 14, 1876.

**considered an established fact:** Hildenbrand, 46–48.

**in his perilous position:** *New York Times*, August 26, 1876.

**I may have overestimated the difficulties:** E. F. Farrington, "Report on Cable-Making to the Chief Engineer," Brooklyn Bridge Papers, RPI.

**important point to all concerned:** EWR, Brooklyn Bridge Scrapbook No. 2, *Union*, May 28, 1878, Emily Roebling Papers, RPI.

**every step that is taken on the bridge:** EWR, Brooklyn Bridge Scrapbook No. 1, *Sun*, September 3, 1876, Emily Roebling Papers, RPI.

**the best that ever were made:** E. F. Farrington, "Report on Cable-Making to the Chief Engineer," Brooklyn Bridge Papers, RPI.

**Steel is stronger:** Blockley, 125.

**not be willing to undertake it:** "Specifications for Steel Cable Wire for the East River Suspension Bridge, 1876," Brooklyn Bridge Papers, RPI.

**their house on Columbia Heights:** *Brooklyn Daily Eagle*, May 20, 1877.

**dangerous and difficult tasks:** C. C. Martin to WAR, December 31, 1876, Brooklyn Bridge Papers, RPI.

**made the crossing that very morning:** Ibid., February 12, 1877.

**"Beauty on the Bridge":** Ibid., February 23, 1877.

**finally dismissed:** *New York Times*, April 1, 1880.

**what is the reason in your opinion:** WAR to William Paine, December 3, 1877, John A. Roebling Papers, RPI.

**any two men can test so much wire:** WAR to H. C. Murphy, December 3, 1877, John A. Roebling Papers, RPI.

**The meeting was adjourned:** *New York Times*, September 8, 1876.

**Surely the time was right:** McCullough, 374.

**a thief who tries to get the best of you:** WAR, 35.

**his own personal advantage:** WAR to H. C. Murphy, September 8, 1876, John A. Roebling Papers, RPI.

**investigate a little:** WAR to H. C. Murphy, September 11, 1876, John A. Roebling Papers, RPI.

**indebted to Hewitt:** McCullough, 396.

**gratifying his private revenge:** WAR to A. Schroeder, October 18, 1876, Brooklyn Bridge Papers, RPI.

**He sold all his shares:** WAR to H. C. Murphy, November 2, 1876, Brooklyn Bridge Papers, RPI.

**rigid inspection will serve:** WAR to FWR, undated, Brooklyn Bridge Papers, RPI.

**galvanizing indifferent:** in "Notes and Designs 1856–1896," undated, Brooklyn Bridge Papers, RPI.

**contract to the lowest bidder:** LER, WAR to H. C. Murphy, December 15, 1876.

**same certificate of inspection:** McCullough, 444.

**two "harrowing" years:** WAR to H. C. Murphy, Spring 1878, John A. Roebling Papers, RPI.

**alter the wrapping machines:** WAR to H. C. Murphy, August 6, 1878, John A. Roebling Papers, RPI.

**as he deems proper under the circumstances:** McCullough, 447.

**nothing further takes place:** WAR to H. C. Murphy, August 6, 1878, John A. Roebling Papers, RPI.

**at his own expense:** McCullough, 446.

was revealed to the world: McCullough, 446.
Guardian Angels: *New York Times*, April 13, 1878.
Bell Telephone Company: Landau and Condit, 110.
known and determined beforehand: *Brooklyn Daily Eagle*, October 7, 1878.
hard labor at Sing-Sing: McCullough, 472.
had tried to bribe his guards: *New York Times*, August 7, 1880, and September 19, 1880.
a young miss of 10: *New York Times*, January 8, 1881.

## Chapter Thirteen: "Trust me"

beginning of the end: *Brooklyn Daily Eagle*, May 7, 1878.
remain intact for centuries: *New York Sun*, June 8, 1878.
list of dead and wounded: *Brooklyn Daily Eagle*, June 15, 1878.
catastrophic results: Francis Collingwood in *Brooklyn Daily Eagle*, June 15, 1878; conversation with Donald Sayenga.
verdict of accidental death: *Brooklyn Daily Eagle*, June 20, 1878.
benefit Brooklyn exclusively: *Brooklyn Daily Eagle*, August 12, 1878, and August 14, 1878.
add largely to the cost of the enterprise: *Brooklyn Daily Eagle*, August 13, 1878.
incapable: McCullough, 395.
not returning it to the manufacturer: *Brooklyn Daily Eagle*, May 6, 1879; McCullough, 466.
whether you believed it or not: McCullough, 467.
Tammany men were turfed out: McCullough, 272.
trust me: WAR to H. C. Murphy, November 21, 1876, Brooklyn Bridge Papers, RPI.
Danish and Swedish: RPI, WAR to E. F. Farrington, November 1876.
a matter of trust: WAR, "Annual Report of the Chief Engineer," 1876, Brooklyn Bridge Papers, RPI.
enormous elevation above the river: Shapiro, 66.
in an ice cellar: *New York Sun*, June 15, 1879.
in about 18 months: *New York Times*, December 16, 1879.
have now been surmounted: *Scientific American*, January 15, 1881.
Ten thousand dollars per annum: WAR to JARII, January 5, 1898, Case, III, RUL.
in spite of all my opposition: WAR to H. C. Murphy, December 1, 1881, in "Reports of the Trustees of the NY and Brooklyn Bridge, 1877–82," Brooklyn Bridge Papers, RPI.
successfully completed: WAR to H. C. Murphy, April 15, 1881, Brooklyn Bridge Papers, RPI.
suited to bridge construction: Vogel, 36–37.
stiffening trusses: Ibid.
loses all control over his mind: *New York Star*, July 1882.
get rid of him: WAR to C. C. Martin, July 21, 1882, Brooklyn Bridge Papers, RPI.
Cannot meet the trustees: McCullough, 486.
Five Forks affair: Jordan, 308.
would find in his favor: McCullough, 491.
hands of a sick man: Ibid., 491.
because it pleases me: Ibid., 491.
necessity of such a change: Ibid., 492.
pushing forward the work: *New-York Tribune*, August 22, 1882.
shattered in its defense: *Brooklyn Daily Eagle*, August 23, 1882.
due to the Chief Engineer: McCullough, 493.

his remarks at the meeting: McCullough, 455.

my strength would let me: WAR to C. C. Martin, August 18, 1882, Brooklyn Bridge Papers, RPI.

services in the future: WAR to H. C. Murphy, undated, Brooklyn Bridge Papers, RPI.

for the sake of policy: EWR to William Marshall, undated, Brooklyn Bridge Papers, RPI.

you have shown in me: WAR to Ludwig Semler, September 10, 1882, Brooklyn Bridge Papers, RPI.

cordage business he helped to found: *New York Times*, William Marshall obituary, June 10, 1895.

Until 1950: Andrew J. Sparberg, *From a Nickel to a Token.* (New York: Fordham University Press, 2015), 70; and RPI, "Report of the Chief Engineer and Superintendent," June 1, 1884.

born in Prussia: *Brooklyn Daily Eagle*, Otto Witte obituary, September 25, 1891.

retained in his post: *Brooklyn Daily Eagle*, September 12, 1882.

Here the storm was a fearful one: WAR to C. C. Martin, September 12, 1882, C. C. Martin Collection, RPI.

never set foot on the bridge: McCullough, 515.

used over a river: Ibid., 516.

trotting a horse: *New-York Tribune*, May 17, 1883.

acting president of the bridge trustees: McCullough, 520.

closing the New York Stock Exchange: *New York Times*, May 23, 1883.

the whims of an official: WAR to James T. Stranahan, May 4, 1883, Brooklyn Bridge Papers, RPI.

should not be permitted here: Ibid., May 5, 1883.

constructive world of art: Hewitt in *Opening Ceremonies of the New York and Brooklyn Bridge*, 56–58.

Whatever is, is right: WAR to Abram Hewitt, May 10, 1883, Brooklyn Bridge Papers, RPI.

came from afar: Oration by Richard S. Storrs, in *Opening Ceremonies of the New York and Brooklyn Bridge*, 117.

the moral qualities: Hewitt, 56–58.

New York side you know: McCullough, 523.

he slipped back upstairs: McCullough, 525.

all along the Heights: *Brooklyn Daily Eagle, New York Times, New-York Tribune*, May 25, 1883.

and the first dude: *New York Times*, May 25, 1883.

floor line of the masonry: WAR, 256.

blindly followed: WAR, 251.

from there to the Towers: WAR, 262; thanks to Donald Sayenga for his clear discussion of these changes.

an excellent musician and as a mineralogist: "Biography of Washington A. Roebling," undated, Brooklyn Bridge Papers, RPI.

what I needed most in life: WAR, 165

desire to relinquish: WAR, "Memoranda and Letter Copies," undated, Brooklyn Bridge Papers, RPI.

his life work was the Brooklyn Bridge: *New York Times*, C. C. Martin obituary, July 12, 1903.

world to do yet: *Brooklyn Union*, May 16, 1883.

lace garment: JARII, August 16, 1883, John A. Roebling II Papers, RUL.

surely never have imagined: RPI, EWR, *Troy Telegram*, February 18, 1887.

John was due to graduate: WAR to JARII, May 30, 1888, John A. Roebling II Papers, RUL.

you must read someday: Ibid., November 26, 1893.

$14,110 for the week: *Brooklyn Daily Eagle*, November 17, 1885.

to their jobs in Manhattan: Burrows and Wallace, 1054–55.
stretching out to infinity: Bender, in Latimer, et al., 327.
track had been laid: Burrows and Wallace, 1045.
dynamism and energy of New York City: Bender, in Latimer, et al., 330.
Manhattan's modern skyline: Landau and Cardit, 62.
"Sky Building in New York": Burrows and Wallace, 1050.
true Americanism: Landau and Condit, 197.
sky-scraper: Burrows and Wallace, 1051.
league-long bridges: Henry James, "The American Scene," 1907, in Lopate, 375.
wind produces no effect: WAR to JARII, January 21, 1904, Case, III, RUL.
mausoleum: WAR to JARII, August 28, 1910, Case, III, RUL.
within ten days: WAR to JARII, May 30, 1888, Case, III, RUL.
not any such a load: RPI, *Railroad Gazette* 30, no. 46): 824.
pull up the anchorages: WAR to *Railroad Gazette*, November 21, 1898, Brooklyn Bridge Papers, RPI.
Isn't that long enough: *New York Times*, July 29, 1922.
Let the Bridge alone: WAR to Palmer C. Ricketts, August 7, 1922, Brooklyn Bridge Papers, RPI.

## Chapter Fourteen: *"She goes everywhere and sees everything"*

the plan of an idiot: WAR to W. Hildenbrand, October 13, 1896, Washington A. Roebling Papers, RPI.
well into the twentieth century: Blockley, 288.
I twice that time: EWR to JARII, January 18, 1896, Emily Warren Roebling Papers, RUL.
living through last night: Ibid., July 17, 1898.
nothing the matter: Ibid., May 20, 1894.
hazardous aspect: Phillips, 154.
not paying their debts: WAR to JARII, May 4, 1893, Case, III, RUL.
was indefatigable: *Brooklyn Daily Eagle*, July 24, 1869.
for future expansion: Zink, 98.
3,000 barrels a day: WAR to JARII, Winter 1893–94, Case, I, RUL.
Truth is stranger than fiction: WAR to Reta Roebling, April 16, 1925, Case, I, RUL.
no happier in it: WAR to JARII, May 5, 1894, Case, III, RUL.
Eleven thousand dollars a year: WAR to JARII, June 1903, Case, III, RUL.
ramrod keeping order: WAR JARII, December 1893, Case, III, RUL.
number 333: Zink, 96–97.
competitive vanity: WAR to JARII, September 22, 1893, Case, III, RUL.
feel like killing them: Ibid., WAR to JARII, October 13, 1893.
thought of me as a linguist: EWR to JARII, May 4, 1893, "Addendum" Papers, RPI.
making any such statement: e-mail from Don Sayenga, October 5, 2015.
Mrs. Washington Roebling: *Engineering and Mining Journal*, January 21, 1882, "Addendum" Papers, RPI.
usually has to come back: JARII to EWR, April 21, 1898, Emily Warren Roebling Papers, RUL.
art of naval war: JARII to Department of the Navy, March 18, 1898, John A. Roebling Papers, RUL.

conditions were appalling: Tucker, 93.

out of her own pocket: Walworth, 75.

30 nightshirts: *New York Times*, August 14, 1898.

thanking her for her efforts: Walworth, 105.

am being buried alive: EWR to JARII, May 14, 1894, John A. Roebling II Papers, RUL.

the ruling passion of all women: WAR to JARII, April 26, 1896, Case, III, RUL.

Just think what I missed: WAR to JARII, May 11, 1896, Case, III, RUL.

not a footman: EWR to WAR, May 31, 1896, Washington A. Roebling Papers, RUL.

under a silken canopy: Ibid., May 17, 1896.

don't stop day or night: WAR to JARII: June 1896, Case, III, RUL.

to our low level: WAR to JARII, August 17, 1896, John A. Roebling II Papers, RUL.

do belong to Sorosis: Ibid., April 30, 1896.

Dickens on his American tour: *New York Times*, "On This Day," May 15, 1869.

metropolitan stomachs: Whitman in Burrows and Wallace, 819.

entirely dependent upon: EWR to JARII, July 12, 1894, John A. Roebling II Papers, RUL.

twice as great: Schuyler, 349–52; Zink, 105.

Hildenbrand about the contract: WAR to W. Hildenbrand, January 18, 1900, John A. Roebling's Sons Company Papers, RPI.

quite out of my world: EWR to JARII, December 10, 1899, John A. Roebling II Papers, RUL.

going to tell: Ibid., March 2, 1896.

thick coating of vanity: WAR to JARII, winter 1896–97, John A. Roebling II Papers, RUL.

out of bed: EWR to JARII, January 17, 1897, John A. Roebling II Papers, RUL.

Modern American law: Ruth Crocker, "Nothing More for Men's Colleges," in Walton, 260.

for rights is still denied: Weigold, 123.

give us our diplomas: EWR to JARII, March 26, 1899, John A. Roebling II Papers, RUL.

transition period of society: *New York Times*, March 31, 1899.

responsibility of their essays: EWR to JARII, April 9, 1899, John A. Roebling II Papers, RUL.

Knows nothing about either: WAR to JARII, November 5, 1899, Case, III, RUL.

law of the land: EWR to JARII, September 6, 1899, John A. Roebling II Papers, RUL.

with which she is associated: National Society Daughters of the American Revolution, *American Monthly Magazine* 17 (July–December, 1900): 246.

Cain & Abel variety: WAR to JARII, January 5, 1898, Case, III, RUL.

more than two thousand men went on strike: Zink, 102.

30 percent he had been getting: Zink, 10.

it's not safe: FWR to WAR, February 13, 1899, Ferdinand W. Roebling Papers, RUL.

nothing to live for: EWR to JARII, undated, John A. Roebling II Papers, RUL.

brothers would accept: WAR's draft letter to R. V. Lindabury, February 15, 1899, Ferdinand W. Roebling Papers, RUL.

deal did not go through: Schuyler, 352; Zink, 103.

dinosaur bone: Smithsonian Institution, geogallery.si.edu.

San Diego County: Mineralogical Record, www.minrec.org, biographical archive, Frank B. Schuyler.

every fellow wants what the other has: WAR to R. B. Gage, May 13, 1926, Washington A. Roebling Papers, RUL.

Science in this Country: WAR to G. F. Kunz, October 24, 1917, Case, IV, RUL.

A tiny nugget of gold: information and image from Paul Pohwat, Smithsonian Institution.

the visitor interested only in display: RUL, *Scientific Monthly* (October 1927 [probable]); 314–20.

**Harquahala Mountain in Arizona:** per https://www.blm.gov/az/st/en/prog/recreation/auto tour/harq_summit.html.

**Table Mountain in California:** Smithsonian Institution Annual Report, 1927.

## Chapter Fifteen: "The image of his wife floats before him"

**Paris Exposition in 1900:** *New York Times*, June 14, 1903.

**I certainly grow weaker each day:** EWR to WAR, November 10, 1902, Washington A. Roebling Papers, RUL.

**whom shall I believe:** WAR to JARII, December 9, 1902, John A. Roebling II Papers, RUL.

**she remained very bad:** Emily Warren Roebling, *Journal of the Reverend Silas Constant*, 33.

**contributed to her debility:** RUL, WAR to Reta Roebling, March 11, 1925; EWR to JARII, October 26, 1895, John A. Roebling II Papers, RUL.

**many a delightful gathering:** *Brooklyn Daily Eagle*, March 1, 1903.

**I miss her all the time:** WAR to JARII, February 8, 1904, John A. Roebling Papers, RUL.

**good night, good night:** WAR, "Personal Miscellany," Washington A. Roebling Papers, RUL.

**content and serene:** WAR to JARII, February 28, 1904, Case, III, RUL.

**worked up of later years:** WAR to JARII, undated, John A. Roebling II Papers, RUL.

**businesslike manner:** Ibid., June 16, 1903.

**things I care most about:** Ibid., August 19, 1905.

**got me in a corner:** Ibid., May 17, 1903.

**Poor beggar:** Ibid., June 27, 1904.

**err on the safe side:** *New York Times*, February 22, 1908.

**a woman of charming personality:** *Trenton Evening Times*, March 28, 1908.

**anything less would be unjust:** JARII to WAR, March 10, 1908, Case, III, RUL.

**would never photograph well:** WAR to JARII, March 21, 1908, Case, III, RUL.

**a gift of the bridegroom:** *New York Times*, April 22, 1904.

**miniature representation of the Brooklyn Bridge:** *Brooklyn Life*, April 25, 1908.

**where the sun will always smile:** *Trenton Evening Times*, June 30, 1908.

**renewed vim:** Ibid.

**picked up his pen once more:** Sayenga, xviii.

**like a pinch too much of carbon:** *Trenton Evening Times*, June 30, 1908.

**he had never seen the original:** WAR to James R. Rusling, January 23, 1916, Case, III, RUL.

**built in the right place:** WAR to Henry S. Jacoby, July 24, 1910, Washington A. Roebling Papers, RUL.

**driven to bankruptcy:** *New York Times*, April 12, 1910.

**trying to keep up with me:** WAR to JARII, August 28, 1910, Case, III, RUL.

**Walter Automobile Company:** "Washington A. Roebling II," www.roeblingmuseum.org.

**spits flames in all directions:** WAR to JARII, October 15, 1908, in Zink, 129.

**perpetual universal war:** WAR to JARII, August 5, 1914, Case, IV, RUL.

**murderous aristocracy:** Ibid., May 14, 1915.

**stepped in to fill the gap:** WAR to JARII, October 16, 1915, Case, IV, RUL.

**antennae wire and aviation wire:** WAR to JARII, March 2, 1917, Case, IV, RUL.

**I am not supposed to know:** WAR to JARII, November 14, 1917, Case, IV, RUL.

**the fate of the world:** RPI, "Wire-roping the German submarine," John A. Roebling's Sons Company, RPI.

**My brothers are getting old:** WAR to JARII, February 19, 1914, in Zink, 134.

wire cloth looms: WAR to FWR, May 31, 1878, Ferdinand W. Roebling Papers, RUL.

overwhelmed by the Brooklyn Bridge: WAR to George W. Pepper, March 19, 1922, Edmund Roebling Papers, RUL.

taken out of him: WAR to FWR, June 4, 1873, Ferdinand W. Roebling Papers, RUL.

can't be trusted alone: Ibid., September 17, 1874.

He is very careless: Ibid., November 27, 1874.

doing something all the time: Zink, 78; Schuyler, 234.

waste thousands in experimenting: Zink, 79, Schuyler, 335–36.

without a cent: RUL, WAR to R. V. Lindabury, July 20, 1898, Washington A. Roebling Papers, RUL.

I send you my opinions: RUL, WAR to FWR, March 24, 1880, Ferdinand W. Roebling Papers, RUL.

had already made his will: Warrant W. Foster to WAR, October 12, 1897, Washington A. Roebling Papers, RUL.

had been less than nine hundred thousand dollars: Schuyler, 146.

50 percent of all the wire rope: Zink, 99.

then you are the master: Zink, 111.

microscopic examination of steel: Schuyler, 366.

a baseball diamond: Zink, 125.

stayed for thirty years: Ibid., 109.

manly art of self-defense: Zink, 160: *Blue Center* 1, no. 1 (November 1925): frontispiece.

as usual he succeeded: Zink, 84, Schuyler, 341–42.

$45,000 worth: WAR to JARII, December 17, 1904, Case, III, RUL.

trusses between the wings: McCullough, *The Wright Brothers*, 90.

workers from mosquitos: Zink, 127.

swept away in a few hours: Zink, 129.

two dozen plants in Trenton: *Trenton Evening Times*, May 23, 1914.

directly after the fire started: WAR to JARII, December 17, 1915, John A. Roebling II Papers, RUL.

supply the Allied armies: *New York Times*, January 19, 1915.

peace and contentment: WAR to JARII, April 15, 1915, Zink, 137.

who succeeds in doing that: Zink, 143.

How to fill the place of Charles: WAR to JARII, September 1, 1918, Case, IV, RUL.

between those two sighs: WAR to Dr. Evan M. Evans, October 10, 1918, Washington A. Roebling Papers, RUL.

double strike of fate: WAR to Mrs. Stockton, October 10, 1918, Washington A. Roebling Papers, RUL.

into an investment: WAR to JARII, October 19, 1918, Case IV, RUL.

increase in production of 75 percent: Zink, 148.

he will soon break down too: WAR to JARII, November 20, 1918, Case, IV, RUL.

which I knew long ago: Ibid., September 3, 1920.

house was remodeled: *New York Times*, May 30, 1921.

very satisfactorily: WAR to Reta Roebling, September 20, 1920, Case, IV, RUL.

## Chapter Sixteen: *"You can't desert your job"*

shortly to graduate: *Princeton Alumni Weekly*, January 17, 1936.

next to the cable bands: WAR to Palmer C. Ricketts, November 7, 1925, Washington A. Roebling Papers, RPI.

laying of the cables: Case, IV, RUL.

the longest suspension bridge in the world: Petroski, "Benjamin Franklin Bridge."

risky work: WAR to JARII, March 30, 1921, Case, IV, RUL.

Camden Bridge was finished: WAR to George W. Pepper, August 26, 1922, Washington A. Roebling Papers, RUL.

proposes to indulge in: WAR to JARII, September 23, 1921, Case, IV, RUL.

cable engineer on the Williamsburg: *Brooklyn Daily Eagle*, November 11, 1902.

do not bid on making the cables: JARII to WAR, October 4, 1921, Case, IV, RUL.

how to get out of it: Ibid., March 8, 1922.

half-starved condition: JARII to WAR, March 18, 1922, Washington A. Roebling Papers, RUL.

I have never had any talk: WAR to George W. Pepper, August 26, 1922, Washington A. Roebling Papers, RUL.

cables can only suffer: WAR to Clarence E. Case, September 23, 1923, Case, IV, RUL.

Bear Mountain Bridge over the Hudson: JARII, Biographical Sketch of W. A. R. (c. 1926), John A. Roebling II Papers, RUL.

thirty-two feet longer: Zink, 155.

West Pointers do: RUL, WAR to JARII, March 1924, John A. Roebling Papers, II, RUL.

joy of the celebrations: *New York Times*, May 27, 1923.

with me among them: WAR, "Personal Miscellany:, June 7, 1924, Washington A. Roebling Papers, RUL.

philological accomplishments: WAR to JARII, undated, John A. Roebling Papers, RUL.

have no one to fall back on: WAR to Mrs. H. DeLanoy Sr., July 1925, Washington A. Roebling Papers, RUL.

50 years ago: WAR to JARII, April 21, 1924, Case, IV, RUL.

infirmities of old age: letter and poem sent by Mrs. H. W. Costigan, August 24, 1924, "Addendum" Papers, RPI.

the death of Colonel Roebling: *Brooklyn Daily Eagle*, May 24, 1883.

propriety of it: WAR to Edmund Roebling, February 18, 1924, Case, IV, RUL.

old man of over 70: WAR to Clarence Case, May 16, 1924, Case, IV, RUL.

as Billy Sunday had been: WAR to Reta Roebling, March 11, 1925, Case, IV, RUL.

Great White Way: Ibid., May 22, 1925.

only three such close events: author's correspondence with Martin Rees, Dr. Robin Catchpole, and Dr. Steve Bell, 2016.

correspondence, investments: WAR to JARII, July 13, 1925, Case, IV, RUL.

as early as 1933: *New York Times*, March 13, 1926.

Strains are not certain: WAR notes on *New York Times* article, March 13, 1926, Washington A. Roebling Papers, RUL.

Professor of Mineralogy: Geological Society of America, memorial page for Paul Kerr. http://www.minsocam.org/ammin/AM69/AM69_586.pdf.

I can't come down: WAR to R. B. Gage, May 13, 1926, Case, IV, RUL.

the sleep of ages: WAR to Reta Roebling, May 14, 1926, Case, IV, RUL.

formerly was a part: Thanks to Donald Sayenga for this information.

could pay their respects: *New York Times*, July 24, 1926.

A special train: *New York Times*, July 25, 1926.

runic cross: WAR to JARII, April 1903, John A. Roebling II Papers, RUL.

twenty-nine million dollars: Zink, 161.

three small cases of cut stones: *Bernardsville (NJ) News*, August 5, 1926.

**would mean as much to me:** Cornelia Roebling to JARII, undated, John A. Roebling II Papers, RUL.

**make a violin talk:** George H. Brown to Cornelia Roebling, undated, Case, I, RUL.

**his estate in Florida:** *New York Times*, June 16, 1936.

**seems to have disappeared:** conversation with Sean Rush, who restored Spottiswoode estate in Clearwater, Florida.

**ignition cables:** Zink, 166.

**could not have devised:** *New York Times*, July 22, 1926.

## Chapter Seventeen: *"Time & age cures all this"*

**hoisted white flags:** *New York Times*, August 12, 2014.

**musical prosthetics:** humanharp.org

**when I took charge of that bridge:** *New York Sun*, October 11, 1918.

**Charles Roebling was not an engineer:** RUL, *Christian Science Monitor*, October 18, 1918.

**averts bad decisions:** Blockley, 217–18.

**latest and most extensive:** Rob Hans, "John A. Roebling Bridge Rehabilitation History."

**Steinman, one of the leading:** Trachtenberg, *Brooklyn Bridge, Fact and Symbol*, 45.

**in which Roebling left them:** Latimer, et al., eds., 125.

**the cables themselves remain essentially as they were:** Donald Sayenga, e-mail, October 5, 2016.

**WAR to Henry C. Murphy,** August 6, 1878.

**loads had increased:** Author correspondence with Henry Perahia, March 23, 2016.

**ensured the disaster would happen:** University of Minnesota. "Independent Study Of The I-35 W Bridge Collapse Results Parallel NTSB Report." *ScienceDaily*. www.sciencedaily.com/releases/2008/11/081120122207.htm (accessed December 8, 2016).

**cures all this:** WAR to H. C. Murphy, December 31, 1877, John A. Roebling Papers, RPI.

**matters of public welfare:** *Bernardsville (NJ) News*, February 7, 1952.

**had been swallowed up:** Cumbler, 5.

**a mere 22,400 tons:** Zink, 167.

**first spun it over the Brooklyn Bridge:** *New York Times*, July 10, 1929.

**1,550 miles of electrical wire:** Zink, 173.

**Golden Gate Bridge:** Ibid., 183.

**closed its doors for good:** Sayenga, "Contextual Essay on Wire Bridges."

**loft-style apartments:** James McAvoy, "State Awards $16M Toward Redevelopment of Former Roebling Steel Mill," *(Trenton, NJ) Times*, March 12, 2015.

**"Superfund Site Profile."** EPA. Accessed January 16, 2016. https://cumulis.epa.gov/supercpad/cursites/csitinfo.cfm?id=0200439.

**prospects of their enterprises:** WAR to Raymond Arnott, August 31, 1922, Washington A. Roebling Papers, RUL.

## Epilogue: *Cold Spring*

**pictures of the imagination:** WAR, 112.

# Bibliography

The archives of the Roebling family are divided between Rensselaer Poly-technic Institute in Troy, New York, and Rutgers University in New Brunswick, New Jersey; most professional and engineering material is at Rensselaer, and most of the family's personal papers are at Rutgers, though this division is not exact I have mainly made use of these archives, as well as the Municipal Archives of New York City, the New-York Historical Society, and the Brooklyn Historical Society.

## Books and Articles

Bandura, Albert, ed. *Self-Efficacy in Changing Societies* Cambridge: Cambridge University Press, 1995.

Barber, Alfred Newton. *John A. Roebling: An Account of the Ceremonies at the Unveiling of a Monument to His Memory*. Roebling Press, 1908.

Barga, M. (2013). *The Long Depression (1873–1878)*. Retrieved August 25, 2015, from http://www.socialwelfarehistory.com/?p=8220.

Bender, Thomas. "Metropolitan Culture: Brooklyn Bridge and the Transformation of New York." In Margaret Latimer, et al., eds. *Bridge to the Future: A Centennial Celebration of the Brooklyn Bridge*. New York: New York Academy of Sciences, 1984.

Benke, Arthur C., and Colbert E. Cushing. *Rivers of North America*. Burlington, MA: Elsevier Academic Press, 2005.

Bernardo, Leonard, and Jennifer Weiss. *Brooklyn by Name: How the Neighborhoods, Streets, Parks, Bridges, and More Got Their Names*. New York: New York University Press, 2006.

Billington, David P. "Building Bridges: Perspectives on Recent Engineering." In Margaret Latimer, et al., eds. *Bridge to the Future: A Centennial Celebration of the Brooklyn Bridge*. New York: New York Academy of Sciences, 1984.

Blockley, David. *Bridges: The Art and Science of the World's Most Inspiring Structures*. Oxford: Oxford University Press, 2010.

Board of Trustees of the New York and Brooklyn Bridge. *New York and Brooklyn Bridge Proceedings, 1867–1884*. New York: Board of Trustees of the New York and Brooklyn Bridge, 1884. In the Brooklyn Museum Libraries and Archives, Special Collections, call no. F129 B79 B755p.

Booth, Malcolm A. "Roebling's Sixth Bridge, 'Neversink.'"*Journal of the Rutgers University Libraries* 30, no. 1 (1966) http://reaper64.scc-net.rutgers.edu/journals/index.php/jrul/article/viewFile/1468/2907.

Bourne, George M. *The Home Doctor: A Guide to Health.* San Francisco: San Francisco News Company, 1878.

Bradford, William, ed. *Samuel Eliot Morison, of Plymouth Plantation, 1620–1647.* New York: Knopf, 1991.

Brown, Henry Collins. *Fifth Avenue Old & New, 1824–1924.* New York: Fifth Avenue Association, 1924.

Brown, R. C.*History of Butler County [Pennsylvania], 1895.* Tucson, AZ: Americana Unlimited, 1974.

Burns, Ric, with Drew Gilpin Faust and Paul, Taylor. "Death and the Civil War," PBS, *The American Experience,* 2012.

Burrows, Edwin G., and Mike, Wallace. *Gotham: A History of New York City to 1898,* Oxford: Oxford University Press, 1999.

Butler, W. P. "Caisson Disease During the Construction of the Eads and Brooklyn Bridges: A Review." *Undersea Hyperbaric Medical Journal* 31, no. 4 (2004).

Cole, Cyrenus. *A History of the People of Iowa.* Cedar Rapids, IA: Torch Press, 1921.

Cumbler, John T. *A Social History of Economic Decline: Business, Politics and Work in Trenton.* New Brunswick, NJ: Rutgers University Press, 1989.

Day, Lance, and Ian McNeil, eds. *Biographical Dictionary of the History of Technology.* London: Routledge, 1996.

Dean, Sheldon and T. S. Lee. *Degradation of Metals in the Atmosphere.* Philadelphia: American Society for Testing and Materials, Special Publications, 1981.

Delp, Robert Worley. *Harmonial Philosopher: Andrew Jackson Davis and the Foundation of American Spiritualism.* Washington, D.C.: George Washington University, 1965.

Dickens, Charles. *American Notes.* London: Thomas Nelson and Sons, 1900.

*East River Bridge, Laws and Engineer's Reports, 1868–1884,* Brooklyn: 1885.

Elias, Stephen N. *Alexander T. Stewart: The Forgotten Merchant Prince.* Westport, CT: Praeger, 1992.

*Engineering News-Record.* New York: McGraw-Hill.

Farrington, E. F. *A Full and Complete Description of the Covington and Cincinnati Suspension Bridge with Dimensions and Details of Construction.* Cincinnati: J. P. Lindsay, 1867.

Fox, Cynthia G. "Income Tax Records of the Civil War Years." *Prologue* 18, no. 4 (Winter 1986). http://www.archives.gov/publications/prologue/1986/winter/civil-war-tax-records.html.

Foyster, Elizabeth. *Marital Violence: An English Family History, 1660–1857.* Cambridge University Press, 2005.

Franklin, R. W., ed. *The Poems of Emily Dickinson,* Reading Edition. Cambridge, MA: Harvard University Press, 1999.

Fraser, James W., ed. *The School in the United States: A Documentary History,* 2nd ed. New York: Routledge, 2010.

Gandhi, Kirti. "The St. Louis Bridge, the Brooklyn Bridge, and the Feud Between Eads and Roebling."Gandhi Engineering, 2011. http://www.gandhieng.com/resources/MEDIA/St.%20Louis%20Bridge,%20the%20Brooklyn%20Bridge,%20and%20the%20feud%20between%20Eads%20and%20Roebling.pdf.

Gillette, William. *Jersey Blue: Civil War Politics in New Jersey, 1854–1865.* New Brunswick, NJ: Rutgers University Press, 1995.

Golway, Terry. *Machine Made: Tammany Hall and the Creation of Modern American Politics.* New York: Liveright/W. W. Norton, 2014.

Green, Theodore, ed. *John A. Roebling: A Bicentennial Celebration of His Birth 1806–2006.* American Society of Civil Engineers. 2007. http://ascelibrary.org/doi/book/10.1061/97807 84408995.

Greve, Charles Theodore. *Centennial History of Cincinnati and Representative Citizens.* Chicago: Biographical Publishing Company, 1904.

Guentheroth, Nele. "Roebling's Development to Being an Engineer." In "John A. Roebling: A Bicentennial Celebration of His Birth 1806–2006." Edited by Theodore Green. American Society of Civil Engineers, 2007. http://ascelibrary.org/doi/abs/10.1061/40899(244)2.

Hall, John, and Mary Anna Hall. *History of the Presbyterian Church in Trenton, N.J.: from the First Settlement of the Town.* Trenton, NJ: MacCrellish and Quigley, 1912.

Hans, Rob. "John A. Roebling Bridge Rehabilitation History." Kentucky Engineering Center, http://www.kyengcenter.org/ accessed January 15, 2016.

Hawthorne, Sophia, ed. *Passages from the American Notebooks of Nathaniel Hawthorne.* Boston: Houghton, Mifflin, 1883.

Hebert, Walter H. *Fighting Joe Hooker.* Indianapolis: Bison Books, University of Nebraska Press, 1999.

Hecht, Roger W. *The Erie Canal Reader: 1790–1950*, Syracuse, NY: Syracuse University Press, 2003.

Hempel, Charles J. *Complete Works of Friedrich Schiller*, vol. 1. Philadelphia: I. Kohler, 1861.

Hewitt, Abram S., *The Production of Iron and Steel in Its Economic and Social Relations.* United States Commissioner, Paris Universal Exposition, 1867, Reports of the United States Commissioners, Washington, D.C.: U.S. Government Printing Office, 1868.

Hildenbrand, Wilhelm. "Cable-making for Suspension Bridges, with special reference to the cables of the East River Bridge, etc." *Van Nostrand's Science Series*, no. 32. New York: Van Nostrand, 1877.

Hindle, Brooke, and Steven, Lubar. *Engines of Change: The American Industrial Revolution 1790–1860.* Washington, D.C.: Smithsonian Institution Press, 1991.

Holmes, Richard. *Footsteps: Adventures of a Romantic Biographer.* London: Hodder & Stoughton, 1985.

Jackson, Kenneth T. "Technology and the City: Transportation and Cultural Form in New York." In Margaret Latimer, et al., eds. *Bridge to the Future: A Centennial Celebration of the Brooklyn Bridge.* New York: New York Academy of Sciences, 1984.

James, Henry. *Substance and Shadow, or Morality and Religion in their Relation to Life: An Essay upon the Physics of Creation.* Boston: Ticknor and Fields, 1863.

———. *The American Scene.* London: Chapman & Hall, 1907.

Jordan, David M.*Happiness Is Not My Companion: The Life of General G. K. Warren*, Bloomington, IN: Indiana University Press, 2001.

Kaestle, Carl F.*Pillars of the Republic: Common Schools and American Society, 1780–1860.* New York: Hill and Wang, 1983.

Kahlow, Andreas. "Johann August Röbling (1806–1869): Early Projects in Context."Cambridge University, Department of Architecture. http://www.arct.cam.ac.uk/Downloads/ichs/vol-2-1755-1776-kahlow.pdf.

Kaplan, Sidney. *American Studies in Black and White: Selected Essays, 1949–1989*. Boston: University of Massachusetts Press, 1991.

Kemp, Emory L. "Roebling, Ellet, and the Wire-Suspension Bridge." In Margaret Latimer, et al., eds. *Bridge to the Future: A Centennial Celebration of the Brooklyn Bridge*. New York: New York Academy of Sciences, 1984.

Kranzberg, Melvin. "Confrontation or Complementarity?: Perspectives on Technology and the Arts." In Margaret Latimer, et al., eds. *Bridge to the Future: A Centennial Celebration of the Brooklyn Bridge*. New York: New York Academy of Sciences, 1984.

Landau, Sarah Bradford, and Carl, W. Condit. *Rise of the New York Skyscraper 1865–1913*. New Haven and London: Yale University Press, 1996.

Latimer, Margaret, Brooke Hindle, and Melvin Kranzberg, eds. *Bridge to the Future: A Centennial Celebration of the Brooklyn Bridge*. New York: New York Academy of Sciences, 1984.

Latimer, Margaret. *Two Cities: New York and Brooklyn the Year the Great Bridge Opened*. Brooklyn, NY: Brooklyn Educational & Cultural Alliance, 1983.

Lopate, Philip, ed. *Writing New York: A Literary Anthology*. New York: Washington Square Press, 1998.

Lorent, Stefan. *Pittsburgh: The Story of an American City*. Garden City, NY: Doubleday, 1964.

Lowe, Thaddeus S. C. *My Balloons in Peace and War*. Edited by Michael Jaeger and Carol Lauritzen. In *Studies in American History*, vol. 53. New York: Edwin Mellen Press.

Malcolm, Janet. *The Silent Woman: Sylvia Plath and Ted Hughes*. London: Picador, 1994.

Marx, Leo. *The Machine in the Garden: Technology and the Pastoral Ideal in America*. Oxford: Oxford University Press, 1964.

Masur, Louis P. *Lincoln's Hundred Days: The Emancipation Proclamation and the War for the Union*. Cambridge, MA: Harvard University Press, 2012.

Matthews, Alva T. "Emily W. Roebling: One of the Builders of the Bridge." In Margaret Latimer, et al., eds. *Bridge to the Future: A Centennial Celebration of the Brooklyn Bridge*. New York: New York Academy of Sciences, 1984.

McCullough, David. *The Great Bridge*. New York: Simon and Schuster, 1982.

———. *The Wright Brothers*. New York: Simon and Schuster, 2015.

Miers, Earl Schenk, ed., *Washington Roebling's War*, Newark, Delaware: Spiral Press, 1961.

Mintz, Steven. *Huck's Raft: A History of American Childhood*. Cambridge, MA: Belknap Press, 2004.

*Modern Industries: A Series of Reports on Industry and Manufactures, As Represented in the Paris Exposition in 1867 by Twelve British Workmen*. London: Macmillan, 1868.

Morris, Jan. *Fisher's Face*. London: Viking, 1995.

Morton, Nathaniel. *New England's Memorial*. Boston: Congregational Board of Publication, 1855.

Mumford, Lewis. *Sidewalk Critic: Lewis Mumford's Writings on New York*. Edited by Robert Wojtowicz. New York: Princeton Architectural Press, 1998.

*Opening Ceremonies, New York and Brooklyn Bridge, May 24, 1883*. Brooklyn, NY: Brooklyn Eagle Job Printing Department, 1883.

*Paris Universal Exhibition 1867, Complete Official Catalogue*. London: J. M. Johnson and Sons, 1867.

Parkman, Francis. *The Oregon Trail/The Conspiracy of Pontiac*. New York: Library of America, 1991.

Petroski, Henry. "Benjamin Franklin Bridge." *American Scientist*, 90 (September–October 2002), 406–10.

————, *Engineers of Dreams: Great Bridge Builders and the Spanning of America*, New York: Alfred A. Knopf, 1995.

Phillips, John L. *The Bends: Compressed Air in the History of Science, Diving and Engineering*. New Haven and London: Yale University Press, 1998.

Pleck, Elizabeth. *Domestic Tyranny: The Making of Social Policy Against Family Violence from Colonial Times to the Present*. Oxford: Oxford University Press, 1987.

Porter, Roy, Ed. *The Medical History of Waters and Spas*, Medical History, Supplement No. 10, London, Wellcome Institute for the History of Medicine, 1990.

Priessnitz, Vincent. *Hydropathy, or the Effectual Cure of Acute and Chronic Diseases by Cold Water Only, with Directions for Its Application as Practised by the Inventor, Vincent Priessnitz, of Graeffenberg, in Silesia*. Halifax: Nicholson and Wilson, 1842.

Richman, Steven M. *Reconsidering Trenton: The Small City in the Post-Industrial Age*. Jefferson, NC: McFarland, 2011.

Ricketts, Palmer C., *History of Rensselaer Polytechnic Institute, 1824-1934*. New York: John Wiley and Sons, Inc., 1934.

Rittner, Don. *Troy: A Collar City History*. Charleston, SC: Arcadia Publishing, 2002.

Roe, Sue, *The Private Lives of the Impressionists*. London: Vintage, 2007.

Roebling, Emily Warren. "A Wife's Disabilities." *Albany Law Journal* n (April 15, 1899).

————. *Richard Warren of the Mayflower and Some of His Descendants*. Boston: David Clapp, 1901.

————. *The Journal of the Reverend Silas Constant, Pastor of the Presbyterian Church at Yorktown*. Philadelphia: J. B. Lippincott, 1903.

Roebling, Washington Augustus. *Early History of Saxonburg*. Butler County [Pennsylvania] Historical Society, 1924.

————. *Washington Roebling's War*. Edited and with an introduction by Earl, Schenk Miers. Newark DE: Spiral Press, 1961.

————. *Pneumatic Tower Foundations of the East River Suspension Bridge*. New York: Averell & Peckett, 1873.

Sayenga, Donald, ed. *Washington Roebling's Father: A Memoir of John A. Roebling*. Reston, VA: American Society of Civil Engineers Press, 2009.

Sayenga, Donald. "Contextual Essay on Wire Bridges." *Historic American Engineering Record*, 1999. Library of Congress archive website: http://cdn.loc.gov/master/pnp/habshaer/nj/nj1700/nj1749/data/nj1749data.pdf.

Schuyler, Hamilton. *The Roeblings: A Century of Engineers, Bridge-builders and Industrialists* Princeton: Princeton University Press, 1931.

Shambaugh, Benjamin F., ed. The Iowa Journal of History and Politics 17 (1919). Iowa City: State Historical Society of Iowa. https://archive.org/details/iowajournalofhis17stat.

Shapiro, Mary J.*A Picture History of the Brooklyn Bridge*. New York: Dover Publications, 1983.

Skrabec, Quentin R. *Henry Clay Frick: The Life of the Perfect Capitalist*. Jefferson, NC: McFarland, 2010.

Smith, Andrew H. "The Effects of High Atmospheric Pressure including the Caisson Disease, by Andrew H. Smith, M.D., Surgeon to the New York Bridge Company." Alumni Association of the College of Physicians and Surgeons, New York, 1878.

Smithsonian Institution, *Annual Report of the Board of Regents, Year Ending June 30, 1927*. Washington, D.C.: U.S. Government Printing Office, 1928.

Sommers, Christina Hoff. *Who Stole Feminism?* New York: Simon and Schuster, 1994.

Sparberg, Andrew J.*From a Nickel to a Token: The Journey from Board of Transportation to MTA*. New York: Fordham University Press, 2015.

*Stanford Encyclopedia of Philosophy*. http://plato.stanford.edu/entries/hegel/.

Stiles, Henry R.*A History of the City of Brooklyn, Including the Old Town and Village of Brooklyn, etc*. 2 vols. Brooklyn, 1867–1870.

Stone, William L.*The Centennial History of New York City, from the Discovery to the Present Day*. New York: R. D. Cooke, 1876.

Stuart, Charles B. *Lives and Works of Civil and Military Engineers of America*. New York: D.Van Nostrand, 1871.

Tarr, Joel A., ed. *Devastation and Renewal: An Environmental History of Pittsburgh and Its Region*. Pittsburgh: University of Pittsburgh Press, 2003.

Taylor, Emerson Gifford. *Gouverneur Kemble Warren: The Life and Letters of an American Soldier 1830–1882*. Boston: Houghton Mifflin, 1932.

Thulesius, Olav. *The Man Who Made the Monitor: A Biography of John Ericsson, Naval Engineer*. Jefferson, NC: McFarland, 2007.

Thurston, George H.*Pittsburgh's Progress, Industries and Resources, etc*. Pittsburgh: Anderson, 1886.

Tolzmann, Don Heinrich. "Roebling Heritage Tour: A Guide to Sites Related to John A. Roebling (1806–69) and His Bridge on the Ohio at Covington, Kentucky." *Bulletin of the Kenton County Historical Society* 5 (November–December 2013), 8, 11.

Townsend, Jan. *The Civil War in Prince William County*. Prince William County Historical Commission, 2011. http://www.pwcgov.org/government/dept/planning/Documents/Hist-Comm_Book_The_Civil_War_in_PWC.pdf.

Trachtenberg, Alan. "Brooklyn Bridge as a Cultural Text." In Margaret Latimer, et al., eds. *Bridge to the Future: A Centennial Celebration of the Brooklyn Bridge*. New York: New York Academy of Sciences, 1984.

———. *Brooklyn Bridge, Fact and Symbol*. 2nd ed. Chicago: University of Chicago Press, 1979.

Trager, James. *The New York Chronology*. New York: HarperCollins, 2003.

Twain, Mark. *The Complete Short Stories*. Stilwell, KS: Digireads, 2008.

Tucker, Spencer C. *The Encyclopedia of the Spanish–American and Philippine–American Wars: A Political, Social and Military History*, vol. 1 Santa Barbara, CA: ABC-CLIO.

Urban, Wayne J., and Jennings, L. Wagoner Jr. *American Education: A History*. 4th edition. London: Routledge, 2009.

*John A. Roebling: An Account of the Ceremonies at the Unveiling of a Monument to His Memory*. Trenton, NJ: Roebling Press, 1908.

———. "Designing Brooklyn Bridge. In Margaret Latimer, et al., eds. *Bridge to the Future: A Centennial Celebration of the Brooklyn Bridge*. New York: New York Academy of Sciences, 1984.

Walton, Andrea, ed. *Women and Philanthropy in Education*. Bloomington, IN: Indiana University Press, 2005.

Walusinski, O. "Pioneering the Concepts of Stereognosis and Polyradiculoneuritis: Octave Landry (1826–1865)." *European Neurology* 70, no. (2013), 5–6.

Walworth, Ellen Hardin. *Report on the Women's National War Relief Association, Organized for the Emergency of the Spanish–American War, March 1898 to January 1899*. New York: Women's National War Relief Association, 1899.

*Washington Post. Civil War Stories: A 150th Anniversary Collection*. New York: Diversion Books, 2014.

Weigold, Marilyn E. *Silent Builder: Emily Warren Roebling and the Brooklyn Bridge.* Port Washington, NY: Associated Faculty Press, 1984.

Weiss, Harry B., and Howard, R. Kemble. *They Took the Waters: The Forgotten Mineral Springs Resorts of New Jersey and Nearby Pennsylvania and Delaware.* Trenton, NJ: Past Times Press, 1962.

Werner, Göran. "John August Roebling, Niagara Railway Suspension Bridge." http://www .arct.cam.ac.uk/Downloads/ichs/vol-3-3315-3332-werner.pdf.

West, John B. "Torricelli and the Ocean of Air: The First Measurement of Barometric Pressure." *Physiology* 28, no. 2 (March 1, 2013), 66–73. http://physiologyonline.physiology. org/content/28/2/66.

Westhofen, Wilhelm. *The Forth Bridge.* Centenary edition. Edinburgh: Moubray House, 1989.

Winschel, Terrence J, ed. *The Civil War Diary of a Common Soldier: William Wiley of the 77th Illinois Infantry.* Baton Rouge, LA: Louisiana State University Press, 2001.

Whitman, Walt. *Leaves of Grass.* London: Trübner, 1881.

Wright, John D. *Routledge Encyclopedia of Civil War Era Biographies.* New York: Routledge, 2013.

Zink, Clifford W. *The Roebling Legacy*, Princeton, New Jersey: Princeton Landmark Publications, 2011.

## Selected Websites

Erie Canalway National Heritage Corridor
http://www.eriecanalway.org/

*Bulletin of the Kenton County Historical Society*
http://www.kentoncountyhistoricalsociety.org/home

"The Sanitary and Moral Condition of New York City" (August 1869)
http://www.yale.edu/glc/archive/1021.htm

Historic Pittsburgh
http://digital.library.pitt.edu/pittsburgh/

The Canal Society of New Jersey
http://canalsocietynj.org/

Troy, New York
http://www.visittroyny.com/aboutTroy/history/history.aspx

New Jersey State Archives
http://www.nj.gov/state/archives/

Warren Family Genealogy
http://net.lib.byu.edu/fslab/AGBI/Warren.pdf

New York State Library, Gouverneur Kemble Warren papers
http://www.nysl.nysed.gov/msscfa/pr/sc10668.pdf

Civil War Trust
http://www.civilwar.org

# Newspapers

The *New York Times*, the *Brooklyn Daily Eagle*, the *Brooklyn Union*, the *New-York Tribune*, the *New York World*, *Harper's Weekly*, *Scientific American*, the *Trenton Times*, the *Bernardsville (NJ) News*; and others noted.

# Index

Note: page numbers in italics refer to images.

Adams, Julius W., 16, 164–65
Allegheny River Bridge (Pittsburgh)
  alterations to, 190
  Brooklyn Bridge committee's trip to, 12, 165
  construction of, 80–81, 82
Allen, Horatio, 164
American Steel and Wire Company, 260–61
Ammann, Othmar, 247, 299, 308
Amnesty Act of 1872, 187–88
Anderson, Robert, 84–85, 86
Arthur, Chester A., 234, 235, 237
Aspinwall, Lloyd, 222
aviation, Roebling's Sons wire and, 283, 301–2

balloon reconnaissance, Civil War, 104–6
Barrows, Cornelia, 131
Bear Mountain Bridge, 294–95
Bell, Alexander Graham, 279
Beni-Barde, Alfred, 192
Benjamin Franklin Bridge, 292–94
Billy Sunday (dog), 291, 293, 295, 298
Black Friday, 8–9
Blondin, Charles, 67
Bourne, George Melksham, 59–60
Brady, Mathew, 123
bridge collapses, in 21st century, 306–7
Bristol Chalybeate Baths, 43
Brooklyn
  growth of, 7, 8, 11–12
  history of, 10–11
  Washington and Emily's home in, 168, 195–96, 205–6, 211
  Washington and Emily's move from, 242
Brooklyn Bridge. See also Chief Engineer for Brooklyn Bridge
  buckling of truss chords (1898), 245
  cable slippage over time, 245–46

experience of crossing, xi–xii
first crossing of, 209
as iconic image, xii–xiii, 243
John Roebling's proposal for, 2
length of, 247
lights for, 234
load increases over time, 245, 246
long-term stability of, Washington on, 246
need for, 7, 11–12
New York and Brooklyn takeover of, 206–7
opening ceremonies for, 234–38
opposition to, 212, 219
public interest in, 168, 185–86, 209
renovations of, 305–6
Roeblings' scrapbook on, 205, 210, 211, 212, 226, 238, 251
and stability, engineering of, xii
state legislature's authorization of, 143
as subject of art and literature, 303–4
as unprecedented structure, 2, 7, 14
utility of, Washington on, 274
Brooklyn Bridge cables
  assembly and hanging of, 206–11, 217, 218, 223
  choice of steel for, 211
  fatal accident involving, 219–20
  manufacturing flaws in wire for, 212, 215–18, 239, 306
  margin of safety in, 216–17, 224
  as still sound, 306
  wire contracts for, 192, 212–15, 218
Brooklyn Bridge caissons, 175
  air pressure system for, 172, 182
  and blowouts, danger of, 175–78
  Brooklyn side, 173–82
  construction and placement of, 173–74
  dangers of working in, 175–78, 183–86

Brooklyn Bridge caissons *(continued)*
  and decompression sickness ("the bends"),
       172, 183–86, 189, 190–91, 248
  depth of sinking, 173, 183, 186
  Eads' patent lawsuit against, 194–95
  fire in, 179–81
  as new technology, 172
  New York side, 182–86, 239
  pay for workers in, 174, 186
  rocks blocking descent of, 174–75, 181
  souvenir piece of bedrock saved by
       Washington, 186–87
  structure and function of, 172, 173, 182
  workers' experience in, 174
Brooklyn Bridge construction
  accidents and deaths during, 175–78,
       178–81, 183, 185–86, 217, 219–20, 307
  anchorages, fixing in place, 208
  completion of towers and anchorages, 196
  congressional authorization of, 12
  cost estimates for, 151, 166, 199, 200
  cost increases and, 200, 203, 220, 230
  and Emily Roebling as engineering partner,
       xvi–xvii, 201–5, 228–30, 252
  floor beams and trusses, 221, 224, 225–26
  health effects on Washington, 180, 181, 187,
       189, 190–91
  as life-changing test for Washington, 171
  Long Depression and, 200
  vs. modern construction environment, 304–5
  planning for future Pullman car use, 225
  and political corruption, 7–8, 151–52, 166,
       197–200, 213–15
  preparations by John Roebling, 12–13
  primary assistants for, 168, 200, 203, 210,
       223, 224, 226–27
  scale models used in, 209–10
  surveying of bridge path, 12–13
  temporary footbridge, public use of, 211–12
  tower construction, 202–3, 206
  Washington on obstructors of, 236
  Washington on personal cost of, 260, 280
  Washington on progress of, 206
  Washington's claiming of credit for, 273–74,
       304
  Washington's health-related absences from,
       191, 195–96, 201–11, 219–22, 226–27, 234
  Washington's other work during, 189–90
Brooklyn Bridge plans
  approval of, 166
  and Bridge committee trip to other Roebling
       bridges, 12, 165–66

  consultants evaluating, 164
  credit to John Roebling for, xii
  critics of, 163–64
  as incomplete at death of John Roebling, 18,
       171
  John Roebling's drawings, 2, 167
  public release of John Roebling's plans, 6–7
  Washington on trustee-made changes to, 245
  Washington's assistance in preparing, 15
  Washington's changes to, 239–40, 305
*Brooklyn Daily Eagle*
  articles in Washington's scrapbook, 205
  on Bridge, need for, 12
  on Bridge cable slippage, 246
  on Bridge construction, 163–64, 165, 166, 174,
       183, 208, 211–12, 214, 217–18, 219–20
  on Bridge opening ceremonies, 237
  on Bridge plans, 2–3, 7
  confusion of Washington and father, 297
  on death of Emily Roebling, 267–68
  on efforts to oust Washington, 228, 230–31,
       233
  Emily's biography of Washington published
       in, 238–39, 240
  on John Roebling's funeral, 15
  publication of Bridge plans, 7
  on Washington's qualifications as Chief
       Engineer, 17–18
Brown, John, 83, 87
Buck, Leffert L., 139, 256
Burnside, Ambrose, 96, 99, 101, 102

caissons. *See also* Brooklyn Bridge caissons;
       Eads, James B., caisson use by
  early uses of, 172, 182, 184
  Washington's research on, 145, 155, 156,
       158–59, 172, 182
Campbell, Allan, 231, 232, 233
Case, Clarence, 292, 294, 297–98
*Centennial History of New York* (Stone), 8
Chief Engineer for Brooklyn Bridge
  anonymity of, xiii
  John Roebling's appointment as, 2, 143
  Martin as, 241
  Washington as
    age at appointment, 1
    anxieties about, 15–16
    appointment as, ix, xii, 169
    credentials for, 14–15, 17–18, 171
    efforts to reduce salary, 224–25
    efforts to replace, 213–14, 221–23, 224–33
    and practical rigor, xiii

resignation, 241
struggle over wire contract and, 213–15
Washington's failing health and, 191,
195–96, 201–11, 219–22, 226–27, 234
civil engineering, as profession, 69
Civil War
and Atlanta, fall of, 126–27
Battle of Five Forks, 128–29
Battle of Fredericksburg, 101, 102
civilian fear during, 106–7
and conscription, 93, 101
death toll in, 87–88
Ellet in, 89
expected short duration of, 86
John Roebling's views on, 83, 85
*Merrimack* and *Monitor* in, 91–92
poor medical care in, 88
start of, 84–85
tensions preceding, 83–84
and Vicksburg, surrender of, 104
Washington in
balloon reconnaissance by, 105–6
and Battle of Antietam, 99–100
and Battle of Chancellorsville, 102–4
and Battle of Gettysburg, 107–11
and Battle of Mine Run, 113–14
and Battle of the Crater, 126
and battle of Williamsburg, 94
battles experienced by, xvii, 3, 87
and bridge construction, 95–96, 100–101,
102, 111–12
and "broken down nerves," 145
on Burnside's generalship, 99
conduct in, 88
on Confederate soldiers, 91, 93–94, 98,
122
cost of to Washington's future, 279
and courtship of Emily Warren, 75–76,
116–18, 119–20, 123–25
enlistment, 3, 85
fortuitous avoidance of injury, 87–88,
103–4, 109, 110, 121
frequent letter-writing, 113
illness, 100
on life in camp, 86–87, 89, 113
on McClellan's generalship, 87, 92, 94
at *Monitor* and *Merrimack* battle, 92
on newspapers' coverage of, 111
and Peninsula Campaign, 93–94
and Pickett's Charge, 109–10
on Pope's generalship, 97–98
on possibility of death, 126

as reconnaissance scout, 97
rise in rank, 3, 88, 89, 96, 121, 129
road-building by, 102
and Second Battle of Bull Run, 97–99
at Shipping Point, Maryland, 89–91
and Siege of Petersburg, 122
on soldiers' sacrifices, 122–23
and suffering, hardening to, 111
transfer to engineering duties, 94–95
on Union devastation of South, 97, 113
on Union morale, 98, 102
units served in, 85, 95, 98, 101–2
visit home, 106–7
and Wilderness Campaign, 120–21, 123
Cleveland, Grover, 234
Collingwood, Francis, 168, 186, 187, 196,
202–3, 245
commuting, development of practice, 8
Cooper, Lucia, 70–71, 144–45, 201
Couper, William, 199, 272–74
Covington-Cincinnati Bridge
Brooklyn Bridge committee's visit to, 12,
165–66
construction of, 3–4, 79, 133–42, 139
dangers of, 135, 138, 140
delays in, 77, 80, 83, 124, 134–35
strain on Washington from, 145
Washington's illness during, 141–42
dimensions of, 136, 142
media accolades for, 142–43
opening of, 142
public interest in, 135, 138, 142
renovations of, 139, 140, 305
Washington's criticisms of design, 140, 142
Cramer, Michael J., 42
Croly, Jane Cunningham, 255

Dalton, E. B., 9–10
Darwin, Charles, 43, 45
Davis, Andrew Jackson, 144, 147–48
decompression sickness ("the bends"), 172,
183–86, 189, 190–91, 248
Dee Bridge (England), collapse of, 63–64
Delaware and Hudson Canal, 54, 56–58, 78
De Witt, William C., 232
Draft Riots (New York City, 1863), 101

Eads, James B.
caisson use by
and decompression injuries, 184–85
patent lawsuit against Washington,
194–95

Eads, James B. *(continued)*
  use of physicians in attendance, 183
  Washington's study of, 172, 182, 184–85
  death of, 300
Eaton, Amos, 69–70
Edson, Franklin, 234
electricity, Roebling's Sons wire and, 283
elevators, Roebling's Sons wire and, 141, 283, 308
Ellet, Charles, 2, 62–64, 64–65, 89
Emancipation Proclamation, 100, 101
engineering schools, establishment of, 69
Ericsson, John, 91, 162
Erie Canal, 31
Ermeling, Ernest C., 282–83
Estabrook, Henry Dodge, xv, 37, 273
Europe, travel to
  by Emily alone, 254–55
  by Washington and Emily, 3, 144–45, 150, 152–59, 191–93
  on Continent, 157–59, 191–93
  cost of, 155
  engineering research during, 153, 154, 155–59, 156, 158, 171, 192
  in John Roebling's native town, 157–59
  in London and Britain, 152–53, 155–56
  and opportunity to invest in new gas engine, 160–63
  in Paris, 153–55, 192
  return journey, 163
  Washington's health and, 145, 191–93

Farrington, Edmond F. "Frank"
  and construction of Brooklyn Bridge, 196, 208, 209–10, 210–11, 226
  and Covington-Cincinatti bridge, publication on, 135–36, 137, 138, 143
  lectures on Brooklyn Bridge construction, 223–24
Farrow, John, 286
Fisher, John "Jacky," ix–xi, *x*
Fisk, "Diamond Jim," 9
French and Indian War, 20

gas engines, opportunity to invest in, 160–63
Gates, John W., 260–61
George Washington Bridge (Hudson River), 247, 299, 307–8
Gill, Washington, 31
Golden Gate Bridge, 308
gold market, and Black Friday, 8–9
Gould, Jay, 8–9, 230

government regulation, and changed environment for engineers, 304–5
Grace, William Russell, 231, 233
Grant, Ulysses S., 41–42, 93, 104, 121, 126, 129, 187–88
*The Great Bridge* (McCullough), xii

Haigh, J. Lloyd
  bankruptcy and imprisonment of, 218
  defective wire provided by, 212, 215–18, 239, 306
  and political corruption, 214–15
  as supplier of Brooklyn Bridge wire, 215, 217–18
Hawthorne, Nathaniel, 61–62
Hegel, George Wilhelm, 26, 42, 146
Herting, Adelaide (grandmother), 39, 40
Herting, Ernest (grandfather), 22–23
Hewitt, Abram S.
  and Brooklyn Bridge opening celebrations, 235–37
  and Brooklyn Bridge wire contract, 213–15
  Couper statue of, 272
  on Emily Roebling and Bridge, 252
  as rival of Roebling's Sons, 214–15
  and Universal Exposition, 153–54, 157
Hildenbrand, Wilhelm, 168, 208–9, 247, 256, 270, 290
Hooker, Joseph "Fighting Joe," 99, 101–4, 107
Howe, Elias, 46–47
Humphrey, A. A., 166

income tax, introduction of, 93
Iowa, John Roebling's farmland in, 77–79

Jackson, Andrew, 34
Jacoby, Henry S., xv, 274
James, Henry, Sr., 42, 46
John A. Roebling Bridge. *See* Covington-Cincinnati Bridge
John A. Roebling's Sons, 277
  aging management and, 277
  and Bear Mountain Bridge, 294–95
  and Benjamin Franklin Bridge, 292–94
  bridge projects concurrent with Brooklyn Bridge, 169
  Brooklyn Bridge trustees' attacks on, 221
  and Brooklyn Bridge wire contracts, 212–15, 218
  closing of, 308
  and Covington-Cincinnati Bridge, 139
  establishment in Trenton, 55

family conflict over management of, 280–81
family conflict over sale of, 260–62, 279
and George Washington Bridge, 247, 299, 308
and growing competition, 281
growth of, 39, 249, 275–76, 279, 281, 283, 285
John Roebling's will and, 14
labor unrest and, 283–84
management of, after Ferdinand and Charles' deaths, 284, 286–87, 290–95
opening of New York City store, 143
sale of, 307
Swan and, 14
and Washington as salesman, 133
Washington's income from, 260
Washington's training at, 80, 81
Washington's work at, after Brooklyn Bridge, 249, 256, 289–90
and Williamsburg Bridge, 139, 256, 289–90

Kelly, John Henry, 101, 220
Kenly, John Reese, 100–101
Kingsley, William C.
and Adams-Roebling feud, 164
and Bridge opening ceremony, 234
death of, 300
as driving force behind Brooklyn Bridge project, 166
efforts to undermine Washington's Brooklyn Bridge construction, 224
and effort to oust Washington, 233
as leading Brooklyn contractor, 16
and New York political corruption, 199
proposal for Brooklyn Bridge, 164–65
Washington's dislike of, 16, 166
Kinsella, Thomas, 17, 165
Kirkwood, James Pugh, 164

Langen, Eugen, 160–62, 300
Lee, Robert E., 99, 100, 102, 104, 106, 107, 108, 110, 187–88
Long Depression, 200
Low, Seth, 224, 227, 231–34, 237, 243, 300
Lowe, Thaddeus S. C., 105

Manhattan Bridge, 283
Marshall, Marguerite M., 290–92, 294, 296
Marshall, William, 230, 231–32, 233
Martin, C. C.
as Chief Engineer, 241
and construction of Brooklyn Bridge, 168, 208, 211, 212, 226–27, 230–31

efforts to oust Washington and, 228–29, 233
and John Roebling's funeral, 16
McAlpine, William Jarvis, 164
McClellan, George B., 87, 92–94, 99–101, 126
McCullough, David, xii, xvi, 160, 169
McLeod, David, 53
McNulty, George, 168, 208, 234, 296, 300, 307
McPherson, James, 109
Meade, George Gordon, 88, 107–9, 111, 113–14
Meigs, Montgomery, 85, 94–95, 101
*Merrimack* (Confederate ship), 91–92
Methfessel, Anton (brother-in-law), 83, 135
Miller, Abraham B., 212
Modjeski, Ralph, 292–94
*Monitor* (Union ship), 91–92
Monongahela Bridge (Pittsburgh), 52
Morison, George S., 247
Morse, Samuel F. B., 46
Mumler, William H., 144
Murphy, Henry Cruse
and Brooklyn Bridge construction, 143, 200, 203, 204, 212, 214–16, 225–26, 307
and effort to oust Washington, 214, 222, 224–25, 229, 231, 233
and John Roebling's funeral, 16
Kingsley and, 166–67
and political corruption, 198–99

Native Americans
European settlement and, 19, 20, 67–68, 78
John Roebling on, 30
New York Bridge Co. *See also* Murphy, Henry Cruse
bridge plan submitted to, 6–7
and Chief Engineer job, 14–15, 143, 169
dissolving of, 206–7
and political corruption, 166, 198–99
New York City
architecture, Washington on, 244
economic inequality in, 8–10
19th century growth in, 7, 8, 152, 243
political corruption in, 7–8, 151–52, 166, 197–200, 213–15
skyline, vertical rise of, 243–44
Niagara Falls Suspension Bridge
bids for, 62–63
Brooklyn Bridge committee's visit to, 12, 166
Ellet and, 63, 64
John Roebling's construction of, 64–67, 65
public reactions to, 66–67
and safety concerns, 63–64
Washington's restoration of, 244–45

Ordish, Roland Mason, 155, 158
Overman, Frederick, 147, 148–49

Paine, William
    in Civil War, 110
    and construction of Brooklyn Bridge, 168,
        196, 212, 216, 218, 221
    death of, 300
    and surveying of bridge path, 12–13
Panic of 1837, 34, 46–47
Panic of 1857, 77, 79, 80, 83
paper currency, introduction of, 93
patent system, US, 61
Pennsylvania. See also Saxonburg
    state legislature, Washington's visit to, 58
    Western, European settlement of, 19–20
Pittsburgh
    fire of 1845, 52
    Washington as father's assistant in, 80–82
    Washington's boarding schools in, 50–52
Pittsburgh Aqueduct, 12, 28, 47–48, 58
Pope, John, 97–98
Porter, Ray, 45
Priessnitz, Vincent, 43–44, 56, 59
Probasco, Sam, 168

railroads, in 19th-century America, 61–62,
    243
Rapid Transit Act of 1875, 243
Reichenbach, Carl von, 149
Rensselaer Polytechnic Institute
    curriculum, 70, 71–72
    history of institution, 68–69
    programs for women, 69–70
    Washington's departure for, 70–71
    Washington's education at, 3, 67, 70–76
    Washington's fundraising for, 269–70
    Washington's son John at, 73, 242
    Washington's troubled friend at, 74–76, 120
    Washington's unhappiness at, 71, 73, 74
Rhule, David, 100
Riedel, Ed, 145–46
Riedel, Julius, 49–50
Riley, John, 213
Roebling, Carl (uncle), 20, 28, 29
Roebling, Charles (brother), 277
    character of, 39–40, 279
    death of, 284–85, 304
    education of, 13–14, 135
    estate, value of, 285
    father's will and, 279–80
    financial stability of, 155

and Roebling's Sons, family dispute over sale
    of, 260–61, 279
    at Roebling's Sons, 249, 256, 277, 279, 280,
        281, 282, 283, 284
    Trenton residence of, 251
    Washington on, 285–86, 290
Roebling, Christoph P. (grandfather), 25
Roebling, Cornelia W. Farrow (2nd wife)
    courtship, 270
    death of, 301
    marriage, 271–72
    and married life, 298
    son of, 270, 286
    Washington on, 271
    Washington's death and, 300
    and Washington's renewed vigor, 274–75
    in Washington's will, 301
Roebling, Edmund (brother), 14, 147, 181, 190,
    192, 278, 281, 297–98
Roebling, Elvira (sister), 14, 39, 86–87, 114,
    117, 118, 131–32
Roebling, Emily Warren (1st wife), 160. See
    also Europe, travel to
    appearance of, 117
    biographies of Washington, 187, 238–39, 240
    birth of son, 159
    as Brooklyn Bridge engineering partner,
        xvi–xvii, 201–5, 228–30, 252
    and Brooklyn Bridge opening, 205, 234–37
    burial and grave of, 269, 311–12
    character of, 254, 259
    concerns about married life, 124–25, 145
    courtship and correspondence, 75–76,
        116–18, 119–20, 123–25
    and Covington-Cincinnati Bridge, 133–34, 140
    death of, 267–68
        impact on Washington, 268–69, 270
    distinguished ancestry of, 115, 124, 259
    editing of Journal of . . . Silas Constant, 267
    education of, 116
    first meeting of Washington, 114–15
    health decline in later life, 259–60, 266–67
    on husband's hypochondria, 248
    introduction of Roebling family, 116, 118–19
    marriage to Washington, 131–32
    on marriage to Washington, 248
    opposition to Washington's later-life work,
        256, 261, 288–89
    plaque to, on Brooklyn Bridge, xvi
    pregnancy, 144, 145, 157, 159
    pre-wedding visits to Roebling family,
        123–24, 127

as sister of G. K. Warren, 115–16, 119–20
and Spanish American War, 252, 253
on taxation, 259
and Trenton mansion, 250–51
visit to spirit medium, 144
"A Wife's Disabilities," 130–31, 258
will of, 269
Woman's Law degree, 130, 257, 258
and women's organizations, 255, 256–57, 268
and women's rights, 130–31, 254–55, 257–58
Roebling, Ferdinand "Ferdy" (brother)
childhood of, 39, 60
death of, 284
on father's spiritualism, 145–46
father's will and, 279–80
financial stability of, 155
and gas engine investment, 161
relationship with Washington, 86, 277–78
and Roebling's Sons, family dispute over sale of, 260–61
at Roebling's Sons, 13, 133, 135, 143, 169, 221, 249, 277, 280
Trenton residence of, 251
Roebling, Ferdinand Jr. (nephew), 284, 286, 307–8
Roebling, Frederike Amalia (aunt), 162
Roebling, Friederike Dorothea (grandmother), 25–26
Roebling, Johanna Herting (mother)
death of, 17, 127–28, 145
husband's abuse of, 37, 38–39, 40, 145
husband's contact with from beyond the grave, 144, 145–49
husband's extreme frugality and, 40–41, 42
and husband's relationship with Cooper, 71
husband's water cure treatments and, 45
limited education of, 125
marriage, 22, 31
sad life of, as concern to Emily Warren, 124, 125, 145
Saxonburg, Pennsylvania and, 34, 54, 55
Washington on, 71, 128
Washington's early separation from, 50
Washington's last visits with, 107
Roebling, John A. II (son), 160
birth of, 159
career choice for, 73
children of, 248–49, 286, 288, 301
death of father and, 264, 300
education of, 204
and engineering, lack of interest in, 242
and father's remarriage, 270–71

marriage, 248
poem celebrating Brooklyn Bridge, 241
poor health of, 242
at Rensselaer Institute, 242
and Spanish American War, 252–53
and Washington's biography of father, xv, xvi
Washington's relationship with, 242–45
Roebling, John Augustus (father), 32–34, 35.
See also John A. Roebling's Sons;
Saxonburg, Pennsylvania
bizarre dietary notions of, 59–60, 81–82
bridge-building experience, 2, 12
character of
caution in business, 34–35
charitable giving, 14, 16, 36, 38
childhood influences and, 41
extreme frugality of, 35, 38, 40, 52, 57
on illness as mental weakness, 100
limited self-awareness in, 46
public vs. private versions of, 36
relentless drive of, 65, 273
tyranny over family, 37–40, 59–60, 273
volatility, 28, 72–73, 78–79, 81, 140–41
Washington on, xvi, 13, 65
childhood of, 19, 25–26, 41
children of, 13–14
concealment of wealth from, 41
brutality and, 37, 38, 39
dislike of doctors and, 42
and water cure treatments, 42, 45
and Civil War, 94, 106
death of, ix, xii, 1, 3–6, 169
and death of son Willie, 147
and death of wife, 127–28, 145
and doctors, animosity toward, 4–5, 41, 42
early years, 20–21
education as engineer, 1, 21, 26, 28
emigration to US, 1, 20–21, 28–31
engineering career in Germany, 27–28
engineering career in US, ix, 31–32, 52, 54, 56–58 (See also Allegheny River Bridge; Covington-Cincinnati Bridge; Niagara Falls Suspension Bridge; Pittsburgh Aqueduct)
funeral of, 15, 16–17
and gas engine investment, 160–63
genius of, xii, xiii, 1
land investments in Iowa, 77–79
and Lucia Cooper, 70–71, 144–45, 201
marriage of, 22, 31
metaphysical interests, 26–27, 146, 147–50
and New York political corruption, 199

Roebling, John Augustus *(continued)*
  and organized religion, 41–42
  and Panic of 1837, 34–35
  and Pennsylvania oil investment, 250
  pocket notebooks of, 12–13
  politics of, 57–58
  remarriage of, 145
  settler's experience of, Washington on, 249–50
  and spiritualism, 144, 145–49
  statue of in Trenton, 37, 199, 272–74
  "Truth of Nature" treatise by, 146
  and US Capitol construction, 85
  Washington as assistant to
    after father's injury, 56–58
    on Allegheny Bridge, 80–82
    conflicts in, 72–73, 78–79, 81, 140–41
    on Covington-Cincinnati Bridge, 3–4,
      133–42
    efforts to break from, 160–63
    and qualifications as Brooklyn Bridge
      Chief Engineer, 17, 171
  Washington on enduring influence of, 242
  Washington's biography of, xiv–xvii, 23,
    249–50
  on Washington's engagement, 118–19
  and water cure, faith in, 4, 42, 43–46, 56,
    59–60, 127–28
  wealth of, xii, 14, 16, 41
  will of, 14, 16, 36, 279–80
  as wire cable manufacturer, xii, 14, 23–24,
    32–34, 35
    development of process for, 58–59
    and move to New Jersey, 54–55
    patent for, 33, 48
Roebling, Josephine (sister), 14, 39, 297
Roebling, Karl (nephew), 284, 286
Roebling, Laura (sister), 14, 39, 83, 123–24, 297
Roebling, Margaret S. McIlvaine "Reta"
    (daughter-in-law), 248–49, 287, 300
Roebling, New Jersey, 39, 281–82
Roebling, Washington A. II (nephew), 274–75
Roebling, Washington Augustus, x, 277–78.
    *See also* Brooklyn Bridge construction;
    Civil War, Washington in; *other specific
    topics*
  as ahead of his time, 247–48
  awards in later life, 295
  bridge-building experience of, 3–4
  and Brooklyn Bridge opening celebrations,
    235, 237
  character of
    directness and candor, xv, 88

    flaws in, xiii
    generosity and kindness, 54
    as private man, xvii
    strength of feeling in, xvi
    tenacity and drive, xiii, 72, 240
    Washington on, 72
  charitable work and giving, 269–70
  childhood of
    abuse from lunatic roommate, 53
    as assistant to his father, 56–58
    at boarding schools, 50–53
    father's frugality and, 35, 52
    father's violent tyranny over family and,
      37–40, 42, 45, 54
    grandfather and, 22–23
    impact on character, 54
    as privileged, 53
    in Saxonburg, Pennsylvania, 22–25,
      33–34, 48–49
  death of, 299–300, 302
  education of, 50–53
    as engineer, at Rensselaer Institute, 3, 67,
      71–76
    at home as child, 48–49
    interruptions of to assist father, 56–58
    in New Jersey, 56–57, 60
  engineering work after Brooklyn Bridge,
    241, 244–45, 247, 289–90, 292–95
  as father's assistant
    after father's injury, 56–58
    on Allegheny Bridge, 80–82
    conflicts in, 72–73, 78–79, 81, 140–41
    on Covington-Cincinnati Bridge, 3–4,
      133–42
    efforts to break from, 160–63
    and qualifications as Brooklyn Bridge
      Chief Engineer, 17, 171
  and father's bizarre diets, 59–60, 81–82
  and father's Iowa farming project, 78–79
  final years, daily life in, 297–300
  funeral and burial of, 300–301
  grandchildren, concerns about, 288
  grave of, 311–12
  health of
    abdominal attacks (1858), 81–82, 193–94
    health-related absences from Brooklyn
      Bridge construction, 191, 195–96,
      201–11, 219–22, 226–27, 234
    impact of Brooklyn Bridge construction
      on, 180, 181, 187, 189, 248
    improvement after Brooklyn Bridge
      completion, 244

in later life, 193, 244, 263, 266, 292, 294
psychological factors in, 193, 247–48
trips abroad for, 145
and wife as engineering partner, 201–5
later-life preference for home, 256, 257, 262
on longevity, secret of, 296
mineral collection of, 251, 256, 262–65, 300
money worries of, 155, 190, 250, 256
as musician/music lover, 24, 81, 83, 301, 303
as native German speaker, 49, 51, 157–58
and New York political corruption, 199–200
pocket notebooks of, xiv, 168
religious beliefs of, 269, 291
remarriage of, 270–72
reputation of, Eads lawsuit and, 195
and Roebling's Sons, family dispute over sale
    of, 260–61, 279
and Roebling's Sons management after
    brother's deaths, 286–87, 290–95
siblings of, 13–14
and son, choice of career for, 73
and spiritualism, 144, 147, 149
will of, 301
Roebling, Willie (brother), 13, 144, 146–47, 149
Rusling, James F., 109, 113, 128, 274
Russell, Isaac Franklin, 257, 258

Saxonburg, Pennsylvania
    founding by Roebling brothers, 1, 20, 21–22
    as German settlement, 19, 23, 24, 49
    John Roebling as farmer in, 21, 22
    John Roebling's sale of parcels in, 37
    John Roebling's temper and, 40
    religious controversy in, 49–50
    Roebling family move from, 54–55
    Roebling home in, 37–38
    Washington's childhood in, 22–25, 33–34,
        48–49
Schuyler, Hamilton
    on Brooklyn Bridge caissons, 172
    on Charles Roebling, 39–40
    on Emily Roebling, xvii
    as Roebling family biographer, xiv–xv
    on Washington at Rensselaer, 71
    on Washington's disposition, 54
Second Bank of United States, 34
Semler, Ludwig, 228–29, 230, 233
Sheridan, Philip, 126, 128–29
Sherman, William T., 124
Slocum, Henry, 102–3, 221–22, 233
Smith, Andrew, 183, 184, 185, 189, 190–91
Sorosis, 255, 257

Spanish–American War, 252–53, 281
spas, in early America, 43
Specie Circular of 1836, 34
spirit photography, 144
spiritualism
    John Roebling and, 144, 145–49
    19th century interest in, 144
steel
    Brooklyn Bridge cables as, 211
    choice of for Bridge floor beams and trusses,
        221, 224, 225–26
    replacement of iron as construction metal,
        154
    Washington's research on, 159
Stephens, Lemuel, 53
Stephenson, Robert, 64
Strong, George Templeton, 109, 126–27
Stuart, Charles B., 62
suspension bridges
    John Roebling's early interest in, 27–28
    John Roebling's experience with, 2, 12
    mid-19th century technology and, 2
    rigidity as issue in, 73–74
    and safety concerns, 63–64, 74
    Washington's study of, 73–74
Swan, Charles
    and death of Johanna Roebling, 128
    and effort to oust Washington, 233
    and gas engine investment, 161, 163
    as John Roebling's partner, 14, 48
    and John Roebling's stashing of gold, 80
    and Trenton wire mill, 55, 56, 141
    Washington and, 67, 80, 81

Tammany Hall, 151, 197–98, 220
technology, 19th century advances in, 61–62,
    243, 283
Thomas, M. Carey, 258
*Titanic* (ship), 275
Townsend, Robert, 47
transportation, 19th century advances in, 8,
    243. *See also* railroads, in 19th-century
    America
Trenton, New Jersey
    decline of industry in, 307
    Emily Roebling's dislike of, 269
    Roebling family move to, 54–55
    Roebling family residences in, 251
    Roebling home and wire mill in, 55
    Washington and Emily's mansion in,
        250–51, 259, 301
Trenton Academy, 60

Trenton Iron Works, 55, 307
Troy, New York. *See also* Rensselaer Polytech-
    nic Institute
  history of, 67–68
  Roeblings' move to, 242
Twain, Mark, 67
Tweed, William M. "Boss," 9, 151–52, 166,
    197–99, 217, 220

Uncle Sam, origin of image, 68
Universal Exposition (Paris, 1867), 153–55, 162

Van Rensselaer, Stephen, 68–69

Warren, Edgar Washburn, 131
Warren, G. K.
  as brother of Emily Roebling, 115–16,
    119–20
  in Civil War, 98, 102, 108, 113, 120–21, 123,
    128–29
  death of, 227
  and Eads bridge investigation, 194
  later life of, 226
Washington, DC, Washington's visit to, 85
water cure
  history and methods of, 43–44
  and Johanna Roebling's death, 127–28

John Roebling's belief in, 4, 42, 43–46, 56,
    59–60, 127–28
  Washington on, 45
  Washington's spa trips and, 191–93, 263
Webb & Bell, shipbuilders, 173
Williamsburg Bridge, 139, 256, 289–90
Wilson, Samuel, 68
wire cables
  for bridges, fabrication process for, 47–48,
    52, 81, 137, 139–40
  John Roebling as manufacturer of, xii,
    23–24, 32–34, 35 (*See also* John A.
    Roebling's Sons)
  development of process for, 58–59
  and move to New Jersey, 54–55
  patent for, 33, 48
Witte, Otto, 232–33
women's rights
  Emily Roebling and, 130–31, 254–55,
    257–58
  in late 19th century, 252, 257–58
  Washington on, 258
World's Columbian Exposition (Chicago
    1893), 251, 252
World War I
  Roebling's Sons and, 275–76, 286
  Washington's views on, 275

# A Note on the Author

Erica Wagner is the author of *Ariel's Gift: Ted Hughes, Sylvia Plath, and the Story of "Birthday Letters,"* the novel *Seizure*, and the short story collection *Gravity*; she is the editor of *First Light: A Celebration of Alan Garner*. Literary editor of *The Times* (London) for seventeen years and twice a judge of the Man Booker Prize, she is now a contributing writer for the *New Statesman* and consulting literary editor for *Harper's Bazaar*. Her work has appeared in the *Guardian*, the *Economist*, the *Financial Times*, and the *New York Times*, among others. She was the recipient of the Eccles British Library Writer's Award in 2014, and she is a lecturer in creative writing at Goldsmiths, University of London. She was born in New York City and lives in London with her husband and son.